3rd International Symposium on Bifurcations and Instabilities in Fluid Dynamics 2009

Journal of Physics: Conference Series Volume 216

Nottingham, UK
10-13 August 2009

ISBN: 978-1-61738-431-8
ISSN: 1742-6588

Printed from e-media with permission by:

Curran Associates, Inc.
57 Morehouse Lane
Red Hook, NY 12571

Some format issues inherent in the e-media version may also appear in this print version.

Printed by Curran Associates, Inc. (2010)

For permission requests, please contact the Institute of Physics
at the address below.

Institute of Physics
Dirac House, Temple Back
Bristol BS1 6BE UK

Phone: 44 1 17 929 7481
Fax: 44 1 17 920 0979

techtracking@iop.org

Additional copies of this publication are available from:

Curran Associates, Inc.
57 Morehouse Lane
Red Hook, NY 12571 USA
Phone: 845-758-0400
Fax: 845-758-2634
Email: curran@proceedings.com
Web: www.proceedings.com

3rd International Symposium on Bifurcations and Instabilities in Fluid Dynamics 2009

Journal of Physics: Conference Series Volume 216

Nottingham, UK
10-13 August 2009

ISBN: 978-1-61738-431-8
ISSN: 1742-6588

IOP Institute of Physics

TABLE OF CONTENTS

Forward

Hydrodynamic stability is of fundamental importance in fluid dynamics and is a well-established subject of scientific investigation that continues to attract great interest of the fluid mechanics community. Bifurcations and instabilities are observed in all areas of fundamental and applied fluid dynamics and remain a challenge for experimental, theoretical and computational studies. Hydrodynamic instabilities of prototypical character are, for example, the Rayleigh-Bénard, the Taylor-Couette, the Bénard-Marangoni, the Rayleigh-Taylor, and the Kelvin-Helmholtz instabilities. A fundamental understanding of various patterns of bifurcations such as identifying the most dominant mechanisms responsible for the instability threshold is also required if one is to design reliable and efficient industrial processes and applications, such as melting, mixing, crystal growth, coating, welding, flow re-attachment over wings, and others. Modeling of various instability mechanisms in biological and biomedical systems is currently a very active and rapidly developing area of research with important biotechnological and medical applications (biofilm engineering, wound healing, etc). The understanding of breaking symmetry in hemodynamics could have important consequences for vascular biology and diseases and its implication for vascular interventions (grafting, stenting, etc).

The collection of papers in this volume is a selection of the presentations given at the Third International Symposium on Instability and Bifurcations in Fluid Dynamics, University of Nottingham, UK, 10–13 August 2009. With more than 100 invited and contributed papers the symposium gave an overview of the state-of-the art of the field including experimental, theoretical, and computational approaches to problems related to convection, effects of magnetic fields, wake flows, rotating flows, and many others. The complete program can be found at the conference website.

The symposium was the follow-up of two previous symposia held in 2004 in Madeira, Portugal and in 2006 in Copenhagen, Denmark. The number of participants in these symposia has been steadily rising. The next symposium in the series will take place in Barcelona, Spain, 18–21 July 2011. See the conference website for further details.

P Z Bar-Yoseph, M Brøns, K A Cliffe, A Gelfgat and A Oron
Editors

All papers published in this volume of *Journal of Physics: Conference Series* have been peer reviewed through processes administered by the proceedings Editors. Reviews were conducted by expert referees to the professional and scientific standards expected of a proceedings journal published by IOP Publishing.

The Third International Symposium on Bifurcations and Instabilities in Fluid Dynamics IOP Publishing
Journal of Physics: Conference Series **216** (2010) 012001 doi:10.1088/1742-6596/216/1/012001

Dynamics and formation of localized states in flowing thin films: Bound states of solitary waves

A. Bergeon[1,2] and E. Knobloch[3]

[1] Université de Toulouse; INPT, UPS; IMFT; Allée Camille Soula, F-31400 Toulouse, France
[2] CNRS; IMFT; F-31400 Toulouse, France
[3] Department of Physics, University of California, Berkeley CA 94720, USA.

E-mail: knobloch@berkeley.edu

Abstract. The flow of a thin liquid film down an inclined heated substrate generates solitary waves in the form of traveling holes. These waves in turn form a variety of different bound states in which the constituent waves all travel with a common speed. The formation and properties of these states are investigated as function of the system parameters.

1. Introduction

The study of flow of thin liquid films down an inclined substrate has a long history [12, 4]. The problem has been studied both experimentally [5] and numerically using a variety of different thin film equations [15] as well as the primitive equations [14]. These studies all reveal the formation and steepening of waves that may lead to shock-like structures or the formation of solitary waves. The latter result when the steeping of the wave due to nonlinearity is balanced by spatial dispersion. Similar phenomena take place on inclined heated substrates in which the temperature dependence of the surface tension is of paramount importance [16].

Heated substrates are of both practical and theoretical interest. On a horizontal substrate the primary instability as the Marangoni number increases is subcritical and leads to the breakup of the film into arrays of drops [7, 10, 1]. These drops are separated by nominally dry regions, or by an ultrathin liquid film if the substrate is wetting. In the latter case the resulting multidrop state undergoes a slow coarsening process since the drops are able to communicate via the intervening film. In a recent paper Thiele and Knobloch described a general construction of multidrop solutions [18] leaving aside their realizability via the solution of an initial value problem. In addition, they argued that on an inclined substrate the breakup of the liquid film may not proceed to completion owing to the presence of a dynamically generated thin film between the drops. This film is present even in the absence of an explicit disjoining pressure describing the wetting properties of the substrate, and the resulting solution is self-consistent in the sense that the resulting multidrop states all slide down the substrate unless trapped by wetting inhomogeneities or other types of surface roughness [19]. Thiele and Knobloch [18] examined sliding solutions of constant form subject to periodic boundary conditions together with their stability. The speed of these states solves a nonlinear eigenvalue problem, and increases with increasing inclination. In addition they identified the presence of intervals in the spatial period with no stable periodic solutions of constant form.

2. Formulation

In this paper we examine these intervals in order to identify the solutions present and describe their properties. We study the lubrication or longwave equation [18]

$$h_t = -\left[\frac{h^3}{3}(h_{xxx} + \alpha - h_x) - \frac{h^2}{2}\text{Ma}\ T_x\right]_x , \tag{1}$$

where $h(x,t)$ is the film thickness and $T(x,t)$ is the surface temperature given by

$$T_x = -\frac{\text{Bi}\ h_x}{(1+\text{Bi}\ h)^2}. \tag{2}$$

The equations have been nondimensionalized using the unperturbed film thickness h_0 for $h(x,t)$ and the capillary length $\ell \equiv \sqrt{\sigma_0/\rho g} \gg h_0$ for the downstream coordinate x, where σ_0 is the isothermal surface tension, ρ is the density and g is the acceleration due to gravity. Thus Bi and Ma are the Biot and Marangoni numbers that measure the importance of heat transfer to the ambient air and the temperature dependence of the surface tension, respectively. Owing to the choice of ℓ the Bond number Bo = 1 [18]. Equations (1)-(2) describe a thin liquid film with a free upper surface on an inclined homogeneously heated substrate, with (scaled) inclination α. The corresponding physical inclination is sufficiently small that the film is primarily supported hydrostatically. Newton's law of cooling is assumed at the liquid-gas interface with passive overlying gas. The flow is sufficiently slow that all inertial effects are negligible [2, 13].

The critical Marangoni number for the onset of instability of the flat film $h = 1$ on a periodic domain of period L is

$$\text{Ma}_n = \frac{2(1+\text{Bi})^2}{3\text{Bi}}\left(1 + k_n^2\right), \tag{3}$$

where $k_n = 2\pi n/L$, $n = 1, 2, \ldots$. At each Ma_n a branch P_n of periodic states with wavenumber n bifurcates from the $h = 1$ solution. These states take the form $h(x,t) \equiv h_n(x - ct)$ and hence are "rotating waves". Thiele and Knobloch [18] used numerical continuation to compute the branches P_n and the associated solution profiles $h_n(\xi)$, $\xi \equiv x - ct$, as a function of the spatial period L for several different choices of the Marangoni number Ma and inclination α. In addition, they used numerical continuation to compute the principal eigenvalues describing the solution stability, computing any secondary rotating wave branches so identified. However, no time-stepping computations were performed. Here we use a spectral code with N modes in the x-direction to solve Eqs. (1)-(2) on periodic domains of length L, focusing on intervals in L with no stable solutions of constant form.

3. Results

The flat film $h = 1$ is linearly unstable provided $\text{Ma} > 2(1+\text{Bi})^2/3\text{Bi}$. The primary bifurcation is supercritical when α is large but subcritical for small α [18]. In the following we describe typical behavior in each of these cases, starting with the supercritical case. In all cases we take Ma = 3.5, Bi = 0.5 and explore the dependence of the results on the domain length L. In Fig. 1 we show solutions obtained by time-stepping for $\alpha = 0.4$, Ma = 3.5 and Bi = 0.5 in the form of space-time plots for a succession of imposed spatial periods L. The dark region indicates that the solution is a depression or a hole in an otherwise flat film; the hole is always on the downstream side of several smaller oscillations in the film height that are also visible in the figure (see also Fig. 2). In each case transients have been discarded. When $L = 50$ Fig. 1(a) reveals a single solitary wave of depression (hereafter a "hole") that propagates through the periodic domain at constant speed. When $L = 100$ one finds a pair of such holes that form a bound state of two holes traveling at constant speed; this speed is almost identical to the speed of a

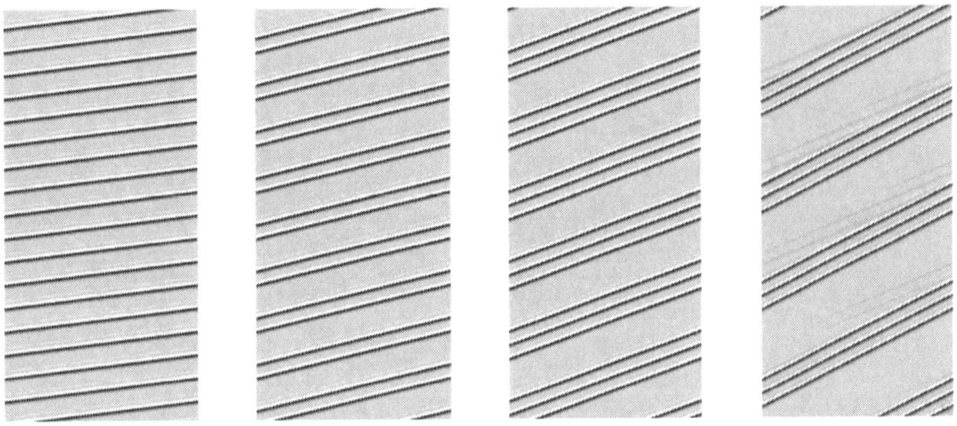

(a) (b) (c) (d)

Figure 1. Isovalues of $h(x, t)$ in the (x, t) plane with $0 \leq x \leq L$ (horizontal axis) and $t \in [25000, 26250]$ (vertical axis). Parameters are $\alpha = 0.4$, $\text{Ma} = 3.5$, $\text{Bi} = 0.5$ and (a) $L = 50$, (b) $L = 100$, (c) $L = 150$, and (d) $L = 200$. Resolution is $N = 128$, starting from $h(x, 0) = 1 + 0.1\sin(2\pi x/L)$.

single hole. When $L = 150$ one finds a bound state of three equally spaced holes traveling at constant speed. This state persists for $L = 200$ but is now embedded in a background of waves that travel faster: the waves grow in amplitude in the downstream direction and are absorbed by the bound state on the upstream side causing negligible perturbation. Solutions of this type are no longer "rotating" waves, i.e., waves that are steady in an appropriately moving reference frame. Figures 1(a,b,c) reveal the presence of n holes in a domain of length $n\lambda$, where λ is the wavelength of the pattern in Fig. 1(a). Thus these states only differ in the *distribution* of the three holes within the spatial period 3λ. Since the speed of these states is almost the same the frequency is also almost the same, as can be seen from the time series $h(\cdot, t)$ shown in Fig. 2. These time series moreover reveal the hole-like structure of the instantaneous profiles. In contrast, the frequency of the bound state in Fig. 2(d) is substantially lower.

Figure 3 shows similar results for $\alpha = 0.5$. The holes move faster, but the bound states seen in Fig. 1 recur, although the small amplitude background traveling waves are present already when $L = 100$. The bound states now have a more complex internal structure. For example, when $L = 150$ the three-hole bound state has a visibly larger separation between the leading and middle holes than between the middle and trailing holes. When $L = 200$ we find a bound state of five holes, again with a monotonically increasing separation between adjacent holes in the downstream direction. A different bound state of five holes is stable when $L = 250$, and bound states of seven and eight holes are found when $L = 300$ and $L = 350$ (not shown), respectively. Figure 4 confirms the trend: the figure shows a stable bound state of 15 holes for $\alpha = 0.8$ and $L = 500$, starting from an unstable P_1 state. Figure 5 shows that the transitions between the different spatially periodic structures shown in Fig. 3 as L increases occur via an interval of quasiperiodicity.

The situation is dramatically different for smaller inclinations α for which the film is closer to the horizontal film and hence closer to potential rupture. In Fig. 6 we show that this regime is characterized by the intermittent formation of deep holes. These evolve slowly from small

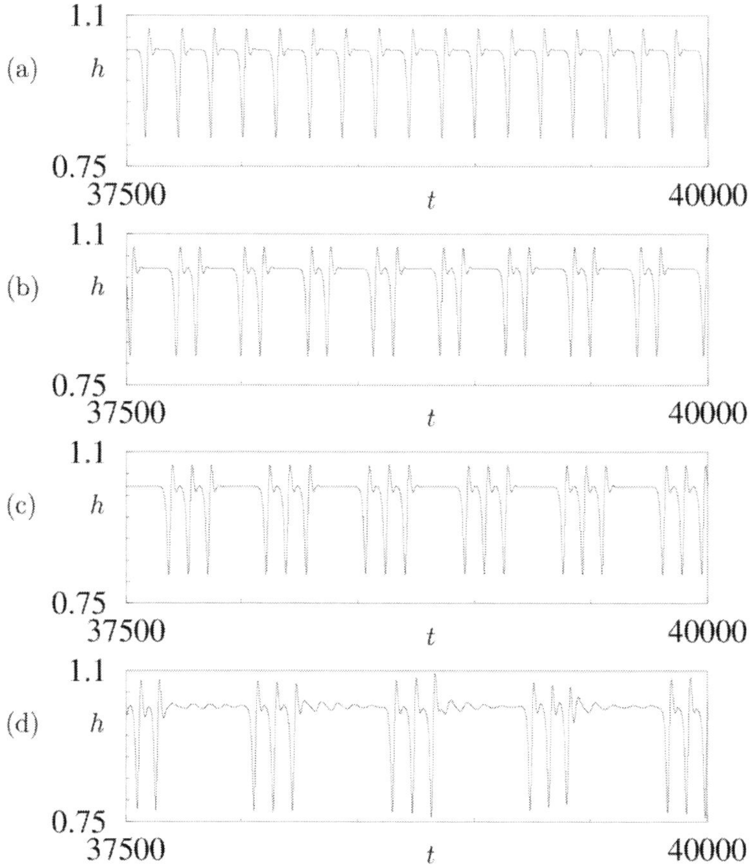

Figure 2. Time evolution of $h(\cdot, t)$ at an arbitrary point in $[0, L)$ for comparison with Fig. 1. Parameters are $\alpha = 0.4$, $\mathrm{Ma} = 3.5$, $\mathrm{Bi} = 0.5$ and (a) $L = 50$, (b) $L = 100$, (c) $L = 150$, and (d) $L = 200$. Note the presence of background traveling waves in (d).

amplitude holes, last for a brief time, before subsiding into the same small amplitude state only to regrow again. These events are best thought of as attempts at film rupture that are thwarted by the flow of the film. The figure shows that this attempted rupture process is approximately periodic albeit with a characteristic frequency that is much lower than the frequency of the small amplitude holes. However, this frequency increases significantly with increasing L, as shown in Fig. 6, and this trend extends to $L = 28$ (Fig. 7) and $L = 53$ (not shown).

Figures 8 and 9 show the details of the transition at fixed α from a small amplitude state with relatively large downstream speed and a larger amplitude state characterized by deep holes and a lower speed as Ma increases for modest spatial period ($L = 16$). Both states consist of a single hole per spatial period L. At $\mathrm{Ma} = 3.6$ the solution takes the form of a periodic train of holes. At $\mathrm{Ma} = 3.7$ this train of holes experiences periodic disruption which becomes more and more frequent as Ma increases further and leads to the establishment of slugs of large amplitude holes within the original train of small amplitude holes. By the time $\mathrm{Ma} = 3.86$ the deep holes predominate and the observed state resembles slugs of the small amplitude holes within a train of deep holes. The frequency of the appearance of the small amplitude holes decreases rapidly with increasing Ma until at $\mathrm{Ma} = 3.9$ the solution takes the form of a periodic train of deep holes; this state persists when Ma is reduced to $\mathrm{Ma} = 3.86$ indicating the presence

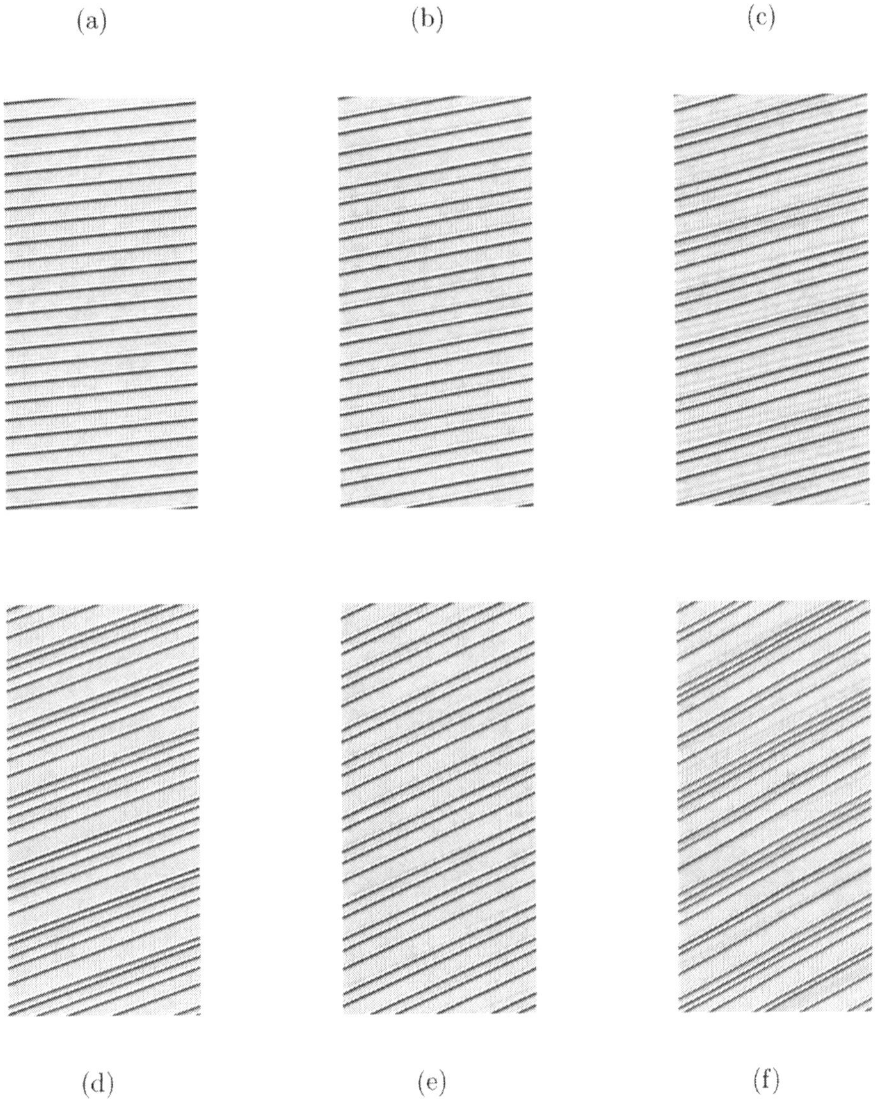

Figure 3. Isovalues of $h(x,t)$ in the (x,t) plane with $0 \leq x \leq L$ (horizontal axis) and $t \in [25000, 26250]$ (vertical axis). Parameters are $\alpha = 0.5$, Ma $= 3.5$, Bi $= 0.5$ and (a-f) $L = 50, 100, 150, 200, 250, 300$. Resolution is $N = 128$, starting from $h(x,0) = 1+0.1\sin(2\pi x/L)$.

of hysteresis. This type of behavior is another example of a quasiperiodic transition between two different spatially periodic states present because of the subcriticality of the $n = 1$ branch: these two states have quite different frequencies, with the amplitude of the small amplitude state dominating at smaller values of Ma and the amplitude of the deep hole state dominating at larger values of Ma. No frequency locking takes place in consequence of translation invariance of the system: in a frame moving with the small amplitude holes the deep hole state is a periodic wave train that has nothing to lock to, and similarly in the frame moving with the deep hole state. States of this type correspond to secondary branches connecting different parts of the $n = 1$ branch and are known to occur at other parameter values [18]. The results shown in Figs. 8 and 9 indicate that these states may be stable.

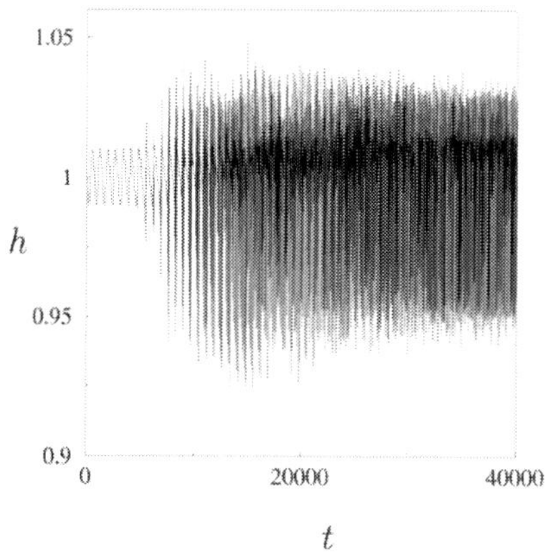

(a)

(b)

Figure 4. Instability of an $n = 1$ traveling wave when $\alpha = 0.8$, Ma $= 3.5$, Bi $= 0.5$. (a) The film height $h(\cdot, t)$ at a fixed location as a function of time. (b) Isovalues of $h(x, t)$ in the (x, t) plane with $0 \le x \le 500$ (horizontal axis) and $t \in [3800, 4000]$ (vertical axis). Resolution is $N = 256$.

4. Discussion and conclusions

In this preliminary account of our time-stepping results we have provided evidence for the stability of secondary branches that bifurcate from the primary branches P_n. There are two types of such secondary branches: branches that connect branches with different values of n (as in Fig. 5) and branches that connect different parts of the same branch (as in Fig. 9). The work of Thiele and Knobloch [18] shows that such secondary branches are common in this problem. We believe, moreover, that the bound states consisting of n holes also originate in secondary bifurcations from the branches P_n. Plots of $h(\xi)$ as a function of ξ and of $h(\xi)$ vs $h'(\xi)$ are useful for understanding the evolution of the solution profile along the secondary branches originating

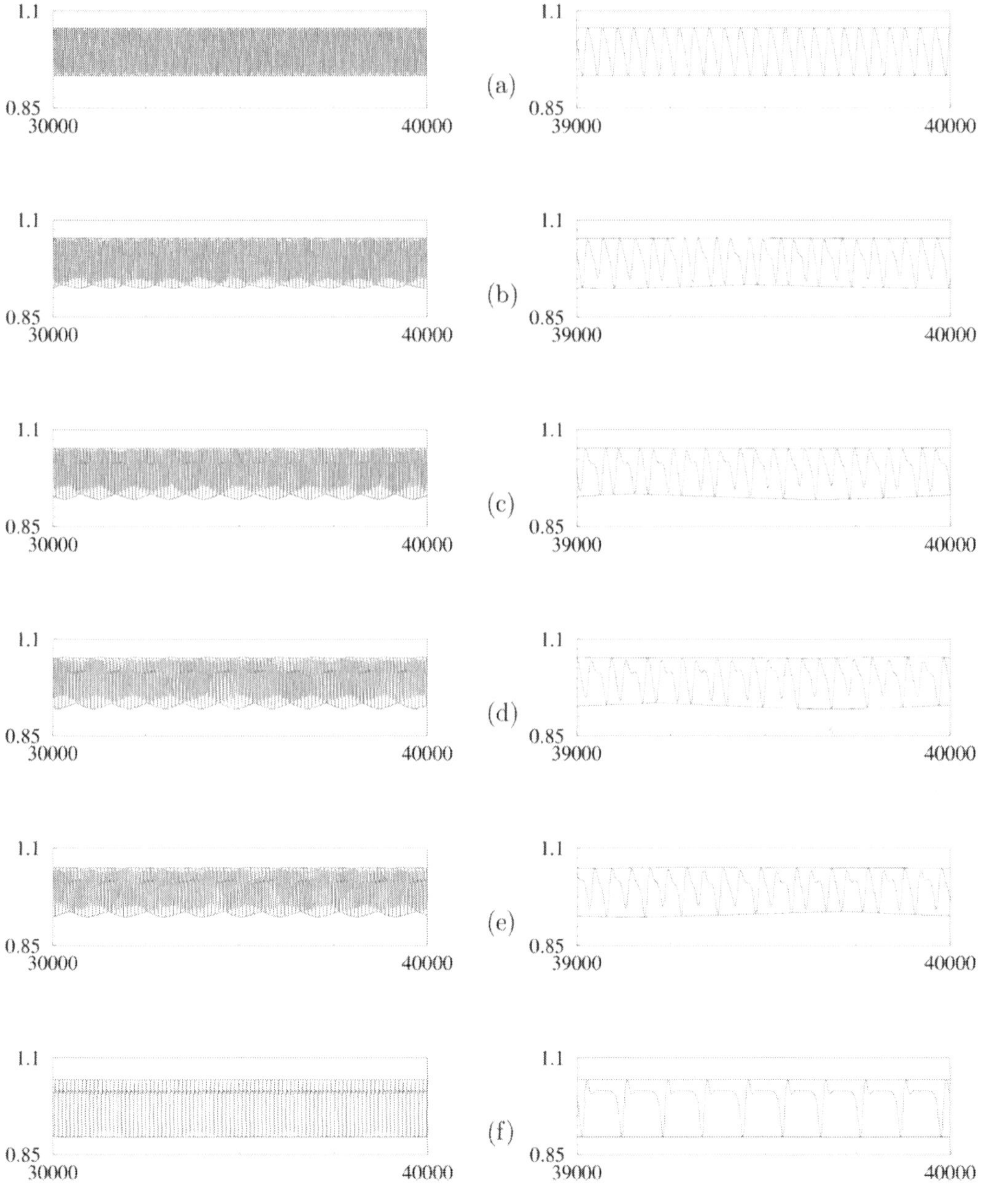

Figure 5. Quasiperiodic transition between two spatially periodic states, showing $h(\cdot, t)$ for $L = 40, 42, 44, 46, 48, 50$ from top to bottom over two different time intervals. Parameters are $\alpha = 0.5$, $\mathrm{Ma} = 3.5$, $\mathrm{Bi} = 0.5$. Resolution is $N = 128$, starting from $h(x, 0) = 1 + 0.1 \sin(2\pi x/L)$.

Journal of Physics: Conference Series **216** (2010) 012001 doi:10.1088/1742-6596/216/1/012001

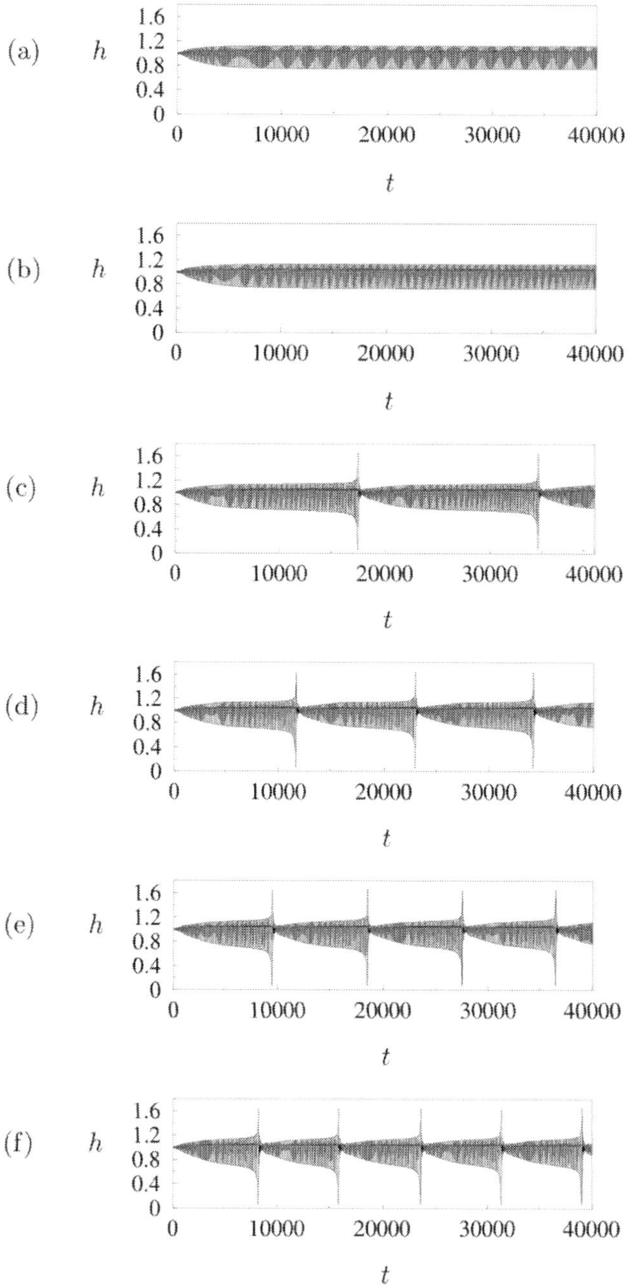

Figure 6. Time evolution of $h(\cdot, t)$ together with $\max_x(h(x,t))$ and $\min_x(h(x,t))$ for x in $[0, L)$. Parameters are $\alpha = 0.3$, Ma $= 3.5$, Bi $= 0.5$ and from top to bottom $L = 25.1, 25.2, 25.3, 25.4, 25.5, 25.6$. Resolution is $N = 128$, starting from $h(x, 0) = 1+0.1\sin(2\pi x/L)$.

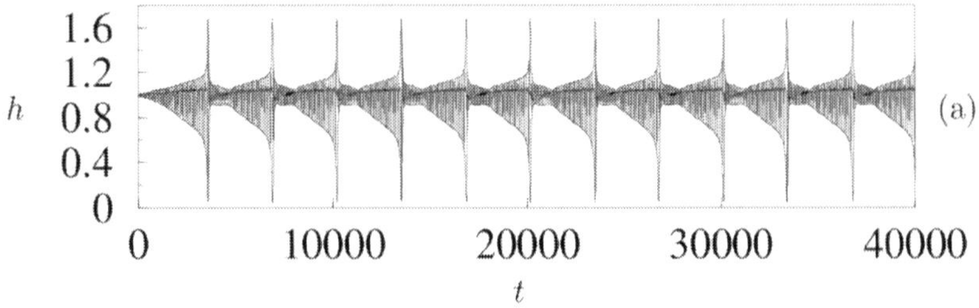

(a)

(b)

Figure 7. (a) Time evolution of $h(\cdot, t)$ together with $\max_x(h(x,t))$ and $\min_x(h(x,t))$ for x in $[0, L]$. Parameters are $\alpha = 0.3$, $\mathrm{Ma} = 3.5$, $\mathrm{Bi} = 0.5$ and $L = 28$. (b) Isovalues of $h(x,t)$ in the (x,t) plane covering approximately one period, and starting from $h(x,0) = 1 + 0.1\sin(2\pi x/L)$. Resolution is $N = 128$.

in these bifurcations. Such plots for $n = 2$ show that the primary P_2 branch consists of a pair of equidistant structures that spread apart as L increases and begin to take on the form of holes. The secondary bifurcations from P_2 lead to nonequidistant structures that start out equidistant (on P_2) but become increasingly nonequidistant as one follows the secondary branch. A typical example is provided in Fig. 17 of [18]: among five secondary branches that bifurcate from P_2 of this type three extend to large L (Figs. 17(b,c,f) of [18]) and hence differ from the branches responsible for the quasiperiodic structures described above. Of these branch (b) remains at small amplitude and takes the form of a flowing, slightly undulated film. In contrast, branches (c,f) develop into well-defined bound states of holes. Thiele and Knobloch did not compute the stability of these secondary branches although naive eigenvalue counting suggests that both of these branches can acquire stability at the final saddle-node bifurcation at which these branches turn towards large L. Moreover, phase space projections indicate that the bound

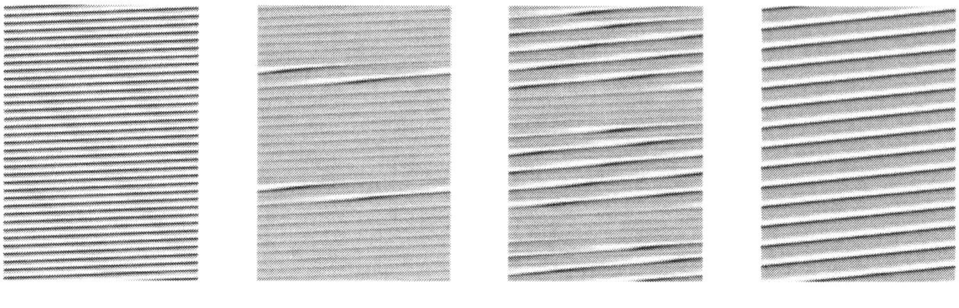

(a) (b) (c) (d)

Figure 8. Isovalues of $h(x,t)$ in the (x,t) plane with $0 \leq x \leq 16$ (horizontal axis) and $t \in [1800, 2000]$ (vertical axis) for Ma = 3.6, 3.7, 3.86 and 3.9, showing the details of the transition from the small amplitude hole state to the large amplitude hole state with increasing Ma. Parameters are $\alpha = 0.3$, Bi = 0.5. Resolution is $N = 128$.

states initially form via spatial period-doubling followed by, with increasing L, an approach to a homoclinic orbit. In the limit of large L this orbit starts from the flat state $h = 1$, executes a pair of oscillations, before returning to the flat state. This type of solution is quite distinct from the 2-pulse solutions that are associated with the Shiln'nikov bifurcation. In the latter the solution undergoes an excursion from the flat state, returns to the vicinity of the flat state, before undergoing a second excursion from the flat state and then returning to it. We believe that this new type of homoclinic orbit is associated with the $n:1$ resonance, cf. [9], since the period-doubled states are triggered by the excitation of an $n = 1$ perturbation of the primary $n = 2$ periodic oscillation. Evidently, the same mechanism is involved in generating bound states involving n holes ($n > 2$). However, since the state $h = 1$ is unstable at these parameter values these solutions *must* lose stability with increasing L (i.e., as the holes move further apart), resulting in a succession of transitions to bound states of ever increasing complexity, in qualitative correspondence with the time-stepping results presented here. The details of these transitions are not fully understood but should also involve quasiperiodic structures since the states involved are still periodic, albeit with a larger period. Similar transitions in systems with reflection symmetry typically involve intervals of chaos [3, 6]. However, there is a further possibility arising from the frequent return of the solution to the special state $h = 1$. This state plays an essential role in the states shown in Figs. 2 and 6 despite the lack of resemblence between our hole states and the solitary elevation waves computed for a related problem in Ref. [8]. A clear example is shown in Fig. 10. A possible mechanism generating complex near-homoclinic structures arising from the close approach to the $h = 1$ state in a system with the same symmetry properties as the present system is identified in [17] and may play a role here as well.

Bound states of n pulses have been observed in a reaction-diffusion system involving both two [11] and three species [20]. Both systems are excitable in the sense that finite amplitude perturbations produce a finite amplitude traveling wave (TW) resembling a spatially periodic train of steadily moving pulses. Since these systems are invariant under spatial reflection the direction of propagation is selected by the initial condition. In the regime of interest these wavetrains are unstable to perturbations with wavelength $n\lambda$, where λ is the TW wavelength and $n > 1$, that evolve on a timescale that is long compared to the time on which the initial structure forms. In systems in which the pattern-forming instability is supercritical these Eckhaus instabilities usually lead to wavelength change mediated by the formation of defects at which the wave amplitude drops to zero. However, in excitable systems the periodic pulse train coexists with a stable homogeneous state. As a result the evolution of the Eckhaus instability conserves phase and the instability leads to *clumping*, i.e., the formation of a bound state of n

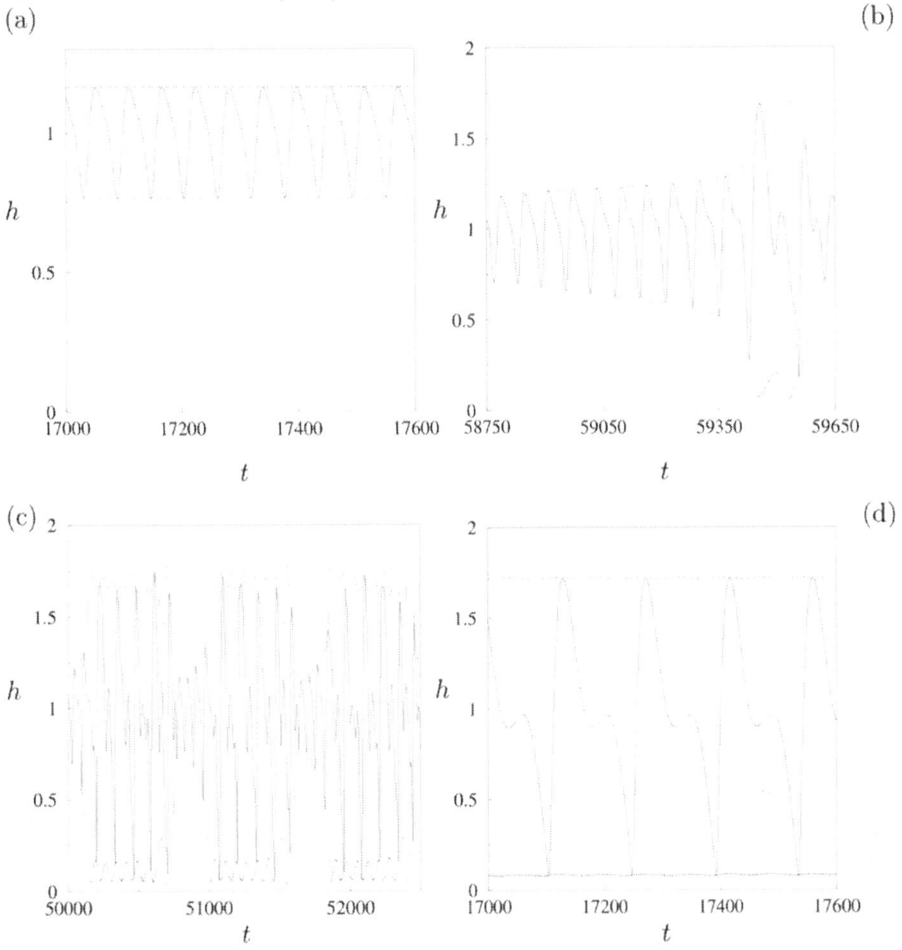

Figure 9. Time evolution of $h(\cdot, t)$ at a fixed point in $[0, L]$ with $L = 16$ and Ma = 3.6, 3.7, 3.86 and 3.9, showing the details of the transition from the small amplitude hole state to the large amplitude hole state with increasing Ma. Parameters are $\alpha = 0.3$, Bi = 0.5. Resolution is $N = 128$.

pulses whenever the domain period $L = n\lambda$. Similar bound states occur in activator-inhibitor systems in the presence of advection [21]; in these systems the advection selects the direction of drift, much as the inclination selects the direction of drift of the holes in the present system. In [11] the formation of these structures is related to the nonmonotonic behavior of the dispersion relation $c(d)$ that determines the speed c of a periodic train of pulses as a function of their separation d.

It is tempting to try to interpret the results of the present work in terms of a similar picture. However, the bound states shown in Fig. 1 and elsewhere are not found in the excitable regime for our system, i.e., the regime in which the primary instability is subcritical. On the contrary, the bound states in Fig. 1 are found for values of L much beyond the threshold for primary instability; moreover, computations in Ref. [18] do not reveal evidence for significant nonmonotonicity in the dispersion relation along the relevant secondary branches that bifurcate from P_n. The possibility remains therefore that the bound states of holes found here are created by a distinct mechanism, perhaps one that involves slow drift along a branch of traveling waves followed by a fast jump to the homogeneous state $h = 1$ and back again. This mechanism is characteristic of slow-

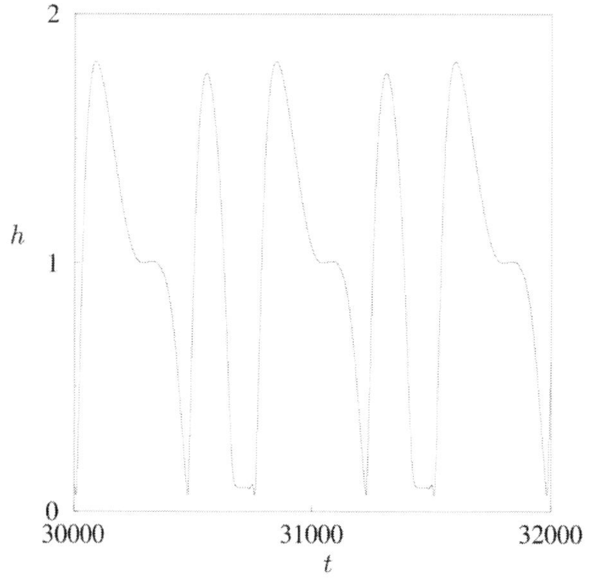

Figure 10. Time evolution of $h(\cdot, t)$ at a fixed point in $[0, L)$. Parameters are $\alpha = 0.135$, Ma $= 3.5$, Bi $= 0.5$ and $L = 42.5$. Observe also the appearance of structure arising from the close approach of the solution to $h = 0$; for $L > 42.5$ we observe rupture of the film (negative values of h), at least with the resolution $N = 256$ used here.

fast systems, and is also capable of generating finite groups of pulses; these may be arranged periodically or otherwise, much as found in [6].

Acknowledgments

This work was supported by the Berkeley-France Fund and the National Science Foundation under grants DMS-0605238 and DMS-0908102.

References

[1] Boos W and Thess A 1999 *Phys. Fluids* **11** 1484
[2] Burelbach J P, Bankoff S G and Davis S H 1988 *J. Fluid Mech.* **195** 463
[3] Coombes S and Osbaldestin A H 2000 *Phys. Rev.* E **62** 4057
[4] Craster R V and Matar O K 2009 *Rev. Mod. Phys.* **81** 1131
[5] Liu J, Paul J D and Gollub J P 1993 *J. Fluid Mech.* **250** 69
[6] Higuera M, Knobloch E and Vega J M 2005 *Physica* D **201** 83
[7] Joo S W, Davis S H and Bankoff S G 1991 *J. Fluid Mech.* **230** 117
[8] Kalliadasis S, Demekhin E A, Ruyer-Quil C and Velarde M G 2003 *J. Fluid Mech.* **492** 303
[9] Kent P and Elgin J 1992 *Nonlinearity* **5** 899
[10] Krishnamoorthy S, Ramaswamy B and Joo S W 1995 *Phys. Fluids* **7** 2291
[11] Or-Guil M, Kevrekidis I G and Bär M 2000 *Physica* D **135** 154
[12] Oron A, Davis S H and Bankoff S G 1997 *Rev. Mod. Phys.* **69** 931
[13] Oron A and Rosenau P 1992 *J. Phys. II France* **2** 131
[14] Ramaswamy B, Chippada S and Joo S W 1996 *J. Fluid Mech.* **325** 163
[15] Ruyer-Quil C and Manneville P 2000 *Eur. Phys. J.* B **15** 357
[16] Ruyer-Quil C, Scheid B, Kalliadasis S, Velarde M G and Zeytounian R Kh 2005 *J. Fluid Mech.* **538** 199
[17] Rodriguez J. D. and Schell M 1990 *Phys. Lett.* A **146** 25
[18] Thiele U and Knobloch E 2004 *Physica* D **190** 213
[19] Thiele U and Knobloch E 2006 *Phys. Rev. Lett.* **97** 204501
[20] Yochelis A, Knobloch E, Xie Y, Qu Z and Garfinkel A 2008 *Europhys. Lett.* **83** 64005
[21] Yochelis A and Sheintuch M 2009 *Phys. Chem. Chem. Phys.* **11** 9210

Journal of Physics: Conference Series **216** (2010) 012002 doi:10.1088/1742-6596/216/1/012002

Sliding fluids: Dewetting experiments reveal the solid/liquid boundary condition

Oliver Bäumchen, Matthias Lessel, Renate Fetzer, Ralf Seemann and Karin Jacobs

Dept. of Experimental Physics, Saarland University, D-66041 Saarbrücken, Germany

E-mail: k.jacobs@physik.uni-saarland.de

Abstract. Nanoscale liquid polymer films are ideal candidates to probe the solid/liquid boundary condition: Prepared on hydrophobized Si wafer, the films are not stable, they dewet. The dewetting induces a flow without applying an external force. Probing the dynamics of the dewetting film and the morphology of the liquid front, we can deduce the slip length. A variation of the type of hydrophobic coating (silane or Teflon®) of the Si wafer enables us to tune the boundary condition from a no-slip to a nearly full-slip condition. For a short introduction to the topic, we offer a phenomenological approach and supply multimedia files.

The solid/liquid boundary condition is of great importance for a theoretical description of liquids in confined geometries. Usually it is assumed, that the velocity profile of the liquid extrapolates to zero at the solid/liquid boundary. This is called 'no-slip' boundary condition. At a gas/liquid interface, a 'full-slip' boundary condition is usually applied, describing a continuity of tangential stress across a fully mobile interface. To quantify the amount of slippage, a parameter is introduced, the slip length b. It is defined as the extrapolation length of the velocity profile of the liquid 'below' the interface. Technically, the full slip condition (b infinite) or a very large b often is desirable since effects such as die swelling using extruder techniques can be reduced. Moreover, the throughput in pipes and channels can be substantially enhanced at the same applied pressure.

A number of different methods have been used to probe the boundary condition at various systems as discussed in recent review articles [1, 2, 3]. Most of them, like colloidal probe microscopy or particle imaging between two shearing plates, apply an external force to create the flow. For the studies presented here, a thin liquid film is observed as it beads off a solid surface, driven by internal forces only. We record the dewetted distance as a function of time, usually with an optical microscope. This gives us information about the dynamics of the dewetting film. Moreover, we characterize the shape of the liquid front as it retracts using atomic force microscopy (AFM). Both the dynamics and the front of the dewetting film are sensitive to the solid/liquid boundary condition.

Let us first look into how dewetting proceeds. An everyday dewetting experiment, the dewetting of honey from a buttered bread, can be performed on a breakfast table. A snapshot of a layer of honey dewetting from a buttered surface is shown in Fig. 1. A video can be found in the supplementary material. Initially, the liquid must be forced into a thin film. At the breakfast table, this is done with a knife, in the video we used a wooden stick. For a polymer melt, that blade technique is rarely used as it usually cannot generate films thinner than several

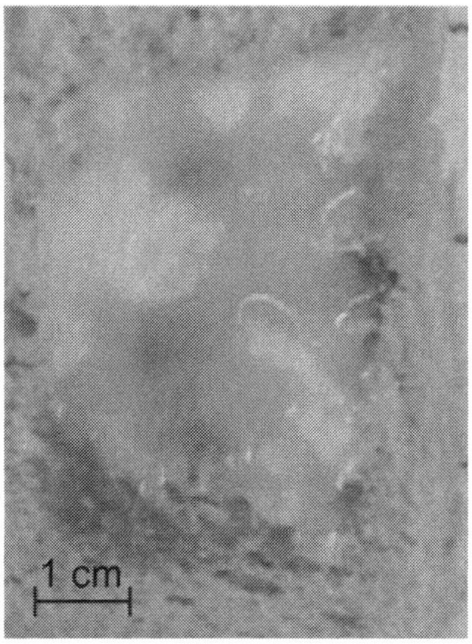

Figure 1. Honey dewetting from a buttered toast surface. A patch of honey, placed on the butterd surface, was forced into a thin film with the help of a stick. Immediately, the honey started dewetting. The movie in the supplementary material is in real time.

micrometers. For the preparation of films in the nm-range, the substrate is placed on a chuck that can spin from several hundreds up to ten thousands rounds per minute. A drop of polymer solution is then placed on the spinning sample. As the solvent evaporates, a smooth film of uniform film thickness is left behind. The film thickness depends on polymer concentration, spinning speed, acceleration, type of solvent and type of polymer and can be measured by e.g. ellipsometry.

On the breakfast table, the layer of honey immediately starts to bead off the buttered surface and dry patches ('holes') appear and grow in size. In case of a polymer instead of honey, it needs to be heated above the glass transition temperature to become liquid, because the polymer film is usually glassy as soon as the solvent has evaporated.

For honey as well as for many polymer films mass conservation is valid, at least for the usual times of observation, minutes to hours. This means that all material that was once on the surface as a flat film must stay on the surface. The material that was inside a hole is accumulated along its perimeter, forming a 'rim'. As the hole grows, also its surrounding rim increases. Dewetting ends when holes meet. Their rims merge and form a common ribbon, a liquid cylinder with a certain contact angle. Due to the Rayleigh-Plateau-instability, the ribbon decays into droplets.

Why does a film dewet? Dewetting occurs if the film can gain energy by minimizing its interfacial area with the substrate (see [4, 5] and references therein). Dewetting of thin liquid films on solid substrates is a result of internal forces, which can be inferred from the effective interface potential Φ [6, 7]. In case of several films of dielectric materials, Φ is the sum of the van der Waals potentials involving gas layers, liquid layers and substrate and a repulsive short range potential, representing the steric repulsion. In case the substrate consists of different layers, the influence of each layer on the van der Waals potential has to be taken into account [5, 7, 10]. If Φ, which is a function of film thickness h, has a global minimum at a finite film thickness, the film can dewet. In the region of Φ where the second derivative of the potential with respect to film thickness, $\Phi''(h)$ is negative, there is a driving force that promotes thinning of the film. In this case spinodal dewetting takes place [6, 11]. Most exiting feature of this case is that there is a wavelength in the capillary spectrum of the liquid film that is amplified

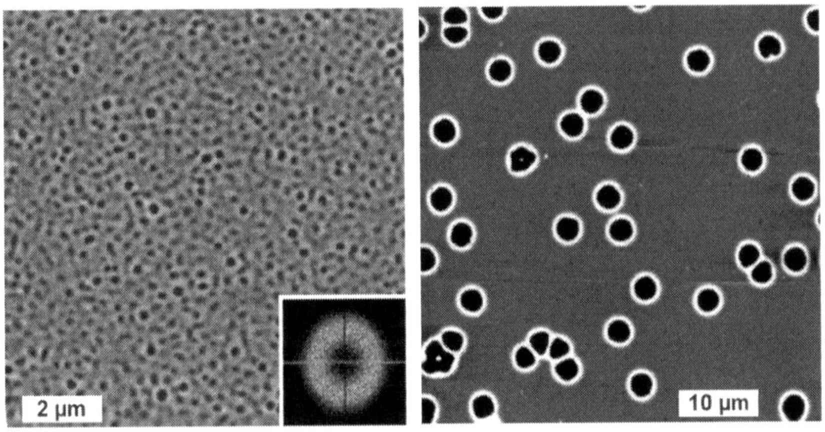

Figure 2. Left: Spinodal dewetting 4 nm thick PS(2k) film on a Si wafer with a thick (191 nm) silicon oxide layer as seen by AFM. Height scale from black to white is 10 nm. Inset: Fourier transform of the AFM image. Right: AFM scan of a 6.6 nm thick PS (2k) film on a Si wafer with a thin (2.4 nm) silicon oxide layer; height scale 20 nm. Holes are randomly distributed and therefore generated by nucleation.

fastest. This wavelength can be observed experimentally, e.g. by an atomic force microscope (AFM). Recording the spinodal wavelength as a function of liquid film thickness can give access to the effective interface potential [7, 10] in the experiment. For a 4 nm thin PS film of 2 kg/mol molecular weight (a short-chained polymer, roughly 20 monomer units) on a Si wafer covered with a thick (191 nm), amorphous silicon oxide layer, spinodal dewetting can be observed, as depicted in the left image of Fig. 2. The inset shows a Fourier transform of the AFM image. Clearly, a preferred wavelength (415 nm $\pm$ 20 nm) can be observed.

To yield information about the dynamics of spinodal dewetting, *in situ* AFM scans can be performed. In order to not disturb the liquid surface, an especially soft, so-called 'non-contact' mode of scanning is used [12]. To get a reasonable time resolution below one minute, only a small area in the μm^2 range is scanned. In Fig. 3, an AFM image of a spinodally dewetting polystyrene (PS) film is shown. The area is $(1.3\ \mu m)^2$, the height scale from black to white is 12 nm. Since the area is so small, a Fourier transform will not yield reliable results. Hence, the image has been analyzed by Minkowski measures and compared to numerical simulations [13, 15]. The surface undulations in both cases can be described by a Gaussian random field, which one theoretically would also expect for spinodal dewetting. By Minkowski measures, also the morphology of the pattern in the experiments and the simulations can be compared and have shown to match perfectly. The time scale, however, when e.g. first valleys of the undulation reach the solid surface, do not match in simulation and experiment. In the experiments, holes appear faster as compared to the simulations, yet in the simulations the coarsening of the rims and ribbons is quicker than in the experiments. The reason is not, as first suggested, a false assumption of the viscosity for the simulations, but rather the neglect of the thermal noise in them. Including this thermal noise, a perfect match of the dynamics of dewetting in experiments and simulations could be achieved [17].

If there is a global minimum of $\Phi(h)$ at finite h, but $\Phi''(h)$ is positive at the prepared film thickness h, the system needs to overcome a potential barrier to reach the minimum in $\Phi(h)$. A nucleus - whatever it may be (e.g. a chemical heterogeneity, a dust particle) - can locally change $\Phi(h)$. There, the film thins, a hole is generated and dewetting starts. If not intentionally prepared, nuclei are randomly distributed and so are the holes. An example is shown in Fig. 2, right. The analysis of the sites of holes by Minkowski measures confined a random distribution [13, 14].

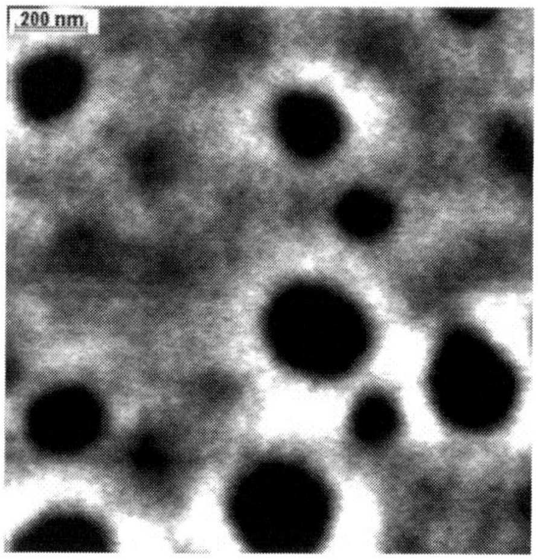

Figure 3. Snapshot of a series of AFM images of a spinodally dewetting polystyrene film (2k) with a thickness of 4 nm on a Si wafer with 191 nm silicon oxide on top. The entire series can be found in the supplementary material. Height scale from black to white is 12 nm.

In everyday life, where films of some 100 nm to some 100 μm thickness are dominant, dewetting is usually initiated by nuclei since the time for the spinodal instability to grow in amplitude scales with h^5 [16]. Spinodal dewetting can therefore only be observed in films of a few nm thickness. For the dewetting experiments evaluating the solid/liquid boundary condition, the dewetting mechanism is not important. Dewetting is used as a tool to initiate a flow without applying an external force. Dewetting - and that is the main advantage - is only driven by internal, capillary forces. Viscous dissipation within the liquid (which mainly occurs at the three-phase contact line [8]) and friction at the solid/liquid interface (slippage) counteract these driving forces. For an overview about the basic models and experimental results concerning especially sliding polymeric melts, the reader is referred to [9], where also concepts that aim to explain slippage on the molecular scale are summarized and discussed.

Fig. 4 shows a sequence of micrographs of a 130 nm thick PS(13.7k) film at T = 120 °C from two different hydrophobic surfaces. The only difference of the two films is the type of hydrophobic layer on top of the Si wafer. In Fig. 4 left, a silanized wafer was used, in Fig. 4 right, a Teflon®-coated wafer [18]. Judging from the contact angle only (which corresponds to the depth of the global minimum in $\Phi(h)$ and, thus, to the driving force), one expects the polymer melt to dewet faster from the substrate with the larger contact angle (Teflon®, 88(2)°, compared to silanized wafer, 68(2)°). The experimental results, however, contradict this expectation: A snapshot of the experiment is shown in Fig. 4. In the supplementary material, a video sequence is given which shows the dynamics of the dewetting experiment: Both films are 130 nm thick and dewet at T = 120 °C. In both cases, the time when the first tiny hole can be detected by optical microscopy is set to t = 0. Clearly, dewetting proceeds faster on the left, the silanized wafer. Concerning silanization, two different types of molecules were used: Octadecyltrichlorosilane (OTS) consisting of a backbone of 18 carbon atoms and Doedecyltrichlorosilane (DTS), with 12 carbon atoms as backbone. Both are forming self-assembled monolayers on Si wafers covered with a native SiO_2-layer, leading to water contact angle above 110 °. Both hydrophobic layers have comparable surface roughness, surface tension and water contact angle hysteresis [21]. Data for the hole radius R as function of time is shown in Fig. 5 left.

A detailed study reveals that on the silanized surfaces, slip lengths in the range of micrometers

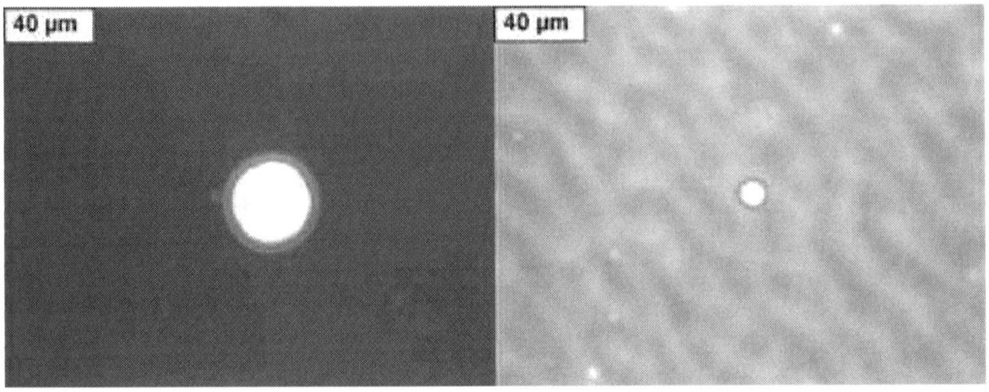

Figure 4. Dewetting of a 130 nm thick PS(13.7k) film at T = 120 °C from two different surfaces as seen by reflective light microscopy. Left: Si wafer as a substrate, hydrophobized by dodecyltrichlorosilane (DTS). Right: Teflon®-coated Si wafer as substrate. The striped pattern in the surrounding of the hole stems from slight nm-sized height differences on a lateral scale of tens of μm in the Teflon®-coating and do not affect dewetting.

can be achieved for unentangled polystyrene films. On the Teflon®-coated Si wafers, however, slip lengths below 100 nm are found [21]. The results for all three surfaces, Teflon®, OTS and DTS, are shown in Fig. 5 right. Slip lengths are determined by either recording the dewetting velocity as a function of time [22] or by analyzing the rim profile as it 'decays' into the unperturbed film [23, 21]. The largest slip lengths range up to 10 μm (at the film thickness of approx. 100 - 130 nm).

The relatively small slip lengths in case of the Teflon®-coated Si wafers are in good agreement with molecular dynamic simulations for a simple Newtonian liquid that predict slip lengths exceeding 30 times the molecular diameter of the liquid molecules [24]. For polystyrene chains like PS (13.7k) with a radius of gyration $R_G \approx 3$ nm, one would expect slip lengths up to roughly 100 nm. This raises the question of the underlying molecular mechanism of slippage. The fact that slip is substantially different on OTS-, DTS- and Teflon®-coated Si wafers is not obvious and various explanations are being discussed. One explanation is that an alignment of

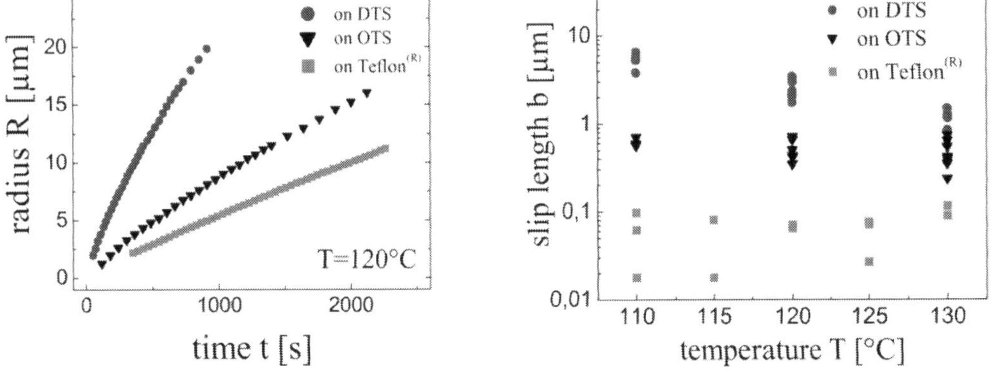

Figure 5. Left: Radius versus time of a growing hole on a 130 nm thick PS(13.7k) film at T = 120 °C from different hydrophobic substrates measured by optical microscopy. Right: Slip length plotted versus dewetting temperature on DTS-, OTS- and Teflon®-coated surface. Both graphs adapted from [21].

the silane molecules takes place during the dewetting process. Another explanation blames an interpenetration of PS-chains into the silane monolayer. Both hypotheses imply a change of the solid/liquid interface. Recently performed neutron scattering experiments indeed corroborate a diffusion of PS chains into the silane layer [25], yet the exact molecular mechanism of the high slip lengths on the silane surfaces is still unclear. A third explanation is revealed with our newest experiments: Entangled polymer melts on hydrophobic, 'no slip', Teflon®-coated Si wafers showed a reduced number of entanglements of polymer chains at the solid/liquid interface and create by that a slip effect [20], which corroborates the hypothesis of a polymer interphase with properties different from the bulk. In this case, the slip lengths characterize apparent slip.

In summary, we have provided a short overview on dewetting experiments revealing the solid/liquid boundary condition, providing a 'dynamic' insight by including multimedia files of dewetting liquids. For thin polymer films on hydrophobic substrates, viscous dissipation as well as slippage are relevant. Future experiments will focus on the molecular origin of slippage and on the characterization of the lubrication layer.

We like to thank S. Grandthyll, M. Klos and S. Howell for their help with the movies. The work is funded by the German Research Foundation DFG, by grant No. Ja905/3, Se1118/2 and by the Graduate College 1276.

References

[1] C. Neto *et al.*, Rep. Prog. Phys. **68**, 2859 (2005).

[2] E. Lauga *et al.*, in *Springer Handbook of Experimental Fluid Mechanics*, edited by C. Tropea, A.L. Yarin, and J.F. Foss (Springer, Berlin Heidelberg New York, 2007).

[3] L. Bocquet and J.-L. Barrat, Soft Matter **3**, 685 (2005).

[4] G. Reiter, Phys. Rev. Lett. **68** 75 (1992).

[5] K. Jacobs, R. Seemann, and S. Herminghaus, in *Polymer Thin Films*, edited by O.K.C. Tsui and T.P. Russell (World Scientific, New Jersey London Singapore, 2008).

[6] A. Vrij, Discuss. Faraday Soc. **42** 23 (1966).

[7] R. Seemann, S. Herminghaus and K. Jacobs, Phys. Rev. Lett. **86**, 5534 (2001).

[8] F. Brochard-Wyart *et al.*, Langmuir **10**, 1566 (1994).

[9] O. Bäumchen and K. Jacobs, J. Phys.: Condens. Matter **22**, 033102 (2010)

[10] R. Seemann, S. Herminghaus and K. Jacobs, J. Phys.: Cond. Mat. **13**, 4925 (2001).

[11] S. Herminghaus, K. Jacobs, K. Mecke, J. Bischof, A. Fery, M. Ibn-Elhaj, and S. Schlagowski, Science **282**, 916 (1998).

[12] S. Herminghaus, A. Fery and D. Reim, Ultramicroscopy **69** 211-217 (1997).

[13] K. Mecke, Int. J. Mod. Phys. B **12**, 861 (1998).

[14] K. Jacobs, K. Mecke, and S. Herminghaus, Langmuir **14**, 965 (1998).

[15] J. Becker, G. Grün, R. Seemann, H. Mantz, K. Jacobs, K. R. Mecke, R. Blossey, Nature Materials **2**, 59 (2003).

[16] M. Sferrazza *et al.*, Phys. Rev. Lett. **81**, 5173 (1998).

[17] R. Fetzer, M. Rauscher, R. Seemann, K. Jacobs, and K. Mecke, Phys. Rev. Lett. **99**, 114503 (2007).

[18] A Teflon coating on top of the silicon is achieved by directly spin-coating a solution of a prepolymer ('AF 1600') from a perfluorinated solvent, resulting in an approx. 20 nm thick, amorphous coating.

[19] R. Seemann, S. Herminghaus and K. Jacobs, Phys. Rev. Lett. **87**, 196101 (2001).

[20] O. Bäumchen, R. Fetzer and K. Jacobs, Phys. Rev. Lett. **103**, 247801 (2009).

[21] O. Bäumchen, K. Jacobs and R. Fetzer, Proceedings of 1st conference on microfluidics in Bologna 2008 Dez 10-12, ISBN 2-906831-76-X

[22] R. Fetzer and K. Jacobs, Langmuir **23**, 11617-11622 (2007).

[23] R. Fetzer, A. Münch, B. Wagner, M. Rauscher, and K. Jacobs Langmuir **23**, 10559-10566 (2007).

[24] J.-L. Barrat, L. Bocquet, Phys. Rev. Lett. **82**, 4671-4674 (1999).

[25] Phillip Gutfreund, personal communication (2009)

Journal of Physics: Conference Series **216** (2010) 012003 doi:10.1088/1742-6596/216/1/012003

Detecting Bifurcations in Numerical Simulation of Fluid Flow

A Salhi[1] and J M Brooke[2]

[1,2] School of Computer Science, Kilburn Building, University of Manchester, Oxford Road, Manchester M13 9PL, UK

E-mail: [1]salhia1@cs.man.ac.uk [2]john.brooke@manchester.ac.uk

Abstract. We present a general method for detecting bifurcations in numerical simulations of dissipative fluid flow whose solutions possess attracting sets for certain values of the parameters of the system. We describe the method as 'fast-slow' steering. The parameters governing system behaviour are slowly steered within a chosen range of interest, as the system evolves in time. The parameter changes are continuous and are 'slow' relative to the 'fast' rate of the time stepping of the system. This allows the simulated system to evolve continuously through a series of states sufficiently close to a trajectory of attractors of the system held fixed at each successive value on the trajectory. We present the method in its general form which can be applied to any suitable dynamical system. We then present a detailed unfolding of a bifurcation sequence in the numerical solutions of magnetohydrodynamic equations modelling dynamo acting in a toroidal volume. Torus-shaped discs have been observed in astrophysical contexts around collapsed objects (e.g. black holes). Our results conform to those obtained using the traditional computational approach of 'quasi-static' steering, and are also compatible with the mathematical analysis.

1. Introduction

When moving through a bifurcation point, a dynamical system may undergo a transition from one stable state to another, via a series of intermediate states which may be unstable. From a geometric viewpoint these stable states are represented by geometric structures in state space. Such structures can be simple, e.g. points or limit cycles, or more complex, such as tori or strange attractors. Currently most software for following solution branches and detecting bifurcations is designed for bifurcations involving the simpler zero or one dimensional structures in phase space. For example, continuation software can follow certain types of solutions (fixed points, periodic orbits, homo/heteroclinic connections, etc) and indicate their bifurcations, but they are generally only capable of dealing with zero or one dimensional equilibria. The use of scientific visualisation can be employed to gain a qualitative understanding of changes in geometry (and hence topology) involving attractors of higher and/or fractal dimension. Scenarios such as the Ruelle-Takens-Newhouse [1, 2] route to chaos, and symmetry breaking due to on-off intermittency [3, 4, 5], involve interesting changes in the topologies of the solutions in phase space which can be detected via visualisation. Computational steering has developed as a means of investigating the changes in behaviour of numerical simulations as key control parameters are steered. Generally, the visualisation is of the development of the solutions in three dimensions of space, for example in studying development of different structures in multi-phase fluids [6, 7]. However, we are concerned here with visualisation and steering of structures in the state space of the solutions, since the theory of dynamical systems is related to behaviour in state space. Visual methods for studying behaviour are very useful when an analytic or purely numerical approach is difficult or impossible. However, a major challenge that we face is that the state space of the solutions of partial differential is infinite dimensional, and, even when some form of discretization reduces this to a finite dimension, it is very large involving thousands or even millions of dimensions that represent the degrees of freedom of the system.

In this paper we describe a novel method for studying bifurcations and illustrate it by examining the solutions of a magnetohydrodynamic dynamo where physical effects of turbulence and rotation are parameterised, giving the control parameter of the system, and a nonlinearity is introduced which enables the existence of a variety of stable solutions which evolve as the control parameter is varied. The structure of the paper is as follows. In section 2 we present various methods for following changes of state in dynamical systems and discuss their different areas of application. In section 3 we present a novel method for studying bifurcations, which we call *fast-slow steering*. In section 4 we present the system we choose to illustrate the use of this method. In section 5 we present our results and in section 6 we present conclusions and suggestions for future work.

2. Methods for detecting bifurcations

2.1. Numerical continuation

A widely used class of tools for bifurcation analysis are based on numerical continuation [8, 9], which itself relies fundamentally on the knowledge of the model equations of the system. To detect bifurcations during the continuation of stable or unstable branches of solutions, the characteristic values of the system, such as eigenvalues or Floquet multipliers, are monitored. Bifurcations happen when these values cross certain manifolds on the complex plane. In this way, continuation software [8, 10, 11, 12] such as AUTO [9] and MATCONT [13] offer a software solution for scientists and engineers who need a systematic way for bifurcation analysis.

Continuation has been thus far used to deal with mathematical problems defined as a system of differential equations. Sometimes however, the same global behaviour can be reproduced using local microscopic rules applied to a set of nodes, agents or particles which constitute the model, such as predator-prey models approximated by cellular automata modelling techniques [14, 15]. Conclusions about the mathematical model can not be easily extended to the computational one, and *vice versa*, due to the complexity of nonlinear phenomena and the discrepancies between the modelling approaches. The loss of mathematical rigour in the computational approach is compensated for through the richer insight that this experimental framework provides, because it captures aspects which are not easily mathematically defined.

Besides its dependence on the existence of the model equations, continuation tools use a large number of numerical methods, and require an advanced understanding of numerical stability analysis and bifurcation theory. Consequently, most existing continuation software are hard to use by non-specialists. Continuation algorithms do not rely on visualisation, and as such can be used with systems of high order without the necessity of model reduction, nonetheless, they do become prohibitively expensive for such systems. Furthermore, continuation software provide close to no functionality to deal with more complex topologies, such as tori, fractal or chaotic attractors, which has to be dealt with using visual approaches.

2.2. Computational steering

There has been an increasing use of non-conventional modelling techniques, where the governing equations of the system are not represented as partial differential equations. Examples of such systems, which lack a macroscopic mathematical model, are mesoscale simulations [16, 17], cellular automata models [18] and design optimisation [19, 20].

In such cases the use of computational approaches to match parameter configurations to system behaviour, became an indispensible scientific tool. Traditionally, such maps have been constructed by *task farming*; that is by performing large numbers of simulations concurrently, each corresponding to a set of different values of the control parameters. However, task farming is not explorative, as it does not allow any intuition and is not user driven, because it relies on post-processing of the data. Consequently, using this brute-force approach may lead to wasted computational power, and inefficient use of valuable resources. Many investigative scenarios involve a process of exploration of the behaviour of solutions which can be made more efficient if human know-how is utilised. Task farming methods have been used in various domains, for example in identifying interesting features in large-scale molecular dynamics simulations [21] or design steering and optimisation [22].

An alternative to task farming is the exploration of parameter space by a human agent, who intuitively *steers* the parameter values towards regions where the solutions display some desired behaviour, or simply for exploration. Such methods are termed computational steering [23, 24]. Like task farming, steering can be used to build maps between points in parameter space and the corresponding dynamical behaviour in solution space [25]. However, steering utilises *on-line* visualisation as a basis for user interaction and steering decisions support [26, 27, 28]. Steering proved well suited to high-order systems where theory alone falls short of describing the system behaviour, and systems which lack a macroscopic mathematical model, such as mesoscale simulations [16, 17], cellular automata models [18] and design optimisation [19, 20].

Steering starts with a point in state space (often, but not necessarily, a point on a stable attractor), and an initial direction in parameter space. A subsequent simulation is started at an adjacent point in parameter space, using the previous convergent solution for initialisation, until eventually a numerically unstable regime (or a regime with a qualitatively different behaviour) is found. Successive simulations can then be launched to home in on the location of the boundary between stable and unstable behaviour, using an interval bisection algorithm. This method provides an advantage over task farming, as it allows an interactive investigation of the system by allowing choice of the direction in parameter space. However, this conventional way of performing parameter steering — that we shall term 'quasi-static' steering — still suffers from relying on multiple simulation runs or, at least, multiple isolated parameter evaluations and the corresponding induced transient delays. This is because quasi-static steering does not take advantage of system continuability with respect to the parameter vector at hyperbolic neighbourhoods, because it deals with solutions whose asymptotic behaviour approximates the actual attractors, but which are nevertheless different by definition (they are not subject to continuation). Consequently, the adoption of computational steering for stability analysis and bifurcation detection has been limited. Beside the limited application of bifurcation theory outside the realm of mathematical models, which confines the problems of interest to those amenable to continuation, this meagre uptake of computational steering has been partly due to the lack of a sound mathematical basis for its use as a stability analysis tool. It is for this reason that numerical continuation is more widely deployed in the study of bifurcations of dynamical systems, even if this limits the types of dynamical systems that may be studied.

2.3. Topology-based visualisation

In steering, as a human agent must be involved, the representation of the data in the solutions is of paramount importance. Unlike with a machine, humans get overwhelmed with large amounts of data, therefore the data must be presented in a form which emphasises the important aspects. One such aspect is the evolution of structural stability which can be conveyed using *topology tracking* techniques. The topology of a system can be seen via a visual representation, for example as a change of topology of an attractor from a limit cycle (one dimension) to a torus (at least two dimensions). The roots of this approach can be traced back to the works of Smale

[29, 30] who introduced concepts of topological equivalence and their relation to structural stability in dynamical systems, and inaugurated the application of topology to the study of dynamical systems. Topology-based visualisation for tracking singularities in vector fields was practically initiated by Helman and Hesselink 1989 [31]. The approach of topology tracking is now used in a number of fields, most notably in fluid flow studies.

Current topology-based techniques are, again, based on analytical tools, such as space interpolation [32, 33], time interpolation [34, 35], streamline and pathline integration [35], or Poincaré-Hopf index tracking [33]. Furthermore, many are designed to deal with data sets composed of consecutive discrete snapshots of the whole vector field, and in particular *two-dimensional* vector fields. In these low-order fields, the time development can be tracked in a third dimension. The vector field is computed for discrete steps, and then reduced into topological skeletons. Interpolation techniques are subsequently used to deduce the topological changes that took place in the missing time intervals. The more challenging problem, at least in terms of visualisation, lies in developing such methods for higher-order topologies, whether for the vector field as a whole or for individual attractors.

3. Fast-slow steering

To reconcile the experimental approach of computational steering with the mathematical theory of dynamical systems, when dealing with simulations which exhibit rich nonlinear behaviour, extensive analysis has to be carried out, to match results to theory. However, for certain simulated models, which exhibit behaviour characteristic of dynamical systems defined as differential equations, such as continuability and rules governing structural stability, conventional quasi-static steering can be augmented to provide an in-line analysis tool by taking this into consideration. By exploiting the concept of parameter continuation, steering can further distinguish itself from the traditional approach of task farming. In this paper we introduce fast-slow steering as a form of parameter continuation, where parameter updates take place as small steps, while the system evolves in time. Slow time functions can be used to guide and automate the parameter stepping in control space, while the system state is recorded for post-processing, or transformed into some visual output for online monitoring. Evolving the system in this manner produces solutions which do not represent a single isolated fixed configuration, but which travel through a series of consecutive configurations. It is demonstrated that, due to the slow evolution of the parameter vector trajectory, these solutions represent the continuation of the consecutive *frozen* configurations along the created parameter trajectory. This way, fast-slow steering makes the investigated systems natural candidates for applying topology-based visualisation. Tracking dynamical behaviour would then amount to tracking the evolution of the system in a suitable observation space, which can be chosen to geometrically convey and emphasise the topological information about the system. The novelty in this approach lies in incorporating a combination of aspects from steering, continuation, topology-based visualisation, and system control, and as such, it provides a powerful alternative to these traditional approaches for studying the dynamical behaviour of a simulated model. Another major distinction of this method, is that it does not rely on the availability of a mathematical model defined as a system of differential equations, nor does it use the analytical information derived thereof, and thus can be applied to a wider range of problems in computational science.

For the purpose of validating fast-slow steering, we demonstrate its effectiveness for a high-order dissipative system (described in section 4). Such systems have, what can be informally described as 'well-behaved' dynamics, as they have a set number of attractors to which all neighbouring solutions converge. Furthermore, the system chosen should preferably have basins of attraction which divide the parameter space 'nicely', rather than being locally intermingled basins of co-existing attractors, such as in the case of the Hénon map [36, 37]. The system will then have solutions which, when initialised in a particular basin of attraction, would exhibit an easily discernible, deterministic succession of dynamical behaviour in a range of parameter values. This produces results which can be easily validated, through using analytical tools, or using quasi-static steering for some sample parameter evaluations.

4. A bifurcation sequence in the solutions of a mean field dynamo

4.1. Physical processes to be modelled

The study of the process by which the magnetic field is regenerated and sustained within a conductive body through its motion is called dynamo theory. This sustainability of magnetic fields depends on the presence of a highly conducting fluid, such as the earth's molten iron or the ionised gas of the sun, where the flow of these fluids in the interior of the object sustains the magnetic field. Earth's magnetic field for example, is due to the convection of liquid iron within the outer core, which is in turn due to earth's rotation that organises the flow in rolls around the north-south polar axis. Electric currents are induced, which again produces a magnetic field that reinforces the original, and a dynamo that sustains itself is created. Such dynamo action is also believed to be responsible for the large scale fields in stars, including the sun.

We illustrate the method of fast-slow steering by investigating a model derived from mean field dynamo theory, in which the key processes driving the dynamo appear as parameters in the model. We restrict the spatial dimensions from three to two by assuming axisymmetry of the system. The control parameters represent processes on much smaller length scale compared with the mean field. More realistic equations and parameterisations can be derived by combining this approach with direct numerical simulation of the turbulent flow. Our model is solved with the electrically conducting fluid confined to torus surrounded by a vacuum, which is known to exhibit a very rich sequence of bifurcations, ending in chaotic behaviour within a range of parameter values close to the onset of dynamo action. Such bifurcation sequences were observed in dynamos operating in spheres and spherical shells but at much more supercritical parameter values.

The method of 'fast-slow' steering to unfold sequences of bifurcations has not been previously applied to solutions of partial differential equations, therefore it is sensible to begin with a test case whose solutions are known but which is rich enough in its behaviour to demonstrate the full power of the method. The model equations are much simpler than those being used in current dynamo theory which are fully three dimensional and which have much more sophisticated representations of the interaction of the turbulent fluid motions and the growing magnetic field. We use these equations because they are sufficiently rich to allow us to demonstrate the exploration of nonlinear behaviour in fluids, whilst being simple enough to explore a sufficiently large region of parameter space in a computationally feasible time.

4.2. The model equations

Our model is represented by the following equation which has been nondimensionalised by an appropriate choice of length and time scales so that there is a single dimensionless parameter α:

$$\frac{\partial \mathbf{B}}{\partial t} = \nabla \times (\mathbf{u} \times \mathbf{B} + \alpha f(\theta, |\mathbf{B}|)\mathbf{B} - \nabla \times \mathbf{B}). \tag{1}$$

- $\mathbf{B}(\mathbf{s}, \theta)$ is the magnetic field vector.
- $\mathbf{u}(\mathbf{s}, \theta)$ is the fluid velocity.
- α is the dimensionless parameter representing the effect of turbulence and rotation as a term for field generation. This effect is necessary for a dynamo to operate.
- $f(\theta, |\mathbf{B}|)$ is a scalar function that represents both the antisymmetry of the parameterisation of the turbulence in the rotating torus and also the back reaction of the magnetic field to quench the turbulence hence to give the nonlinear feedback. In the results presented here $f(\theta, |\mathbf{B}|) = \sin(\theta)/(1 + |\mathbf{B}|^2)$. This is just about the simplest form for these effects.

Further details of the model and the exact mathematical representation of the differential operators in toroidal geometry is described in [38, 39]. The expressions are given in terms of a set of coordinates s, θ, ϕ where s is a coordinate that determines (but is not equivalent to) the minor radius of the torus. Surfaces of constant s represent tori when the other coordinates range through their full set of values. θ represents the angle around the torus starting from the equator, thus $\theta = 0, \theta = \pi$ represent the points where the cross section of the torus cuts the equator. ϕ is the azimuthal angle along the major axis of the torus, since we assume the system is axisymmetric about the central axis of the torus, this coordinate does not appear in the equations and is therefore ignored.

We choose α as the parameter which will be slowly varied as the model (1) is solved via an explicit time-stepping scheme (Dufort-Frankel). Thus, α changes its value at each time step but sufficiently slowly that the system appears to evolve through the trajectory of attractors. Bifurcations along this trajectory are detected as major changes in the topology of the attractor. In accordance to the language of dynamical systems theory we call a solution where the fields are time invariant a fixed point solution (often called a steady dynamo). A single period solution is a limit cycle. A doubly periodic solution is called a solution on a torus being quasi-periodic if the two periods are incommensurate.

4.3. Projection method

(a) (b)

Figure 1. Poincaré section with control and sampling points, 1(a) one hemisphere only, 1(b) north and south hemipheres (symmetrical sampling).

When we solve equation (1) via discretization of time and space we represent the solutions as values at points on a grid, essentially a matrix. In the current case we are using a finite difference scheme for discretization with a grid of typically 100 values for s and 400 values for θ. This gives a state space of order 10^5 dimensions. To visualise we must find a method of projecting from this space onto something that can be visualised, i.e. using three spatial dimensions and some other visual clues such as colour for extra dimensions. We now describe this method.

We sample the solutions as they progress in time via the Poincaré map. This is obtained by choosing a suitable hyperplane in phase space and plotting the points at which the orbit intersects this hyperplane. To provide graphical insight this hyperplane is projected onto two-dimensional plane. The hyperplane is better chosen not tangential or near tangential to the solution, as it might give results that are difficult to interpret. We do this via a physically motivated method. The dynamo solutions are known to propagate as waves travelling clockwise in the northern and anticlockwise in the southern hemisphere. We choose a small number of points on the solution grid as in figure 1. The Poincaré section is produced by monitoring values of the field at a two chosen control points A and C to check when these values pass through a zero in the north and south hemisphere. Each zero represents the passing of the wave, since they are alternately opposite in magnetic sign. Strictly, for a Poincaré section one should only record crossings in one direction. Here we record both, negative to positive or *vice versa*. This is essentially to aid the visualisation as will be shown later, we treat positive and negative regions of the field as individual waves whereas a strict interpretation would be that they represent together a single wave. To project the Poincaré section to a subspace low enough for visualisation we only record data values at two data points, B and D, which are recorded as tuples of three elements. Since the magnetic field $\mathbf{B}$ is a vector in axisymmetric geometry it has two components. The first is the poloidal component, the magnitude of the field in the meridional plane, the second is the toroidal representing the field in the azimuthal direction. The third value is the time recorded when the values of the field at A (for B) and C (for D) are zero. Note that because the zero crossing was monitored in both directions, the attractors obtained are symmetric with respect to the origin whenever symmetry is preserved. Therefore, we observe that whenever there is a periodic solution, the period is an even number, whenever the solution is doubly periodic (torus) the two loops are commensurate, etc. It is important to note that both control and data points are symmetric with respect to the equator. Just after the value of α at which the dynamo waves start to appear, solutions are symmetric about the equator, this continues until symmetry is broken in a bifurcation. Our choice of points allows our visualisation to be sensitive to this symmetry breaking effect.

5. Results using α as the steering parameter

5.1. Bifurcation sequence

For the investigated model, fast-slow steering proved a powerful method for investigating the way bifurcations unfold as the parameter passes through the critical values. The parameter α was steered in the interval [2,6] where most of the dynamics of the system related to this parameter are observed. Figure 2 represents the time development of the toroidal component of the Poincaré projection. The figure conveys the bifurcation sequence, through which the solution travels as the parameter α is steered in the range of interest for ascending then descending values. One immediately interesting result is the existence of hysteresis; the bifurcation sequence is different in the ascending and descending ranges of α. The area where the red (north hemisphere) data points are observed, indicates the presence of symmetry-breaking and chaos.

Periodic Behaviour (Limit Cycles and Tori): The solution starts as a limit cycle represented by a one dimensional loop, and we observe the evolution of two point solutions representing the intersection of its orbit with the applied hyperplane. This limit cycle develops into a torus, or a two dimensional loop, through a Neimark-Sacker bifurcation (equivalent to a secondary Hopf in state space). The intersection of the torus with the Poincaré projection corresponds to a commensurate pair of closed invariant curves. This bifurcation is observed better for decreasing α values, as for the latter case, there is no hidden unstable solutions — i.e. the limit cycle in the onset of the torus behaviour — because they become stable at the bifurcation point. On the other hand, when the parameter is being increased, the solution keeps following the still existing but now unstable limit cycle, before it suddenly jumps to the emerging torus attractor.

It is possible for periodic orbits to emerge on a torus via saddle-node bifurcations, as the parameter crosses the Arnold tongues bounded by them. This is how we observe the temporary disappearance of the growing torus, and its replacement by a limit cycle, manifested by 2×3 branches on the Poincaré map. Saddle-node bifurcations are dual by nature, as they create a saddle and a node. The steered solution follows the node behaviour as it is attracting. It is

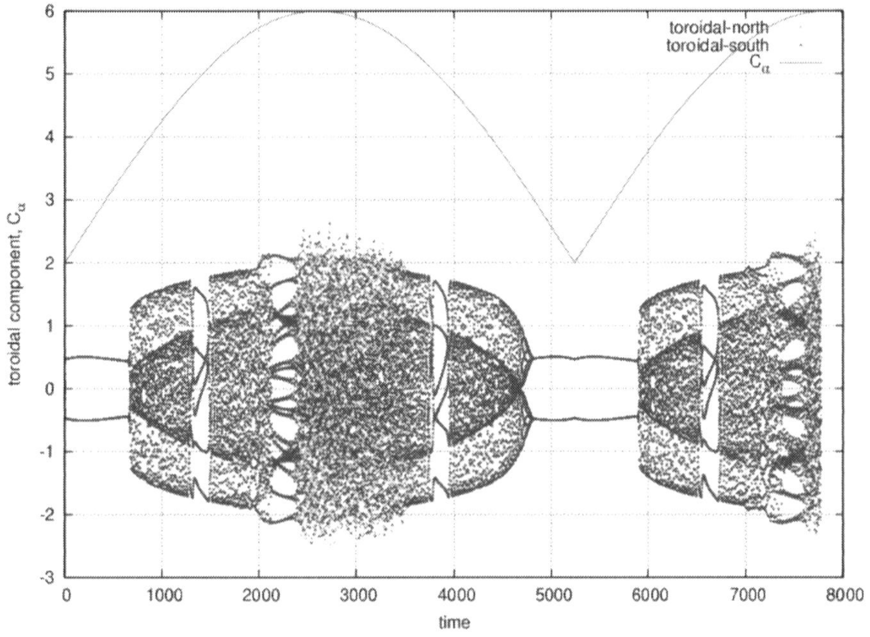

Figure 2. Toroidal component time series for $\alpha \in [2-6]$ using $\alpha = \mid 4sin(\epsilon t) \mid +2$ and $\epsilon = 10^{-6}$ (red curve). The north data points are in red and the south in blue. Where the solutions are symmetric the points coincide and only the south values can be displayed. The symmetry-breaking can be clearly seen in the chaos region where the red (north) data points do not coincide with the blue (south) ones. A hysteresis cycle can also be seen.

observed that the saddle and the node switch roles depending on the direction from which the solution approaches them, i.e. depending on the steering direction. This takes place as part of the global hysteresis taking place throughout the investigated parameter range. Therefore, in actual fact, there is another hidden 2×3 periodic solution in this α range. This fact is backed by two observations. First, the 2×3 node limit cycle disappears for decreasing α values, as the steered solution follows the previously hidden saddle limit cycle, which is now attracting. Second, the examination of the torus attractor near the saddle-node bifurcation, shows that it is denser around 12 points (figure 7(b)), which means that the field is approaching the transition to *two* 2×3 periodic solutions, rather than just one. In both cases, the observed attracting limit cycle is the image of the hidden unstable one by symmetry with respect to the origin. Therefore, despite the 'apparent' breaking of origin symmetry (each of the two observed limit cycles do not show origin symmetry), this symmetry is still present in the set of all existing solutions. North-South Symmetries are discussed in section 5.3.

Chaos and 11 fold Periodic Windows: After the recovery of the torus behaviour, we observe the breaking of the torus, as the solution enters the chaotic regime. There is however, a window of periodic solutions, characterised by 11 fold branches (the case shown in the figure shows 4×11 branches), which appears shortly after the emergence of chaotic behaviour. For other configurations of this model, various 11 fold branches were observed, (e.g. a $3 \times 5 \times 11$ periodic solution is reported in [38]). These orbits can display very delicate structure in phase space. This window of limit cycles does not appear for the decreasing α values, due to the previously mentioned global hysteresis.

5.2. Bifurcation tracking in state space

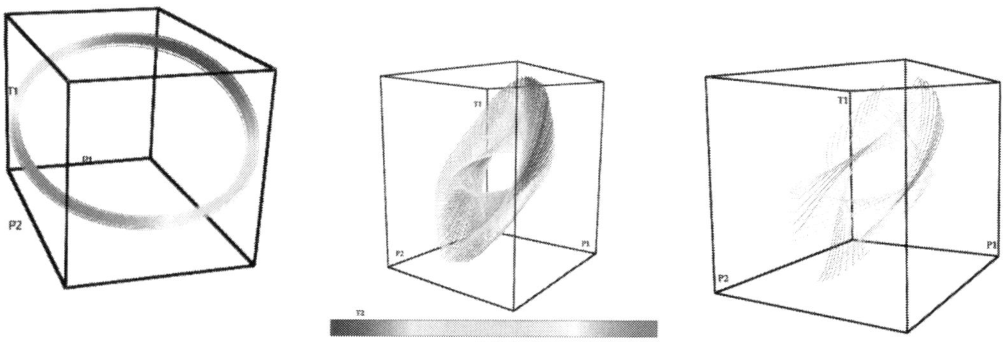

(a) $\alpha \in (2.0, 3.5)$: One-dimensional limit cycle. The loop looks rather like a spiral (growing diameter), due to the continuous steering.

(b) $\alpha \in (3.5, 4.8)$: Solution evolving on a torus topology, this solution is 1M pts in size.

(c) subset of 3(b): Building block pattern of the torus solution, a subset of the previous attractor (100K pts).

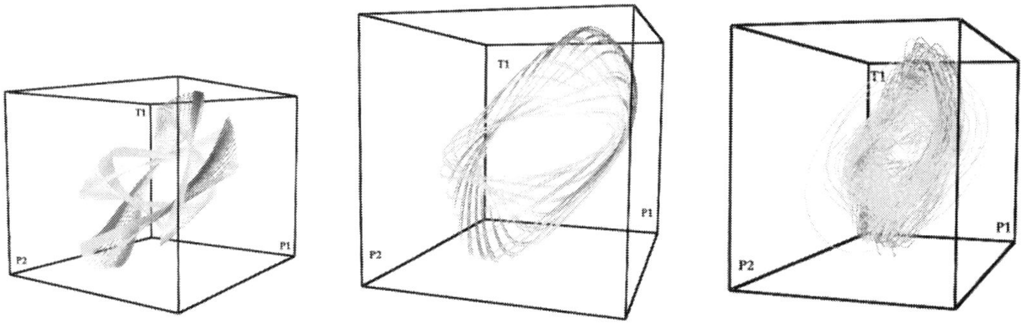

(d) $\alpha \in (4.8, 5.0)$: 2×3 periodic solution forming on top of the torus.

(e) $\alpha \in (5.8, 5.9)$: 4×11 periodic window inside the chaos onset. This solution retains a torus topology.

(f) $\alpha \in (5.9, 6.0)$: Chaotic attractor. The ghost of a torus can still be observed.

Figure 3. We are using *four* data points, two toroidal values and two poloidal. We are visualising the data in a four-dimensional projection of phase space. We used colour to show the second toroidal value. These solutions correspond to the evolution of the system as the parameter is steered in the indicated ranges.

To actually visualise what is happening to the topology of solutions as the system travels through the described bifurcations, the Poincaré section, which served as a dimension reduction technique to show the continuation of the solutions, was removed. The recorded toroidal and poloidal components of the two data points are coupled in a four-dimensional space, with the fourth dimension represented as colour. This enables a more detailed topology tracking, and provides added information about the global dynamical behaviour of the system, which cannot be induced from tracking a single data point. The generic choice of the pair of sampling points allows us to draw conclusions about the global dynamics in the whole spacial domain. In other words, this enables us to investigate whether the dynamics of the system at each grid point are consistent for a given parameter configuration. Results showed that the data points are indeed dynamically synchronous (as will be demonstrated in section 5.3 the grid points exhibit spacio-temporal synchronisation as well).

Figure 3(a) shows the first observed attractor; a limit cycle. Due to the continuous steering of the parameter α in the range [2.0-3.5], the cycle is observed to be slowly growing in diameter as a spiral. For a static configuration in this range, a single line loop that closes on itself is observed. Figure 3(b) shows the appearance of a torus via a supercritical secondary Hopf bifurcation at $\alpha \simeq 3.5$. This behaviour persists through the range $\alpha \in [3.5 - 4.8]$. Two thin edges, where the folding of the torus takes place, are also observed. Figure 3(c) represents the building pattern of the torus. This pattern is obtained by taking a subset of the evolution of the solution in the range $\alpha \in [3.5 - 4.8]$, which is 100K points in size instead of the 1M points sampled to demonstrate the torus behaviour in the whole range. The torus is built by the slow rotation of this pattern. Despite the apparent overall origin symmetry in the torus, the solution develops this symmetry as this pattern rotates. Theoretically, the full symmetry is revealed when this block pattern is dense on the torus. Unfortunately, this requires an infinite integration time. Figure 3(d) represents the saddle-node bifurcation due to the crossing of an Arnold tongue of toroidal/poloidal synchronisation. This bifurcation leads to phase locking and, consequently, the emergence of a limit cycle on the torus. As mentioned before, the loss of symmetry here is only 'apparent', because of the existence of another hidden unstable solution, which becomes stable and hence detectable for the opposite steering direction. The difference between this limit cycle and the pattern in figure 3(c) is a denser edge on the stripe, where the limit cycle is reached. Figures 3(f) (1M pts) represents the chaotic attractor over the range $\alpha \in [5.9 - 6.0]$. The shape of the torus (ghost torus) can still be seen but now the trajectories do not close and we have sensitive dependence on initial conditions. Figure 3(e) (300K pts) illustrates the 4×11 periodic window discussed above. It can be observed that this periodic solution develops on a topology equivalent to the previously broken torus.

It is worth emphasising that to obtain these numerical approximations of the attractors, the solution behaviour was tracked for ranges of parameter values, rather than monitoring the evolution of static configurations of the model. This was possible thanks to the topological equivalence of successive attractors in ranges where there are no stability transitions. Another interesting result, for this particular model, is the fact that the geometrical properties of the attractors were also preserved throughout the explored range, which made easy the task of separating them into sub-solutions of distinct behaviour. This variant behaviour however, developed around a unified topological skeleton; the torus.

5.3. North-south symmetries:

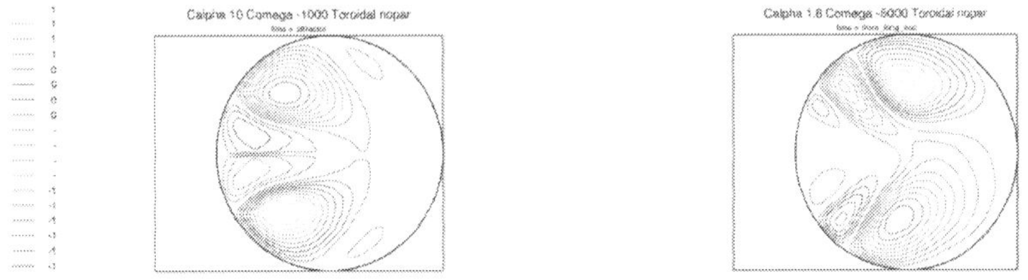

Figure 4. The solutions are travelling waves that migrate from the side nearest the rotation axis to the opposite side. In the case of (anti)symmetric solutions the files are (−) equal about the equator. The solution on the left is symmetrical, while the one on the right is taken after a bifurcation which breaks north-south symmetry.

Figure 5(a) represents solution behaviour throughout the range $\alpha \in [2.0 - 6.0]$. The torus, acting as a topological skeleton, can be seen within the chaotic attractor and within the torus

Journal of Physics: Conference Series **216** (2010) 012003 doi:10.1088/1742-6596/216/1/012003

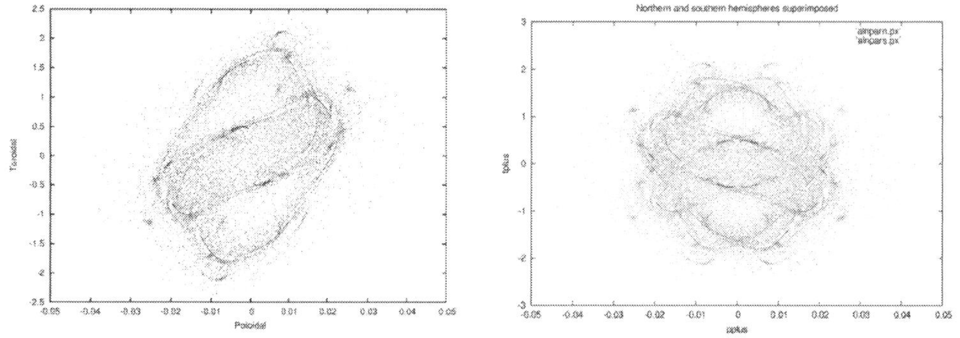

(a) Phase space projection. Attracting solutions for the northern hemisphere in the range $\alpha \in (2.0 - 6.0)$

(b) Northern (red) and southern (blue) attractors super-imposed. Data points chosen symmetrical w.r.t the equator.

Figure 5. Attractors superimposed

regions of higher density. Regions of higher density outside the torus correspond to the 4×11 periodic window, which evolve on a similar but 'bigger' torus. An overall origin symmetry is also conveyed. Figure 5(b) is the result of superimposing the northern and southern attractors. For this figure, the sampling points were chosen symmetrical with respect to the equator. Solutions are demonstrated to exhibit an equator anti-symmetry, i.e. data points in the northern hemisphere, with the same toroidal component as their counterpart in the southern hemisphere, have a sign-inversed poloidal component.

The use of fast-slow steering, in which parameters are coupled with time, enabled us to investigate the developement of these spacial symmetries in more detail by looking at the time series of the solution. Comparison between the two hemispheres time series data, revealed a temporal relationship between the two solutions. The component signals exhibited a constant phase difference between north and south. In particular, the toroidal components were always in phase, and the poloidal ones had a phase difference of $1\ \pi$. This phase coupling persisted through the whole explored steering interval, even in the chaotic region where local symmetries are broken. Amplitudes match in the periodic regions resulting in an exact symmetry (figures 6(a) and 6(b)), and differ in the chaotic region where the local symmetries are broken (figures 6(c) and 6(d)), and only a global symmetry is prevalent. Consequently, solutions in the chaotic range do not show spatial symmetry about the equator at all times. However hemisphere coupling is maintained in phase. This is an interesting result, because even if these symmetries existed at discrete parameter values (e.g. using quasi-static steering), their persistence as the system evolves through a range of parameter values cannot be easily induced, in particular, if that range contains chaotic attractors. Through allowing attractor evolution, fast-slow steering allows the separation of the spacial symmetries into their one-dimensional time components, and hence a simpler way of tracking symmetries and the mechanisms governing symmetry-breaking. This might not be of much benefit when the phase space is of low dimension, however, when phase space is four-dimensional or higher, visualising the spacial symmetries can be more of a challenge. Figures 6(e) and 6(f) convey the correlation between solutions of the two hemispheres in a more compact way. For the toroidal component, the data points representing the north-south correlation function, lie on the line $x = y$ (periodic range) or scattered around it (chaotic range). Similarly, for the poloidal data, the correlation function fits on the curve $x = -y$ due to the phase difference of $1\ \pi$.

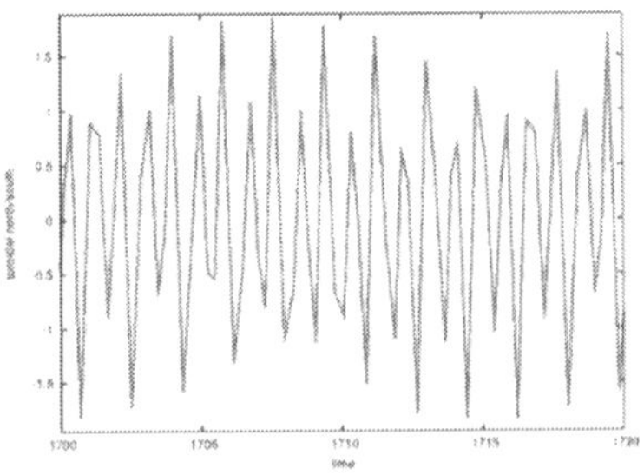

(a) North/south toroidal time series in periodic range: phase and amplitude synchronised, solution has spatial symmetry at all times (complete synchronisation).

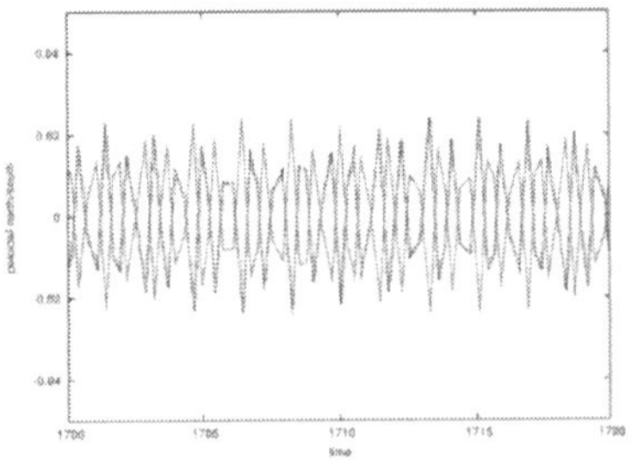

(b) North/south poloidal time series in periodic range: constant phase difference of 1 π, and matching amplitudes (complete synchronisation).

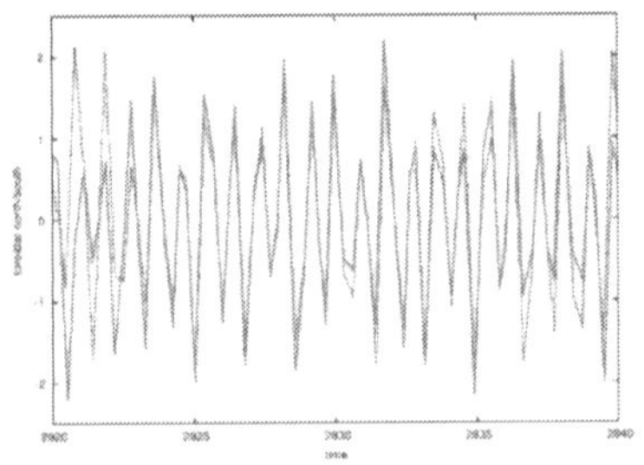

(c) North/south toroidal time series in chaotic range: synchronised in phase but not in amplitude (weak synchronisation).

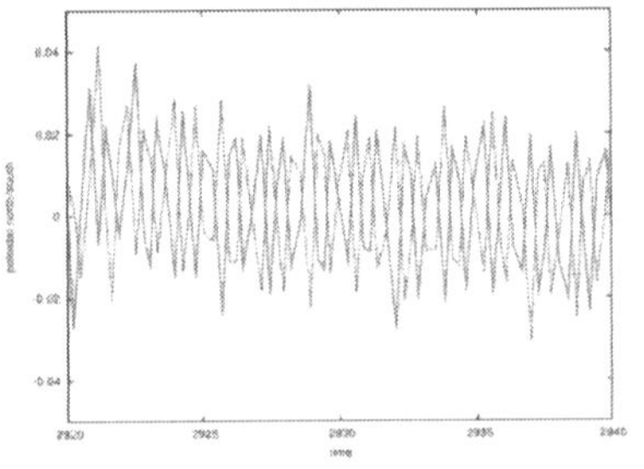

(d) North/south poloidal time series in chaotic range: constant phase difference of 1 π, and mismatching amplitudes (weak synchronisation).

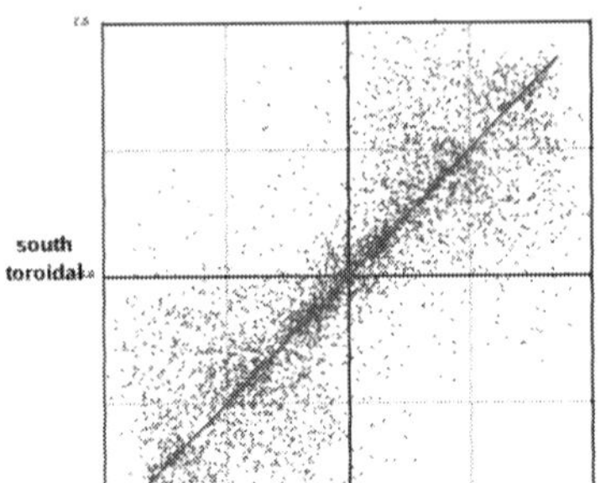

(e) Toroidal: complete(red) / phase-only(blue) synchronisation.

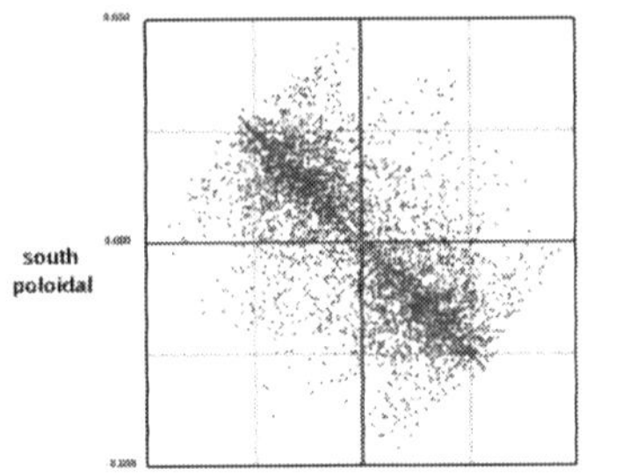

(f) Poloidal: complete(red) / phase-only(blue) synchronisation.

Figure 6. North-south symmetries

5.4. Fast-slow steering advantage

Fast-slow steering can offer a number of advantages over quasi-static steering. Fast-slow steering automates the process of parameter updating, where the user updates parameter trajectories rather than individual parameter values, hence reducing checkpointing frequencies. The user can monitor the system evolution as the bifurcations unfold, which provides more insight on the transition mechanisms during bifurcation occurrence. Furthermore, the slow steering of the

Journal of Physics: Conference Series **216** (2010) 012003 doi:10.1088/1742-6596/216/1/012003

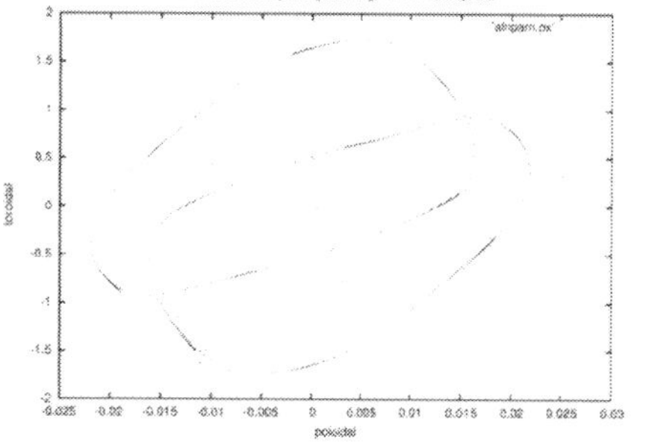

(a) Saddle-node bifurcation giving rise to *two* periodic solutions: a saddle (unstable) and a node(stable) 2×3 limit cycles. This information cannot be directly extracted from this snapshot, because it does not convey the time evolution of the solution.

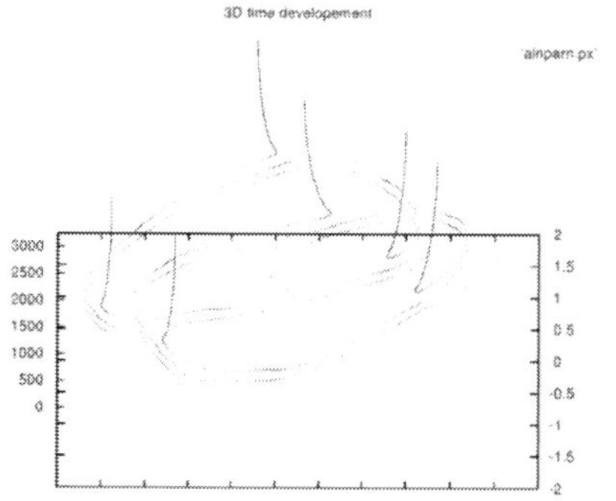

(b) The parameter α is linked to time, thus we unfold the bifurcation. The hidden saddle is observed on the torus before the bifurcation (dense). The appearance of only one of the possible period-3 orbits seems to be linked to the hysteresis. One orbit is selected on the increasing α and the other on the decreasing.

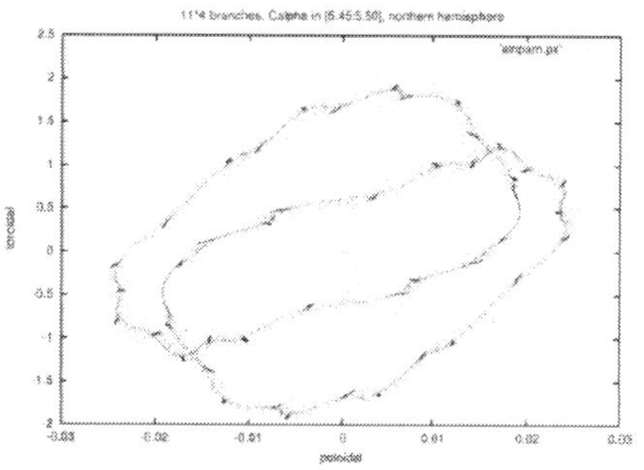

(c) The appearance of a 4×11 periodic window after the breaking of the torus.

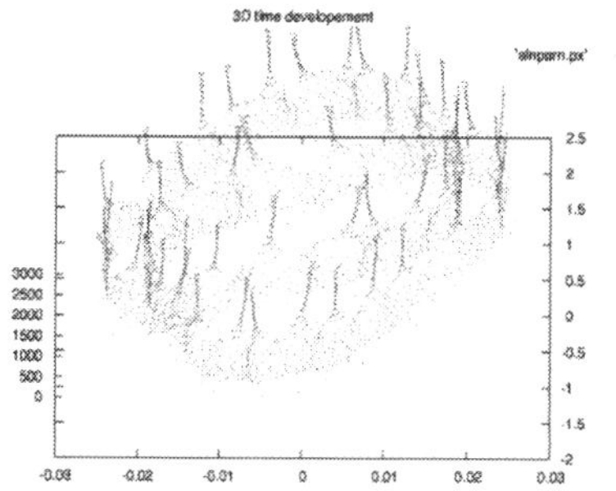

(d) Three-dimensional view describing how the 11 fold periodic solution emerges from the broken torus.

Figure 7. Using fast-slow steering to *unfold* the bifurcation

parameter vector through consecutive attractors cancels the effect of transient delays, caused by repeated initialisation. Comparing it to continuation, fast-slow steering provides a more intuitive tool for studying dynamical behaviour, because it conveys this information in a more conceivable visual format, and offers the user more experimental freedom during the simulation run-time. Fast-slow steering can be easily implemented for already available simulations lacking continuation codes which would otherwise be developed from scratch for the specific simulation. Typically, with small coding effort, simulations can be augmented with steering routines in the context of many available steering environments.

Most importantly, fast-slow steering allows the user to unfold the bifurcation, and investigate the mechanisms by which emerging attractors replace the pre-bifurcation ones. This information is not always easily induced from just looking at static snapshots of the system (figure 7). Furthermore, when the behaviour of orbits is affected by the previous evolution of the system, such as in the case of the hysteresis, steered solutions can be made to reveal results of accumulative nature.

6. Conclusions

We introduced a novel approach to computational steering, which is shown to take advantage of system continuability with respect to the parameter vector, and can be used to unfold bifurcating solutions across a non-hyperbolic points. This steering technique is best suited for dissipative systems, where the notion of an attracting solution is used to track its topological development in a suitable observation space. The emergence of structures at bifurcation points, such as the appearance of strange attractors after the breaking of tori, can be observed as an evolutionary process to gain insight into the transition mechanisms. The fast-slow steering method reveals the dynamic details of the bifurcation sequence that are difficult to see when building up solutions at fixed parameter values. Therefore, we believe that it is a good basis for parameter space exploration in dynamical systems.

The use of controllable trajectories allows an efficient exploration of parameter space, and enables the automation of the steering process. It was also demonstrated how, using this method, spacial symmetries can be decomposed into their time components and hence easily monitored. The concept of fast-slow steering is simple and can be easily implemented in existing simulation codes. Computational scientists who do not have a detailed understanding of the mathematical theory of bifurcations, can use this method for the investigation of dynamical systems using their domain-specific knowledge the dynamics of their systems. and without the need for a deep mathematical grasp of bifurcation theory or continuation methods.

In terms of the actual example studied in this paper, the axisymmetric dynamo in torus geometry, our method reproduces the results obtained with quasistatic steering given in [38] and in [39]. However our method reveals details of the transitions between different solutions that are very difficult, if not impossible, to see via quasistatic steering. This may be due to the ability of the fast-slow method to detect the effects of unstable solution branches on the evolution of stable solutions. In time-stepping methods, unstable solutions are very difficult to detect because the slightest numerical noise destroys them. Our method therefore permits comparison of high-dimensional models with low-dimensional models where methods of analysis exist which can track unstable solutions. This is very important in the study of fluid dynamics and magnetohydrodynamics, where nearly all our solution methods involve some dimensional reduction, the most drastic being from the infinite dimensional state space of partial differential equations to the finite dimensional spaces of numerical discretisations or lattice methods.

In future work we intend to apply more rigorous methods of dimensional reduction from the data produced by the numerical solutions. We have used in our current work, methods based on a knowledge of the behaviour of the dynamo solutions as travelling waves but we would like to use more general methods of model reduction. We are also investigating the use of fast-slow steering in systems where we do not know the governing equations, this is a key area for our method as the conventional continuation methods cannot work with these systems. We will also investigate the extension of our method to fully three-dimensional fluid flows by linking it to simulations on parallel computers. In these cases, efficient methods of dimensionality reduction will be essential for the visualisation used in our methods.

Acknowledgements

We would like to thank the Ministry of High Education and Scientific Research of Algeria for sponsoring this work. We would also like to thank the RealityGrid project for using the RealityGrid steering library.

References

[1] Schuster H G and Just W 2006 *Deterministic Chaos: An Introduction* (Wiley-VCH) ISBN 3527606416, 9783527606412

[2] Newhouse S, Ruelle D and Takens F 1978 *Communications in Mathematical Physics* **64** 35–40

[3] Brooke J M 1997 *Europhysics Letters* **37** 171–176

[4] Hu B and Zhou C 2000 *Phys. Rev. E* **62** 1983–1987

[5] Lai Y C 1996 *Phys. Rev. E* **53** R4267–R4270
[6] Pickles S M, Haines R, Pinning R L and et al 2005 *Royal Society of London Philosophical Transactions Series A* **363** 1843–1853
[7] Chin J and Coveney P V 2004 *Proceedings of the U.K. e-Science All Hands Meeting* (Nottingham, UK: National e-Science Centre)
[8] Goovaerts M J, Gragg W B and Wuytack L 1989 *J. Comput. Appl. Math.* **26** ISSN 0377-0427
[9] Doedel E J, Paffenroth R C, Champneys A R, Fairgrieve T F, Yu, Sandstede B and Wang X 2001 Auto 2000: Continuation and bifurcation software for ordinary differential equations (with homcont) Tech. rep. California Institute of Technology
[10] Khibnik A I, Kuznetsov Y A, Levitin V V and Nikolaev E V 1993 *Proceedings of a NATO advanced research workshop held at the Centre for Nonlinear Phenomena and Complex Systems on Homoclinic Chaos* (Amsterdam, The Netherlands, The Netherlands: Elsevier North-Holland, Inc.) pp 360–371
[11] Govaerts W and Kuznetsov Y A 2007 *Interactive Continuation Tools* (Springer Netherlands) pp 51–75 ISBN 978-1-4020-6355-8 (Print) 978-1-4020-6356-5 (Online)
[12] Khibnik A 1990 *Continuation and Bifurcation: Numerical Techniques and Applications* ed Roose D (Dordrecht: Kluwer Academic Publishers) pp 283–296
[13] Dhooge A, Govaerts W and Kuznetsov Y A 2003 *ACM Trans. Math. Softw.* **29** 141–164 ISSN 0098-3500
[14] He M, Ruan H and Yu C 2003 *International Journal of Modern Physics C* **14** 1237–1249
[15] Farina F and Dennunzio A 2008 *Fundam. Inf.* **83** 337–353 ISSN 0169-2968
[16] Hecht M and Harting J 2009 *Comput. Math. Appl.* **58** 995–1002 ISSN 0898-1221
[17] Chin J, Harting J, Jha S, Coveney P V, Porter A R and Pickles S M 2003 *Contemporary Physics* **44** 417–434(18)
[18] Dattilo G and Spezzano G 2003 *Parallel Comput.* **29** 1403–1418 ISSN 0167-8191
[19] Wenisch P, van Treeck C, Borrmann A, Rank E and Wenisch O 2007 *Int. J. Parallel Emerg. Distrib. Syst.* **22** 275–291 ISSN 1744-5760
[20] Swan JE I, Lanzagorta M, Maxwell D, Kuo E, Uhlmann J, Anderson W, Shyu H J and Smith W 2000 *Visualization 2000. Proceedings* pp 441–444
[21] Beazley D M and Lomdahl P S 1996 *Supercomputing '96: Proceedings of the 1996 ACM/IEEE conference on Supercomputing (CDROM)* (Washington, DC, USA: IEEE Computer Society) p 50 ISBN 0-89791-854-1
[22] Matkovic K, Gracanin D, Jelovic M and Hauser H 2008 *Visualization and Computer Graphics, IEEE Transactions on* **14** 1699–1706 ISSN 1077-2626
[23] Mulder J D, van Wijk J J and van Liere R 1999 *Future Gener. Comput. Syst.* **15** 119–129 ISSN 0167-739X
[24] Brooke J M, Coveney P V, Harting J, Jha S, Pickles S M, Pinning R L and Porter A R 2003 *Proceedings of the U.K. e-Science All Hands Meeting* (Nottingham, UK: National e-Science Centre)
[25] Tweedie L, Spence R, Dawkes H and Su H 1996 *CHI '96: Proceedings of the SIGCHI conference on Human factors in computing systems* (New York, NY, USA: ACM) pp 406–ff. ISBN 0-89791-777-4
[26] Chatzinikos F and Wright H 2001 *VMV '01: Proceedings of the Vision Modeling and Visualization Conference 2001* (Aka GmbH) pp 455–462 ISBN 3-89838-028-9
[27] Wood J D and Wright H 2008 *Concurrency and Computation: Practice and Experience, Special Issue: Selected Papers from the 2005 U.K. e-Science All Hands Meeting (AHM 2005)* **20** 265–276 copyright 2007 John Wiley & Sons, Ltd.
[28] Wright H, Brodlie K and David T 2000 *VIS '00: Proceedings of the conference on Visualization '00* (Los Alamitos, CA, USA: IEEE Computer Society Press) pp 291–296 ISBN 1-58113-309-X
[29] Smale S 1967 *Bull. Amer. Math. Soc.* **73** 747–817
[30] Palis J and Smale S 1970 *Global Analysis, Proc. Symp. Pure Math.* **14** 223–231
[31] Helman J L and Hesselink L 1989 *Computer* **22** 27–36 ISSN 0018-9162
[32] Scheuermann G, Hagen H, Krüger H, Menzel M and Rockwood A 1997 *VIS '97: Proceedings of the 8th conference on Visualization '97* (Los Alamitos, CA, USA: IEEE Computer Society Press) pp 67–74 ISBN 1-58113-011-2
[33] Scheuermann G, Kruger H, Menzel M and Rockwood A 1998 *Visualization and Computer Graphics, IEEE Transactions on* **4** 109–116 ISSN 1077-2626
[34] Tricoche X, Scheuermann G and Hagen H 2001 *Data Visualisation* (Ascona, Switzerland.: Proc. VisSym 01) pp 117–126
[35] Tricoche X, Wischgoll T, Scheuermann G and Hagen H 2002 *Computers & Graphics* **26** 249 – 257 ISSN 0097-8493
[36] Cvitanović P, Gunaratne G H and Procaccia I 1988 *Phys. Rev. A* **38** 1503–1520
[37] Hénon M 1976 *Communications in Mathematical Physics* **50** 69–77
[38] Brooke J M 1997 *Symmetry breaking and chaos in nonlinear mean-field dynamos* Ph.D. thesis
[39] Brooke J M, Pelt J, Tavakol R and Tworkowski A 1998 *Astronomy and Astrophysics* **332** 339–352

Long-scale nonlinear evolution of parametrically excited Marangoni convection

Alexander Mikishev[1], **Alexander Nepomnyashchy**[1] **and Boris Smorodin**[2]

[1]Department of Mathematics, Technion - Israel Institute of Technology, Haifa, 32000 Israel
[2]Perm State University, Perm 614990 Perm, Russia

E-mail: amik@tx.technion.ac.il

Abstract. We consider a horizontal liquid layer with a deformable free upper surface heated from below so that on the bottom the heat flux is periodically changing and the averaged temperature of the layer equals zero. A set of nonlinear evolution equations is derived for the description of the spatiotemporal dynamics of the long-wave Marangoni instability. A bifurcation analysis near the threshold of the convection onset shows the existence of the subcritical as well as supercritical type of bifurcations. The region of supercritical bifurcation regime is found. The evolution equation for the surface deviation in the form of the well-known Cahn-Hilliard equation is obtained in the vicinity of the critical value of the Marangoni number.

1. Introduction

A dependence of the surface tension on the temperature can lead to the development of a thermocapillary (Marangoni) instability in a layer with the free surface. The instability threshold, as well as the nonlinear dynamics, can be significally changed under the impact of different external constraints, like vibration and thermal modulation ([1]-[3]).

This work continues the investigation of the pattern formation and transport phenomena in open liquid layer caused by the instability that is developed under a modulated heat flux due to the thermocapillary effect. This instability has been studied in the framework of a linear stability theory in [4]. By means of an analytical approach, we derive strongly nonlinear equations governing the development of a long-wavelength instability, which is important in some regions of the parameters, as shown in [4]. Near the instability threshold, a weakly nonlinear theory is developed.

2. Derivation of the governing equations

2.1. Basic equations and boundary conditions

We consider a two-dimensional flow in a liquid layer which is infinite in the horizontal direction ($-\infty < x < \infty$) and confined in the vertical direction between a rigid lower plane $z = 0$ and a free deformable upper boundary $z = h(x, t) > 0$ with a temperature-dependent surface tension. In the unperturbed state the layer has a thickness H_0. The prescribed modulated heat flux at $z = 0$ is normal to the lower plane and varies periodically with time. The liquid is Newtonian and incompressible. Under these assumptions the following set of two-dimensional equations describes the problem:

$$\partial_x u + \partial_z w = 0, \tag{1}$$
$$\partial_t u + u\partial_x u + w\partial_z u = -\rho^{-1}\partial_x p + \nu(\partial_{xx} + \partial_{zz})u, \tag{2}$$
$$\partial_t w + u\partial_x w + w\partial_z w = -\rho^{-1}\partial_z p + \nu(\partial_{xx} + \partial_{zz})w - g, \tag{3}$$
$$\partial_t T + u\partial_x T + w\partial_z T = \chi(\partial_{xx} + \partial_{zz})T. \tag{4}$$

This system of equations is supplemented by the non-slip, no-penetration condition for the velocity and the condition of a harmonically varying heat flux on the bottom:

$$z = 0 : u = w = 0, \quad \lambda(\partial_z T) = Q_0 \cos(2\Omega t). \tag{5}$$

The conditions at the free surface $z = h(x,t)$ include a kinematic boundary condition for the free boundary motion, the condition of heat insulation, and the balance of normal and tangential inerfacial stresses:

$$\partial_t h + u\partial_x h = w, \tag{6}$$
$$\lambda \nabla T \cdot \mathbf{n} = 0, \tag{7}$$
$$-p + \frac{2\eta}{N^2}[\partial_x u((\partial_x h)^2 - 1) - \partial_x h(u_z + w_x)] = \sigma\frac{\partial_{xx} h}{N^3}, \tag{8}$$
$$\frac{\eta}{N}[2(\partial_z w - \partial_x u)\partial_x h + (\partial_z u + \partial_x w)(1 - (\partial_x h)^2)] = \partial_s \sigma. \tag{9}$$

Here $\mathbf{v} = (u, w), T$ and p are fields of the fluid velocity, temperature and pressure, respectively; ν, ρ, χ are kinematic viscosity, density and thermal diffusivity; λ is the heat conductivity of the liquid, Q_0 and Ω are the amplitude and frequency of the heat flux oscillations; $\nabla = (\partial_x, \partial_z); N = (1 + h_x^2)^{1/2}, \eta = \nu\rho$ is a fluid viscosity and s is arc length along the interface. We assume that the surface tension varies linearly with the temperature as $\sigma = \sigma_0 - \sigma_T T$.

2.2. Scaling transformation

The linear stability analysis of the problem [4] shows the existence of long-wave instability, i.e. minimum of the neutral stability curve is located at the wave number $k = 0$ under definite conditions. In this case we can assume, that the horizontal spatial scale of flows, L, is much larger than the vertical scale. This assumption permits to simplify the complicated nonlinear problem by means of asymptotic methods using multiscale expansions.

To form the dimensionless set of equations and boundary conditions we choose the thermal skin-layer thickness, $\sqrt{\chi/\Omega}$, as a vertical spatial scale. Then we suppose that the distortions are long-scale and

$$\varepsilon = \frac{2\pi\sqrt{\chi}}{\sqrt{\Omega}L} \ll 1.$$

We restrict our consideration to long-wave motions only. So, we will use the new dimensionless spatial coordinates $X = \varepsilon x\sqrt{\Omega/\chi}$ and $Z = z\sqrt{\Omega/\chi}$. The dimensionless fluid velocity in x- and $z-$ directions are

$$U = \frac{u}{\sqrt{\chi\Omega}}, \quad W = \frac{w}{\sqrt{\chi\Omega}}.$$

Due to the linear analysis we introduce two characteristic times. The first time scale is related to the heat flux modulation ("fast" time), $\tau = t\Omega$, and the second one - to the liquid motion ("slow" time), $T = \varepsilon^2 t\Omega$. In this case the differentiation operator with respect to time looks like

$$\partial_t = \varepsilon^2\Omega\partial_T + \Omega\partial_\tau.$$

After substituting into the governing system we obtain the following scaled system:

$$\varepsilon U_X + W_Z = 0, \tag{10}$$
$$P^{-1}(U_\tau + \varepsilon^2 U_T + \varepsilon U U_X + W U_Z) = -\varepsilon \Pi_X + \varepsilon^2 U_{XX} + U_{ZZ}, \tag{11}$$
$$P^{-1}(W_\tau + \varepsilon^2 W_T + \varepsilon U W_X + W W_Z) = -\Pi_Z + \varepsilon(\varepsilon^2 W_{XX} + W_{ZZ}) - G, \tag{12}$$
$$\Theta_\tau + \varepsilon^2 \Theta_T + \varepsilon U \Theta_X + W \Theta_Z = \varepsilon^2 \Theta_{XX} + \Theta_{ZZ}, \tag{13}$$

and the boundary conditions are given as

$$z = 0: \quad U = 0, \quad W = 0, \quad \Theta_Z = \cos(2\tau). \tag{14}$$
$$z = H: \quad H_\tau = -\varepsilon^2 H_T - \varepsilon U H_X + W, \tag{15}$$
$$\varepsilon^2 \Theta_X H_X - \Theta_Z = 0, \tag{16}$$
$$-\Pi + \frac{2\varepsilon}{N^2}[U_X(\varepsilon^2 H_X^2 - 1) - \varepsilon H_X(U_Z + \varepsilon W_X)] = \tilde{\Sigma}\frac{H_{XX}}{N^3}, \tag{17}$$
$$4\varepsilon W_Z H_X + (U_Z + \varepsilon W_X)(1 - \varepsilon^2 H_X^2) = M\varepsilon(\Theta_X + H_X \Theta_Z)N. \tag{18}$$

Here H is the function of the surface distortion, Π and Θ are pressure and temperature functions respectively, $N = \sqrt{1 + \varepsilon^2 H_X^2}$. The dimensionless parameters here are $P = \nu/\chi$ is the Prandtl number, $G = g\chi^{1/2}/(\nu\Omega^{3/2})$ - Galileo number, $\Sigma = \sigma/(\eta\sqrt{\chi\Omega})$ - inverse capillary number and $M = -\sigma_T Q/(\lambda\eta\Omega)$ - Marangoni number. As we know from the linear problem, the long wavelength instability exists in the range of large inverse capillary numbers, then through the paper we assume that $\Sigma = \tilde{\Sigma}\varepsilon^{-2}, \tilde{\Sigma} = O(1)$.

2.3. Derivation of long wavelength evolution equations

The asymptotic solution of the nonlinear problem (10)-(18) can be constructed by expanding the functions into power series of ε,

$$\begin{aligned} U &= \varepsilon(U_0 + \varepsilon U_1 + \varepsilon^2 U_2 + \cdots), \\ W &= \varepsilon^2(W_0 + \varepsilon W_1 + \varepsilon^2 W_2 + \cdots), \\ \Pi &= \Pi_0 + \varepsilon \Pi_1 + \varepsilon^2 \Pi_2 + \cdots, \\ \Theta &= \Theta_0 + \varepsilon \Theta_1 + \varepsilon^2 \Theta_2 + \cdots, \\ H &= H_0 + \varepsilon H_1 + \varepsilon^2 H_2 + \cdots, \\ M &= M_0 + \varepsilon^2 M_2 + \varepsilon^4 M_4 + \cdots. \end{aligned} \tag{19}$$

At each order in ε we get a linear problem that, according to Fredholm alternative, has solution bounded in space and time, if and only if some solvability conditions are satisfied. These solvability conditions provide the evolution equations for long-scale motions. At leading order of ε we have

$$U_{0,X} + W_{0,Z} = 0, \tag{20}$$
$$P^{-1} U_{0,\tau} + \Pi_{0,X} - U_{0,ZZ} = 0, \tag{21}$$
$$\Pi_{0,Z} + G = 0, \tag{22}$$
$$\Theta_{0,ZZ} - \Theta_{0,\tau} = 0. \tag{23}$$

with following boundary conditions

$$\begin{aligned} &z = 0: \quad U_0 = W_0 = 0, \quad \Theta_{0,Z} = \frac{1}{2}e^{2i\tau} + c.c. \\ &z = H_0(X,t): \quad H_{0,\tau} = 0, \quad \Theta_{0,Z} = 0, \\ &\Pi_0 = -\tilde{\Sigma} H_{0,XX}, \quad U_{0,Z} = M_0 \Theta_{0,X}. \end{aligned} \tag{24}$$

The solution of the system at leading order found by successive integrations is

$$H_0 = H_0(X,T), \tag{25}$$
$$\Pi_0(X,Z,T) = -G[Z - H_0(X,T)] - \tilde{\Sigma} H_{0,XX}, \tag{26}$$
$$\Theta_0(X,Z,\tau,T) = \bar{\Theta}_0 + \tilde{\Theta}_0 = \bar{\Theta}_0(X,T) - \frac{\cosh[\alpha(H_0 - z)]e^{2i\tau}}{2\alpha\sinh(\alpha H_0)} + c.c., \tag{27}$$
$$U_0(X,Z,\tau,T) = \bar{U}_0 + \tilde{U}_0 =$$
$$= \frac{1}{2}(GH_{0,X} - \tilde{\Sigma} H_{0,XXX})z^2 + (M_0\bar{\Theta}_{0,X} - GH_0H_{0,X} + \tilde{\Sigma} H_0 H_{0,XXX})z +$$
$$+ \frac{M_0\cosh(\alpha H_0)H_{0,X}\sinh(\mu z)e^{2i\tau}}{2\mu\sinh^2(\alpha H_0)\cosh(\mu H_0)} + c.c., \tag{28}$$
$$W_0(X,Z,\tau,T) = \bar{W}_0 + \tilde{W}_0 = \frac{(-GH_{0,XX} + \tilde{\Sigma} H_{0,XXXX})z^3}{6} +$$
$$+[G(H_{0,X})^2 + GH_0H_{0,XX} - \tilde{\Sigma} H_{0,X}H_{0,XXX} - \tilde{\Sigma} H_0 H_{0,XXXX} - M_0\overline{\Theta}_{0,XX}]\frac{z^2}{2} +$$
$$+\frac{M_0[\cosh(\mu z) - 1]}{4\mu^2\sinh(\alpha H_0)\cosh(\mu H_0)}\{[\alpha + \frac{2\alpha}{\sinh^2(\alpha H_0)} + \mu\coth(\alpha H_0)\tanh(\mu H_0)](H_{0,X})^2 -$$
$$- \coth(\alpha H_0)H_{0,XX}\}e^{2i\tau} + c.c. \tag{29}$$

Here $\alpha = 1+i, \mu = \frac{1+i}{\sqrt{P}}$ and H_0, $\bar{\Theta}_0$ are functions that do not depend on the "fast" time τ. There evolution in the "slow"timescale T is determined at the next order of expansion. The solutions of the leading order show that functions are separated into averaged part which depends on "slow" time T and pulsating part depending on "fast" time τ.

At the second order of asymptotic expansions we obtain the following system

$$U_{2,X} + W_{2,Z} = 0, \tag{30}$$
$$P^{-1}[U_{2,\tau} + U_{0,T} + U_0U_{0,X} + W_0U_{0,Z}] + \Pi_{2,X} - U_{0,XX} - U_{2,ZZ} = 0, \tag{31}$$
$$P^{-1}W_{0,\tau} + \Pi_{2,Z} = 0, \tag{32}$$
$$\Theta_{2,\tau} + \Theta_{0,T} + U_0\Theta_{0,X} + W_0\Theta_{0,Z} = \Theta_{0,XX} + \Theta_{2,ZZ} \tag{33}$$

with following boundary conditions

$$\begin{aligned}
&z = 0: \quad U_2 = W_2 = \Theta_{2,Z} = 0\\
&z = H_0: \quad H_{2,\tau} = -H_{0,T} - U_0H_{0,X} + W_0,\\
&\quad \Theta_{0,X}H_{0,X} - H_2\Theta_{0,ZZ} - \Theta_{2,Z} = 0,\\
&\quad -\Pi_2 - H_2\Pi_{0,Z} - 2U_{0,X} = \tilde{\Sigma} H_{2,XX} + \tilde{\Sigma} H_2 H_{0,XXZ} - \frac{3}{2}\tilde{\Sigma} H_{0,XX}H_{0,X}^2,\\
&U_{2,Z} + 4W_{0,Z}H_{0,X} + W_{0,X} - U_{0,Z}H_{0,X}^2 + H_2U_{0,ZZ} =\\
&\qquad M_0[\Theta_{2,X} + \frac{1}{2}\Theta_{0,X}H_{0,X}^2 + H_{0,X}\Theta_{2,Z}] + M_2\Theta_{0,X}.
\end{aligned} \tag{34}$$

Here boundary conditions are obtained by the expansion of (15)-(18) in a Taylor series near the point H_0.

2.4. Solvability conditions

Let apply now the solvability conditions to obtain the evolution equations. The first condition reflects the fact that the amount of liquid in the domain does not change in time. Integrating the kinematic equation for the deformation of the surface over a period in τ, we obtain

$$H_{0,T} = -\bar{U}_0 H_{0,X} + \bar{W}_0.$$

The latter relation can be written om the form of the conservation law:

$$H_{0,T} = -\partial_X\left(\int_0^{H_0} \bar{U}_0 dz\right) = \\ = \left(\frac{1}{3} G H_0^3 H_{0,X} - \frac{1}{2} M_0 H_0^2 \bar{\Theta}_{0,X} - \frac{1}{3}\tilde{\Sigma} H_0^3 H_{0,XXX}\right)_X. \quad (35)$$

The pulsating part of the surface distortion depends on τ as $\tilde{H}_{2,\tau} = \tilde{W}_0 - \tilde{U}_0 H_{0,X}$.

The second solvability condition reflects the energy balance. It helps to obtain the evolution of the averaged temperature. For this goal we first average Eqn. (33) with respect to the fast time τ and then across the layer. We obtain the following relation

$$< \bar{\Theta}_{0,T} > = < \bar{\Theta}_{0,XX} > + < \Theta_{2,ZZ} > - < U_0 \Theta_{0,X} > - < W_0 \Theta_{0,Z} > .$$

Here $< \cdots > = \int_0^{H_0} ...dz$. Using boundary conditions, we find that $< \Theta_{2,ZZ} > = \Theta_{2,Z}|_{Z=H_0} - \Theta_{2,Z}|_{Z=0} = \Theta_{0,X} H_{0,X} - H_2 \Theta_{0,ZZ}$.

Finally after all substitutions we obtain the second evolution equation in the form

$$\partial_T(H_0\bar{\Theta}_0) = \partial_X(H_0\bar{\Theta}_{0,X}) - \partial_X \int_0^{H_0} (\bar{U}_0\bar{\Theta}_0 + \tilde{U}_0^+\tilde{\Theta}_0^- + \tilde{U}_0^-\tilde{\Theta}_0^+)dz = \\ = \partial_X(H_0\bar{\Theta}_{0,X} - \int_0^{H_0} \bar{U}_0\bar{\Theta}_0 dz - M H_{0,X} S(H_0, P)). \quad (36)$$

Here $\bar{U}_0$ and $\bar{\Theta}_0$ are averaged parts of horizonthal component of velocity and temperature, respectively, $\tilde{U}_0^+$ and $\tilde{\Theta}_0^+$ are pulsating parts of these variables, which depend on "fast" time as $e^{2i\tau}$, and $\tilde{U}_0^-$, $\tilde{\Theta}_0^-$ are their complex conjugated terms (see (27)-(29)). Function $S(H_0, P)$ will be defined latter.

Eqns. (35) and (36), upon the substitution of the corresponding expressions and calculation of the integral, form a closed set of evolution equations for H_0 and $\bar{\Theta}_0$. We can note, that both of these equations are written in a divergent form.

3. Linear stability analysis

In this section we linearize the equations, which are obtained in the last section, near the base state given by the mean value of the layer thickness and a constant reference temperature, using

$$H_0(X,T) = h_0 + h(X,T), \quad , \quad \Theta_0(X,T) = \theta(X,T), \qquad h(X,T), \theta(X,T) \ll 1.$$

resulting in

$$h_T = \frac{1}{3} G h_0^3 h_{XX} - \frac{1}{2} M_0 h_0^2 \theta_{XX} - \frac{1}{3}\tilde{\Sigma} h_0^3 h_{XXXX}, \quad (37)$$

$$h_0 \theta_T = h_0 \theta_{XX} - M_0 S(h_0, P) h_{XX}. \quad (38)$$

Here

$$S(h_0, P) = \Re\left\{\frac{(1+i)P}{16(1+P)}\left(\frac{1}{\sinh[(1+i)h_0]}((\cot[(1+i)h_0] - \coth[(1+i)h_0])\csc[(1+i)h_0] + \right.\right. \\ + \cot[(1+i)h_0]\coth[(1+i)h_0]\frac{1}{\cosh[\frac{(1+i)h_0}{\sqrt{P}}]}) - \\ \left.\left. - \cot[(1+i)h_0]\coth[(1+i)h_0]\csc[(1+i)h_0]\sec[\frac{(1+i)h_0}{\sqrt{P}}]\right)\right\} \quad (39)$$

For further calculations we will fix Prandtl number to 7, which corresponds to the case of water. Figure 1 depicts function $S(h_0, P)$.

Substituting the normal form for the perturbations

$$h(X,T) = Ae^{\sigma T + ikX} \qquad \theta(X,T) = Be^{\sigma T+ikX},$$

we obtain the following set of equations in terms of amplitudes:

$$\sigma A = -\frac{1}{3}Gh_0^3k^2A + \frac{1}{2}M_0h_0^2k^2B - \frac{1}{3}\tilde{\Sigma}h_0^3k^4A, \tag{40}$$

$$h_0\sigma B = -h_0k^2B + M_0Sk^2A. \tag{41}$$

The solvability condition takes the form of

$$(\sigma + k^2)(\sigma + \frac{Gh_0^3}{3}k^2 + \frac{h_0^3}{3}\tilde{\Sigma}k^4) = \frac{h_0}{2}M_0^2k^4S. \tag{42}$$

In case of the convection onset we take $\sigma = 0$ and obtain for critical Marangoni number

$$M_0 = \pm\sqrt{\frac{2h_0^2}{3S}\left(G + \tilde{\Sigma}k^2\right)}. \tag{43}$$

One can see that the instability takes place within the intervals of h_0 where $S(h_0) > 0$, and the neutral curve $|M| = |M_0(k)|$ has its minimum at $k = 0$. The dependences of the critical Marangoni number$M_c = |M_0(0)|$ on h_0 are shown in Figure 2 for two first intervals of the positiveness of $S(h_0)$. The critical Marangoni number reaches its minimum at $h_c = 3.4017$.

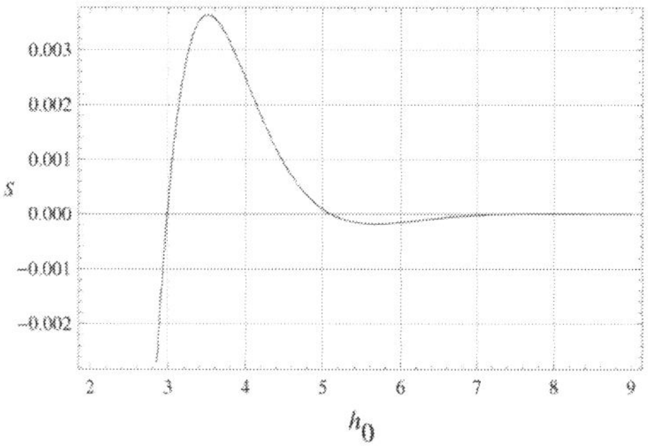

Figure 1. Dependence of function $S(h_0, P)$ on layer thickness h_0 for $P = 7$.

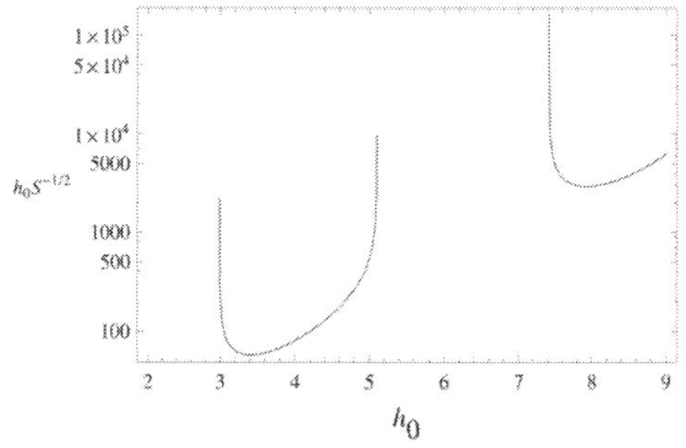

Figure 2. Critical Marangoni number within the first two instability domains for $P = 7$ (water). The minimum is achieved at $h_c = 3.40$.

In the general case, the solution of (42) yields the two modes of growth rate

$$\sigma_\pm = -\frac{(3 + Gh_0^3 + \tilde{\Sigma}h_0^3k^2)}{6}k^2 \pm \frac{k^2}{6}\sqrt{(Gh_0^3-3)^2 + 18M_0^2Sh_0 + 2\tilde{\Sigma}h_0^3(Gh_0^3-3)k^2 + \tilde{\Sigma}^2h_0^6k^4} \tag{44}$$

The analysis of this expression reveals the absence of oscillatory instability: $\Re(\sigma_+) \leq 0$.

Journal of Physics: Conference Series **216** (2010) 012004 doi:10.1088/1742-6596/216/1/012004

4. Nonlinear analysis

4.1. Stationary bifurcations

We present here the analysis of branching in the stationary solutions of (35) - (36) that appear due to the monotonic instability of the base state. The symmetry of the equations permits to set the integration constants equal to zero. After this we have the following nonlinear system

$$0 = \frac{1}{3} G H_0^3 H_{0,X} - \frac{1}{2} M H_0^2 \bar{\Theta}_{0,X} - \frac{1}{3} \tilde{\Sigma} H_0^3 H_{0,XXX}, \tag{45}$$

$$0 = H_0 \bar{\Theta}_{0,X} - \int_0^{H_0} \bar{U}_0 \bar{\Theta}_0 dz - M H_{0,X} S(H_0) = 0. \tag{46}$$

Note here, that we will analyze the periodic solutions of a given wavenumber k, different from the ctitical ($k = 0$). To reduce the system to a simpler form valid near the onset of the instability at fixed wavenumber let use the perturbation analysis. The linear analysis shows that minimum of the primary neutral stability curve is characterized by critical Marangoni number M_0, which is $M_0^2 = 2h_0^2(G + \tilde{\Sigma}k^2)/(3S(h_0))$. In our approach we suppose

$$M = M_0 + \delta^2 M_2 + ..., \quad M_2 > 0, \quad \delta \ll 1, \tag{47}$$

where δ characterizes the amplitude of the bifurcating stationary solution. Asymptotic expansions for surface distortion and temperature functions are

$$H_0 = h_0 + \delta h_1 + \delta^2 h_2 + \delta^3 h_3 + \cdots, \tag{48}$$

$$\bar{\Theta}_0 = \delta\theta_1 + \delta^2\theta_2 + \delta^3\theta_3 + \cdots. \tag{49}$$

At the first order of δ we recover the linear problem with the equation for h_1:

$$h_{1,XXX} + k^2 h_{1,X} = 0. \tag{50}$$

Its solution can be written in the form

$$h_1 = a\cos(kX). \tag{51}$$

After some rearranging the second-order equation is

$$h_{2,XXX} + k^2 h_{2,X} = -\frac{a^2 k}{2h_0\tilde{\Sigma}}(G + \tilde{\Sigma}k^2)(2 - h_0\frac{\partial_{h_0} S(h_0)}{S(h_0)})\sin(2kX) \tag{52}$$

The solution of Eqn. (52) is

$$h_2 = -\frac{a^2}{12h_0 k^2\tilde{\Sigma}}(G + \tilde{\Sigma}k^2)(2 - h_0\frac{\partial_{h_0} S(h_0)}{S(h_0)})\cos(2kX), \tag{53}$$

where the coefficients $C_i, i = 1, \cdots, 4$, are cumbersome expressions which are do not present here. The third-order equation can be written as

$$h_{3,XXX} + k^2 h_{3,X} = C_1\sin(kX) + C_2\cos(kX)\sin(2kX) + C_3\sin(kX)\cos^2(kX) + C_4\sin(kX)\cos(2kX) \tag{54}$$

The solvability condition of this equation is

$$4C_1 + 2C_2 + C_3 - 2C_4 = 0. \tag{55}$$

This condition leads to the stationary Landau expression, in the form

$$\gamma a + \kappa a^3 = 0,$$

which gives us the bifurcation behavior near the instability threshold with fixed wavenumber. In case when $\Re(\kappa) < 0$, then we have a supercritical bifurcation, in case of $\Re(\kappa) > 0$ - a subcritical one.

The calculation gives us that

$$\gamma = \frac{12kM_2}{h_0\tilde{\Sigma}}\sqrt{\frac{2}{3}(G+\tilde{\Sigma}k^2)}.$$

The expression for κ was tabulated and Figure 3 depicts its behaviour as a function of h_0 for $k = 0.1, G = 100, \tilde{\Sigma} = 1000$. The graph is plotted where $S(h_0) > 0$, within the regions of existence of long-wave instability.

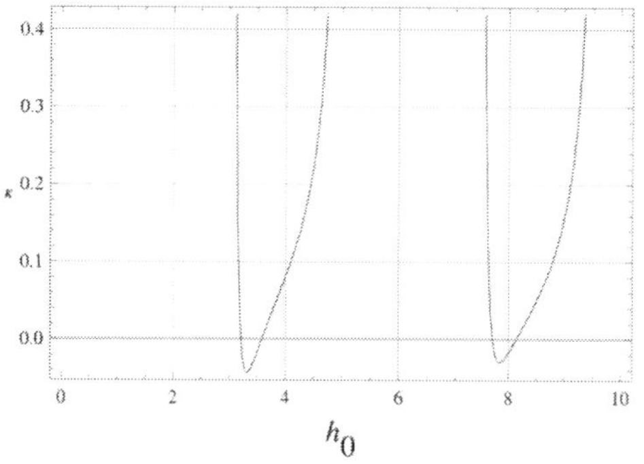

Figure 3. Dependence of κ on the layer thickness h_0 for $k = 0.1, G = 100, \tilde{\Sigma} = 1000, P = 7$.

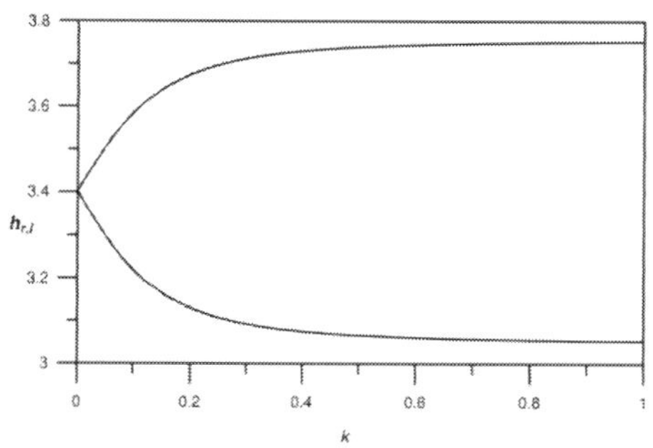

Figure 4. Location of points h_r and h_l as function of k for $G = 100, \tilde{\Sigma} = 1000, P = 7$.

The graph shows the existence both subcritical and supercritical branching of the surface distortion. The size of supercritical branching region depends on considered wavenumber. The next figure represents the location of the right, h_r, and the left ,h_l, intersection points of κ with axis h_0 as function of wavenumber, k, where $\kappa(h_0)$ changes its sign. The region of a supercritical bifurcation $h_l < h_0 < h_r$ shrinks in the limit $k \to 0$ to the point $h_r = h_l = h_c = 3.4017$.

4.2. Evolution near the critical Marangoni number

Here we consider dynamics of surface distortion for the most unstable layer thickness, $h_0 = h_c$. In this region we will apply the multi-scale method again using the deviation of Marangoni number ϵ which measures the declination of Marangoni number from the critical one, M_c at h_c:

$$M = M_c + \epsilon^2 M_2$$

We use Eqns. (35) and (36) as a base equations for the multi-scale analysis. Introduce a new slow spatial coordinate $\xi = \epsilon X$. The evolution occurs on the slow time scale t_s as $t_s = \epsilon^4 T$. According to this scaling we can rewrite the system (35) - (36) as

$$\epsilon^4 H_{0,t_s} = \epsilon^2 G H_0^2 H_{0,\xi}^2 + \epsilon^2 \frac{1}{3} G H_0^3 H_{0,\xi\xi} - \epsilon^2 M H_0 H_{0,\xi} \bar{\Theta}_{0,\xi} - \epsilon^2 \frac{1}{2} M H_0^2 \bar{\Theta}_{0,\xi\xi} -$$

$$-\epsilon^4\tilde{\Sigma}H_0^2H_{0,\xi}H_{0,\xi\xi\xi} - \epsilon^4\frac{1}{3}\tilde{\Sigma}H_0^3H_{0,\xi\xi\xi\xi}, \tag{56}$$

$$\epsilon^4(H_0\bar{\Theta}_0)_{t_s} = \epsilon^2(H_0\bar{\Theta}_{0,\xi})_\xi + \epsilon(\frac{1}{3}\epsilon G H_0^3\bar{\Theta}_0 H_{0,\xi} - \frac{1}{2}\epsilon M H_0^2\bar{\Theta}_0\bar{\Theta}_{0,\xi} -$$
$$-\frac{1}{3}\epsilon^3\tilde{\Sigma}H_0^3\bar{\Theta}_0\bar{\Theta}_{0,\xi}H_{0,\xi\xi\xi} - \epsilon M H_{0,\xi}S(H_0))_\xi. \tag{57}$$

We expand function H_0 around critical thickness h_c as

$$H_0 = h_c + \epsilon h_1(\xi) + \epsilon^2 h_2(\xi) + \epsilon^3 h_3(\xi) + \cdots$$

and temperature $\bar{\Theta}_0$ as

$$\bar{\Theta}_0 = \epsilon\theta_1 + \epsilon^2\theta_2 + \epsilon^3\theta_3 + \cdots.$$

The lowest order of expansion gives relation between h_1 and θ_1 as

$$\frac{1}{3}Gh_c^3 h_{1,\xi\xi} = \frac{1}{2}M_0 h_c^2\theta_{1,\xi\xi}.$$

The next order of expansion leads to the equations for h_2 and θ_2

$$\frac{1}{3}Gh_c^3h_{2,\xi\xi} - \frac{1}{2}M_0h_c^2\theta_{2,\xi\xi} = -\frac{1}{3}Gh_c^2(h_1h_{1,\xi})_\xi = -\frac{1}{6}Gh_c^2(h_1^2)_{\xi\xi}, \tag{58}$$

$$-M_0Sh_{2,\xi\xi} + h_c\theta_{2,\xi\xi} = \{M_0S' - \frac{2G}{3M_0}h_c\}(h_1h_{1,\xi})_\xi =$$
$$= \frac{1}{2}M_0S'(h_1^2)_{\xi\xi} - \frac{1}{2h_c}M_0S(h_1^2)_{\xi\xi}. \tag{59}$$

The solvability condition for this system gives us the relation

$$\frac{S'}{S} = \frac{2}{h_c},$$

which is correct for the minimum of the neutral curve, where $dM_0(h_0)/dh_0 = 0$ at $h_0 = h_c$ (prime is derivative with respect to h_0).

Next order of the expansion contains temporal derivatives

$$\frac{1}{3}Gh_c^3h_{3,\xi\xi} - \frac{1}{2}M_0h_c^2\theta_{3,\xi\xi} = \partial_{t_s}h_1 - \frac{1}{3}Gh_c^2(h_1h_2)_{\xi\xi} + \frac{1}{3}Gh_c^3\frac{M_2}{M_0}h_{1,\xi\xi} + \frac{1}{3}\tilde{\Sigma}h_c^3h_{\xi\xi\xi\xi}, \tag{60}$$

$$-M_0Sh_{3,\xi\xi} + h_c\theta_{3,\xi\xi} =$$
$$= \frac{2Gh_c^2}{3M_0}\partial_{t_s}h_1 - \frac{2G}{9M_0}(h_1^3)_{\xi\xi} + \frac{2Gh_c}{3M_0}(h_1h_2)_{\xi\xi} + \frac{Gh_c^2}{9M_0}(\frac{S''}{S})(h_1^3)_{\xi\xi} + M_2Sh_{1,\xi\xi}. \tag{61}$$

Finally we obtain the following equation for surface deviation:

$$(3 + Gh_c^3)\partial_{t_s}h - \frac{Gh_c^3}{3}(\frac{M_0''}{M_0})(h^3)_{\xi\xi} + 2Gh_c^3(\frac{M_2}{M_0})h_{\xi\xi} + \tilde{\Sigma}h_c^3h_{\xi\xi\xi\xi} = 0. \tag{62}$$

This form of the equation means that in vicinity of the critical Marangoni number the evolution of the surface is governed by the standard Cahn-Hilliard equation.

Acknowledgments

A.M. and A. N. acknowledge partial support from the Israel Ministry of Science through Grant # 3-5799. B.S. acknowledges a support by the Russian Basic Research Foundation (project # 09-01-92472). The research was partially supported by the European Union via FP7 Marie Curie scheme [PITN-GA-2008-214919 (MULTIFLOW)].

Journal of Physics: Conference Series **216** (2010) 012004 doi:10.1088/1742-6596/216/1/012004

References

[1] Gershuni G.Z. and Zhukhovitsky E.M. 1976 *Stability of Incompressible Fluids* (Jerusalem: Keter)
[2] Gershuni G.Z. and Lyubimov D.V. 1998 *Thermal Vibrational Convection* (Chichester: Wiley)
[3] Or A.C. and Kelly R.E. 2002 *J. Fluid Mech.* **456** 161
[4] Smorodin B.L., Mikishev A.B., Nepomnyashchy A.A. and Myznikova B.I. 2009 *Phys Fluids* **21** 062102

Journal of Physics: Conference Series **216** (2010) 012005 doi:10.1088/1742-6596/216/1/012005

Effect of rotation on stability of advective flow in horizontal liquid layer with a free upper boundary

Konstantin G. Shvarts [1], **Abdelaziz Boudlal** [2]
[1] Applied Mathematics Department, Perm State University, Bukireva St., 15, 614960 Perm, Russia
[2] Laboratoire de Mecanique de Lille – UMR CNRS 8107 F59655 Villeneuve d'Ascq, Lille, France

E-mail: kosch@psu.ru
E-mail: abdelaziz.boudlal@univ-lille1.fr

Abstract. A horizontal liquid film of infinite extent is bounded below by a rigid plane and above by free surface, rotating about a vertical axis. Advective flow is set by imposing a constant temperature gradient on both boundaries. In the framework of Boussinesq approximation, we obtain an exact stationary solution of Navier–Stokes equations. The linear theory of stability for such flow is numerically investigated by reducing the initial system to the boundary problem for linear one-dimensional partial differential equations. The neutral stability flow states for normal perturbations are carried out; then the critical flow parameters allow us to get the dependence between the critical Grashof number and the wave number for various Taylor ones at a fixed Prandtl number, Pr=6.7.

Introduction

Advection generated in the liquid layer is due to horizontal heterogeneity in density on the boundaries. The advection usually shows in the form of different flows, the difference from convective currents is perpendicularity of velocity to gravity and buoyancy [1,2]. The advective flows interest is closely related to the geophysical and technological applications. Depending on the values of the Prandtl number, the motion in ocean, the dynamics of the atmosphere and Earth's crust and mantle, the movement of melt in installation for variant horizontal directions of crystallization, the transport processes in shallow basin are among the numerous application fields. When the linear dependence of temperature at the rigid boundary and at the undeformable free surface with the x-coordinate (horizontal) is prescribed, the advective flow can be described analytically.

The theoretical study of advective flow with constant pressure gradient on the boundaries in the absence of rotation was first performed by Birikh [1], considering only one horizontal component of velocity, which is perpendicular to buoyancy force. The buoyancy force is the main cause of motion. This advective flow property remains unchanged for different boundary conditions of velocity. [2–3] made the first investigations on stability of such advective flows. They studied stability of the plane–parallel stationary flows without rotation for either rigid or free upper boundary. Research of stability was conducted with usage of a Galerkin method. It was shown, that the horizontal parallel plate flow is unstable concerning two critical modes of disturbances. At small Prandtle numbers (*Pr*) the

Journal of Physics: Conference Series **216** (2010) 012005 doi:10.1088/1742-6596/216/1/012005

instability has the hydrodynamic nature and is developmental of vortexes on border of counter currents. With growth of a Prandtle number in the field of formation of vortexes there is a steady temperature stratification handicapping development of instability of this type. At moderate and large Prandtle numbers the instability has the Rayleigh nature, it is connected to availability in a flow potentially of unstable zones of distribution of temperature. The stability limit depends on a direction of a wave vector of normal disturbances in relation to speed of the basic flow. Actually this surge represents convective Benard's cells, which one communicates with a fluid flow. Moreover, a disturbance for lower and the high bounds practically do not influence against each other. The flat Rayleigh mode and helical (wave) Rayleigh modes (even and odd) is excreted, if the wave vector is directed bevel way to the basic flow, the arising spatial helical disturbances are more dangerous, than flat one. The researches of stability of advection flow in a layer with flat free high bound and firm lower have shown, that the crisis is produced by flat hydrodynamic mode (for $Pr < 0.075$), or by helical Rayleigh mode (for $Pr > 0.075$). The helical Rayleigh mode will be localized in an instability region near to lower boundary. As well as in case of both firm borders, the hydrodynamic mode with increase of a Prandtle number is stabilized. The flat and helical Rayleigh disturbances developing in the upper unstably stratified zone are less dangerous as contrasted to by localized disturbances near to lower boundary.

Smith M.K. and Davis S.H. [4–5], Hart [6] investigated the stability of the advective thermocapillary flow and the advective flow with thermal insulative boundaries in horizontal layer without rotation.

The other kind of flow, which was analytically described by Ekman [7–9], is generated by the rotation of a horizontal layer of isothermal liquid caused by the friction on the lower boundary and the tangential stress of an external force on the upper free boundary. Ekman flow, which is not an advective one, has the typical spiral, and it is the main boundary flow of the atmosphere and ocean. The stability problem of the Ekman is well studied and basically is resolved (see, for example [10]).

The analytical description of the advective flow in a rotating horizontal layer of liquid was first given by Aristov [11]. In this work, the velocity vector is perpendicular to the buoyancy force too, but it has two horizontal components. The advective nature of this flow is similar to Birikh's one and the corresponding spiral is similar to Ekman's flow. It is possible for the advective flow in rotating layer of liquid to play the same role in the atmosphere and ocean as in Ekman's flow. A large class of advective currents in different media and at different boundary conditions is analytically described [12]. The rate of advective flow grows with an increase of the lateral temperature gradient, and it may become unstable. In such situation the hydrodynamic crisis is accompanied by the crisis of heat transfer. As a result of instability, a secondary flow arises, which, in turn, may loose its stability and the development of turbulence may follow.

The instability of rotating advective flow has been come under research only recently. The flows of this family are used to derive quasi-two-dimensional models of geophysical hydrodynamics [13–15], which are a development of "shallow water" theory for viscous non-isothermal liquids, and it is very important for such models to investigate the stability of advective flows. In the papers [16–19] it was studied the stability of advective flow in rotating liquid layer with rigid boundaries for Prandtl number $Pr=6.7$ (water). This is the simplest situation. Instability has oscillatory character and it develops near the boundaries. Then, to follow the founders [1], we study the stability of advective flow in rotating liquid layer with a free surface and linear temperature on the boundaries. This assumption is used to study the main thermal mechanism in large and mesa scale phenomena in atmosphere and ocean. This is one of the ways to investigate the complicated hydrothermodynamic phenomena with the help of mathematical methods of research. Such boundary conditions for temperature may be realized in laboratory experiment too. And these simple assumptions of boundary conditions are necessary to progress in understanding the complexity of such flows.

1. Mathematical model

In the present investigation, we consider a horizontal liquid layer of infinite extent limited from below and above respectively by a rigid plane and a free upper surface (Fig. 1). The rigid plane rotates with a constant angular velocity Ω_0 about a vertical axis. The rotating rectangular frame of reference $Oxyz$ is chosen as follows: the Oz-axis coincides with the axis of rotation, Oxy lies on the horizontal mid-plane of the liquid layer. We denote: h, the layer half-thickness of the fluid layer; $\vec{v} = (v_x, v_y, v_z)$, the velocity vector; p, the pressure; t, the time; ρ_0 a homogeneous density; T is counted off a homogeneous temperature; β is the coefficient of thermal expansion; $\vec{\Omega} = \Omega_0 \vec{e}_z$, where $\vec{e}_z$ is the unit vector directed vertically upwards; $\vec{r}$ is radius–vector, g is the gravity acceleration, ν is the coefficient of kinematic viscosity and χ is the coefficient of heat diffusivity.

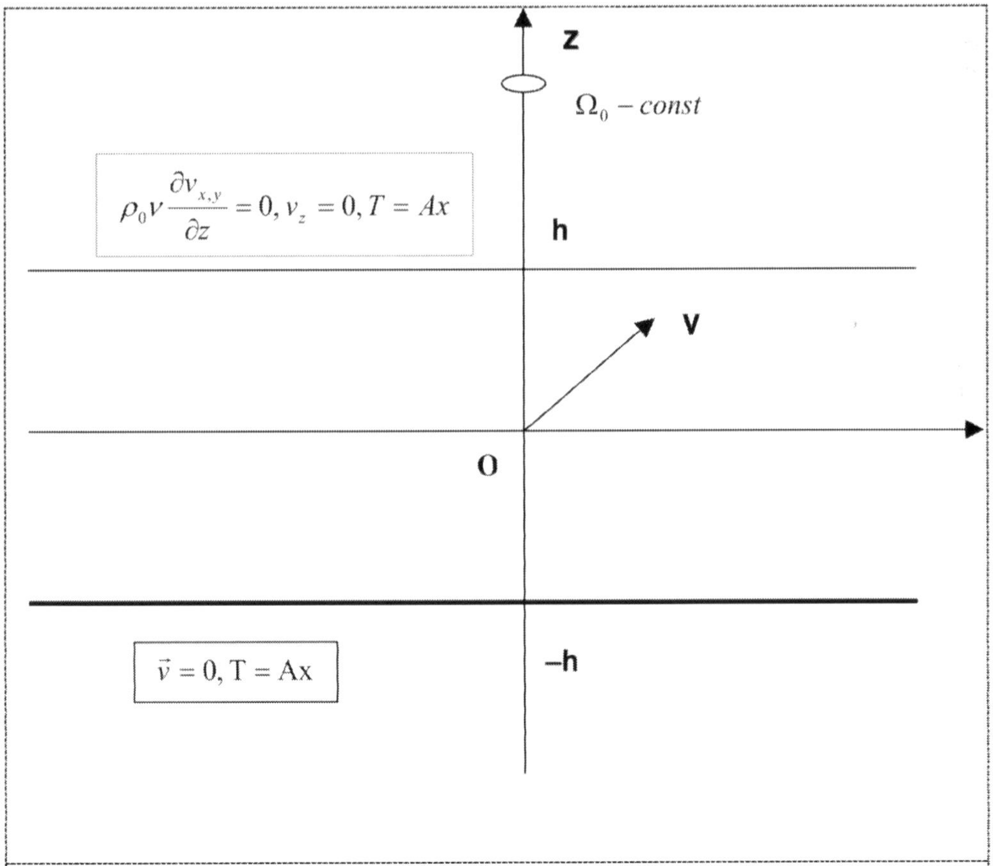

Figure 1. A sketch of the geometry of the infinite horizontal liquid layer.

We assume that T, along the two boundaries changes linearly as a function of x-coordinate. Our aim is to study the flow structure evolutions for an incompressible fluid and analyze its stability in the rotating Cartesian coordinate system $Oxyz$ on the basis of free convective equations in Boussinesq approximation [2, 16, 20]:

$$\rho = \rho_0 (1 - \beta T),$$

$$\frac{\partial \vec{v}}{\partial t} + (\vec{v}\nabla)\vec{v} = -\frac{\nabla p}{\rho_0} + \nu \Delta \vec{v} + g\beta T \vec{e}_z - 2(\vec{\Omega} \times \vec{v}) + \beta T\, \vec{\Omega} \times (\vec{\Omega} \times \vec{r}),$$

Journal of Physics: Conference Series **216** (2010) 012005 doi:10.1088/1742-6596/216/1/012005

$$\nabla\cdot\vec{v}=0,$$
$$\frac{\partial T}{\partial t}+\vec{v}\nabla T=\chi\,\Delta T.$$

We assume that the Froude number [20, 21] $Fr=\Omega_0^2R/g<<1$, i.e., the convective forces, which arises because of the non-homogeneous density in centrifugal field, are negligible with regards to the convective force in gravitational field, The deformations of the free surface caused by rotation and thermocapillary effects are ignored.

The characteristic length, time, velocity, temperature and pressure for scaling are h, h^2/ν, $g\beta Ah^3/\nu$, Ah, and $\rho_0 g\beta Ah^3$, respectively, where, A is a constant temperature gradient on both boundaries. Then, the dimensionless governing equations can be written in the form as follows:

$$\frac{\partial\vec{v}}{\partial t}+Gr\left(\vec{v}\,\nabla\right)\vec{v}+\sqrt{Ta}\left(\vec{e}_z\times\vec{v}\right)=-\nabla p+\Delta\vec{v}+T\,\vec{e}_z, \tag{1}$$

$$\nabla\cdot\vec{v}=0, \tag{2}$$

$$\frac{\partial T}{\partial t}+Gr\,\vec{v}\nabla T=\frac{1}{\Pr}\Delta T. \tag{3}$$

and the boundary conditions:

$$T=x,\ \vec{v}=0\ \text{ at } z=-1;\ \ T=x,\ \frac{\partial v_{x,y}}{\partial z}=v_z=0\ \text{ at } z=1. \tag{4a,b}$$

Here $Gr=\dfrac{g\beta Ah^4}{\nu^2}$, is the Grashof number; $Ta=\left(\dfrac{2\Omega_0 h^2}{\nu}\right)^2$, is the Taylor number; $Pr=\dfrac{\nu}{\chi}$, is the Prandtl number; the Rayleigh number is $Ra=GrPr$. The subscripts x, y in the equation (4,b) mean that the equality holds for both v_x and v_y.

2. Basic flow

An exact steady flow state solution of the system (2)-(4), subject to the boundary conditions (4) can be found in the form :

$$v_x=u_0(z),\ v_y=v_0(z),\ v_z=0,\ T=T_0(x,z)=x+\tau_0(z),\ p=p_0(x,y,z). \tag{5}$$

Notice in this system that the form of the velocity components is admissible, as the boundary conditions do not require lateral variations in v_x, v_y, v_z.

Substituting formulae (5) in the system (1)-(3), we get the following partial differential equations for speed, temperature and pressure distributions:

$$\frac{\partial p_0}{\partial z}=x+\tau_0, \tag{6}$$

$$-\sqrt{Ta}\,v_0=-\frac{\partial p_0}{\partial x}+u_0'', \tag{7}$$

$$\sqrt{Ta}\,u_0=-\frac{\partial p_0}{\partial y}+v_0'', \tag{8}$$

$$Ra\,u_0=\tau_0''. \tag{9}$$

The boundary conditions write:

$$u_0'=v_0'=0,\ \tau_0=0\ \text{ at } z=1, \tag{10}$$

$$u_0=v_0=0, \tau_0=0\ \text{ at } z=\text{-}1 \tag{11}$$

and closed flow conditions read:

$$\int_{-1}^{1} u_0 dz = 0\,,\ \int_{-1}^{1} v_0 dz = 0\,. \tag{12}$$

Solving the system (6)-(12) gives after some tedious manipulations, the exact solution:

$$u_0(z) = \Re\,(M(z)),\quad v_0(z) = \Im\,(M(z)), \tag{14a,b}$$

$$\tau_0(z) = \frac{Ra}{\sqrt{Ta}}\left[v_0(z) - 0.5v_0(1)(1+z) + 0.5\left(1-z^2\right)\Im\left(\frac{D_2}{D}\mu\sinh(\mu)\right)\right], \tag{14c}$$

where,

$$M(z) = \frac{\cosh(\mu(z-1))}{\mu^2 D} - \frac{\sinh(\mu(z+1))}{\mu^3 D} - \frac{\sinh(2\mu)}{2\mu^3 D} + \frac{\sinh(\mu)}{\mu^4 D}(\sinh(\mu z) + \sinh(\mu)) - \frac{z}{\mu^2},$$

$$\mu = \sqrt[4]{\frac{Ta}{4}}\left(1+i\right),\ i = \sqrt{-1}\,,\ D = \frac{\sinh(2\mu)}{2\mu} - \cosh(2\mu)\,,\ D_2 = -\frac{\sinh(\mu)}{\mu^3} + \frac{\cosh(\mu)}{\mu^2}\,.$$

Here, $\Re$ and $\Im$ represent the real and imaginary parts of the corresponding complex quantity, respectively.

For the results presented below, Pr was fixed: Pr=6.7 (water). In the manner of horizontal velocity profiles, plotted respectively in Figs. 2(a), (b), the variation of dimensionless temperature τ_0 is displayed versus z for different values of Taylor number in Figs. 2(c). In general, the flow structure is asymmetric about the horizontal mid-plane and can be divided into two currents moving in opposite directions as it can be checked from the figures. A dynamical boundary layer appears only close to the bottom, whereas a steep temperature gradient develops in the thermal boundary layer near the upper free surface too. Both thicknesses grow when the Taylor number decreases. In the central core, i.e., outside the boundary layers, the component u_0 can be considered as almost constant, whereas v_0 and τ_0, are approximately linear. The intensity of motion is much lower inner the central core of the fluid than at near the boundaries.

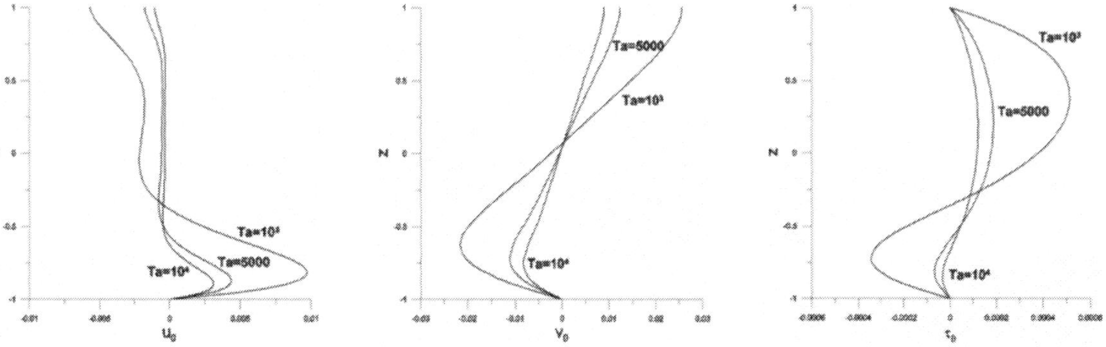

(a) (b) (c)

Figure 2. Dependence of velocity components (a) u_0, (b) v_0, and of (c) temperature τ_0 on vertical z-coordinate for various given Taylor numbers.

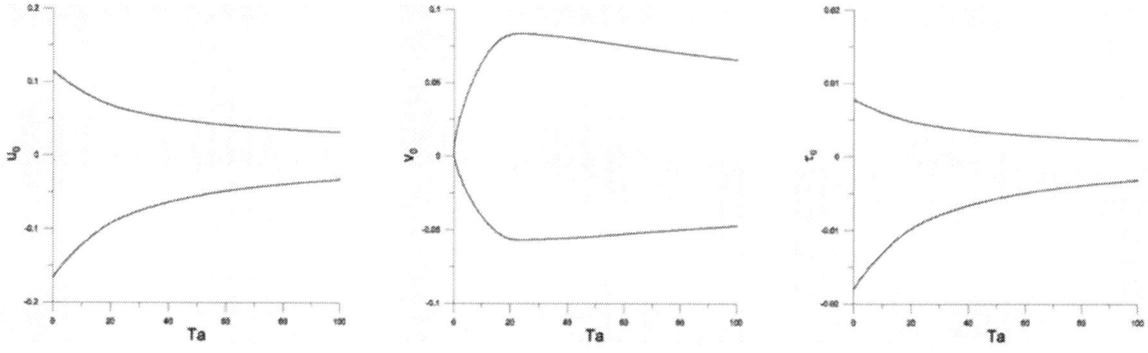

(a) (b) (c)

Figure 3. Minimal and maximal velocity components (a) u_0, (b) v_0, and (c) temperature τ_0 versus Taylor number Ta.

In Figs. 3(a), (b), (c) are displayed the minima (resp. the maxima) of the stationary flow solution $(u_0, v_0, 0)$, τ_0 as a function of Ta. As it can be seen, the minima (resp. the maxima) of τ_0 and u_0 increase monotonically (resp. decrease), as Ta increases. Fig. 3(b) shows that above the horizontal mid-plane, the maxima of v_0, increase with Ta, up to the limit $Ta = 24.5$, then decrease almost proportionally to $1/\sqrt{Ta}$. Below the plane $z = 0$, about the minima of v_0, the tendency is inverted : they decrease with Ta till the value $Ta = 27$, then, they monotonically increase.

3. Linear stability

The stability analysis dealt with herein aims to investigate the growth rate of small perturbations superimposed to a steady flow state solution :

$$\vec{v} = \vec{v}_0 + \vec{V}, \quad \vec{v}_0 = (u_0, v_0, 0), \quad \vec{V} = (u, v, w), \quad T = T_0 + \theta, \quad p = p_0 + \tilde{p}. \tag{15a, b, c}$$

Here $\vec{V}$, θ and $\tilde{p}$ are small perturbations. Linearizing in the usual manner, taking the flow equations for $(\vec{v}_0, T_0, p_0)$ into account, the linearized system (1)-(3) takes the form :

$$\frac{\partial \vec{V}}{\partial t} + Gr\left[(\vec{V}\,\nabla)\vec{v}_0 + (\vec{v}_0\nabla)\vec{V}\right] + \sqrt{Ta}\left(\vec{e}_z \times \vec{V}\right) = -\nabla\tilde{p} + \Delta\vec{V} + \theta\,\vec{e}_z, \tag{16}$$

$$\nabla \cdot \vec{V} = 0, \tag{17}$$

$$\frac{\partial \theta}{\partial t} + Gr\left[\vec{V}\,\nabla T_0 + \vec{v}_0 \nabla \theta\right] = \frac{1}{\Pr}\Delta\theta, \tag{18}$$

$$\vec{V} = 0, \ \theta = 0, \quad \text{at } z = -1, \tag{19}$$

and

$$\frac{\partial u}{\partial z} = \frac{\partial v}{\partial z} = w = 0, \ \theta = 0, \quad \text{at } z = 1. \tag{20}$$

In the general case of spatial normal disturbances, all the variables are proportional to $\exp\left[-\lambda t + i\left(k_x x + k_y y\right)\right]$, where $\lambda = \lambda_1 + i\lambda_2$ is the characteristic decrement, k_x and k_y are the components of the wave vector along $x-$ and $y-$ directions. If the decrement λ is complex, the disturbances oscillate with frequency λ_2 and propagate in the flow like the waves with a phase velocity $c = \frac{\lambda_2}{k}$. Squire's theorem does not apply due to the presence of longitudinal temperature gradient (Gershuni et al., 1989; Gershuni and Zhukhovitskii, 1996). Thus, the disturbance equations cannot be transformed and reduced to the corresponding plane problem. Moreover, we shall take into consideration two more important limiting cases. They are spatial disturbances of the first type in the form of rolls with axes parallel to x-axis and spatial disturbances of the second type in form of rolls with axes perpendicular to x-direction.

3.1. The spatial disturbances in form of rolls with axes perpendicular to x-axis

The problem of linear stability for the advective flow in a horizontal rotating layer of liquid is here solved by reducing it to a boundary problem for a system of linear one-dimensional partial differential equations. The equations for the spatial disturbances of the second type are obtained for $k_y = 0$. All the three components of velocity disturbance and temperature one, are functions of time t and two spatial coordinates x and z. We introduce the stream function $\psi(t,x,z)$: $u = -\frac{\partial \psi}{\partial z}$, $w = \frac{\partial \psi}{\partial x}$ and vorticity $\phi = \frac{\partial u}{\partial z} - \frac{\partial w}{\partial x} = -\Delta\psi$. Looking for all unknown functions F in the form : $F = [F_1(t,z) + iF_2(t,z)]\exp(ik_x x)$, the linearized system (16)-(20) is reduced to the one-dimensional system as follows:

$$\frac{\partial \phi_1}{\partial t} - k_x\, Gr\left[u_0(z)\phi_2 + u_0''(z)\psi_2\right] - \sqrt{Ta}\frac{\partial v_1}{\partial z} = \frac{\partial^2 \phi_1}{\partial z^2} - k_x^2\phi_1 + k_x\theta_2, \tag{21}$$

$$\frac{\partial^2 \psi_1}{\partial z^2} - k_x^2\psi_1 + \phi_1 = 0, \tag{22}$$

$$\frac{\partial \phi_2}{\partial t} + k_x Gr\left[u_0(z)\phi_1 + u_0''(z)\psi_1\right] - \sqrt{Ta}\frac{\partial v_2}{\partial z} = \frac{\partial^2 \phi_2}{\partial z^2} - k_x^2\phi_2 - k_x\theta_1, \tag{23}$$

$$\frac{\partial^2 \psi_2}{\partial z^2} - k_x^2\psi_2 + \phi_2 = 0, \tag{24}$$

$$\frac{\partial v_1}{\partial t} - k_x Gr\left[u_0(z)v_2 + v_0'(z)\psi_2\right] - \sqrt{Ta}\frac{\partial \psi_1}{\partial z} = \frac{\partial^2 v_1}{\partial z^2} - k_x^2 v_1, \tag{25}$$

$$\frac{\partial v_2}{\partial t} + k_x Gr\left[u_0(z)v_1 + v_0'(z)\psi_1\right] - \sqrt{Ta}\frac{\partial \psi_2}{\partial z} = \frac{\partial^2 v_2}{\partial z^2} - k_x^2 v_2, \tag{26}$$

$$\frac{\partial \theta_1}{\partial t} - k_x Gr\left[u_0(z)\theta_2 + \tau_0'(z)\psi_2\right] - Gr\frac{\partial \psi_1}{\partial z} = \frac{1}{\Pr}\left[\frac{\partial^2 \theta_1}{\partial z^2} - k_x^2\,\theta_1\right], \tag{27}$$

$$\frac{\partial \theta_2}{\partial t} + k_x Gr\left[u_0(z)\theta_1 + \tau_0'(z)\psi_1\right] - Gr\frac{\partial \psi_2}{\partial z} = \frac{1}{\Pr}\left[\frac{\partial^2 \theta_2}{\partial z^2} - k_x^2\theta_2\right] \tag{28}$$

subject to the boundary conditions :

$$\begin{aligned} &\psi_{1,2} = \frac{\partial \psi_{1,2}}{\partial z} = v_{1,2} = \theta_{1,2} = 0 \ \text{ at z = -1}, \\ &\psi_{1,2} = \phi_{1,2} = \frac{\partial v_{1,2}}{\partial z} = \theta_{1,2} = 0 \ \text{ at z = 1}. \end{aligned} \tag{29}$$

In the notation $F_{1,2}$, the subscript means that the equalities hold for both F_1 and F_2.

The advective flow stability is characterized by neutral curves, which describe the dependence between the Grashof number and k_x, and separate the zones of stability and instability of this flow. The real part of the characteristic decrement $\lambda_1(Gr,k_x,Pr,Ta)$ is a function of several parameters. It vanishes on neutral curve. In the stability domain, $\lambda_1 < 0$, and in the region of instability $\lambda_1 > 0$. The numerical methodology of computing neutral curves for different values of Taylor number is represented in detail [17–18]. We define the function λ_1 discretely by repeated calculation of the system (21)-(29). This system is solved by finite difference method with the help of implicit scheme for numerical calculations. The implicit schemes for the system (21)–(29) is absolutely stable. The number of points N, used on the $z-$ axis depends on the Taylor number value and the needed precision. It must be noticed that the boundary layer thickness decreases with the growth of Ta. The approximation of schemes is $O(\Delta t + \Delta z^2)$, where Δt is a step of discretization along time axis and $\Delta z = 2/N$ is a step of discretization along z-axis. According to the theory of finite difference method the stability and approximation guaranties convergence. For the values of Taylor number ranged in the interval $0 \le Ta \le 10^6$, N is varying between 101 and 201. It must be increased with the growth of Ta.

Figs. 4 show marginal curves for various values of Taylor number. Below each of these curves the flow is stable, and it is instable above. It is to be emphasized that the critical Grashof number increases with the increasing values of Ta. Stated otherwise, the rotation is a stabilizing factor against this kind of disturbance.

The wave numbers k_x corresponding to the critical Grashof numbers are presented as function of Taylor numbers in Figs. 5. As one can see from the figures, k_x varies with the growth of Ta. At small values of Taylor number: $0 < Ta < 500$, the wave number k_x varies from 3.25 to 4.25. For $0 < Ta < 110$, it monotonically increases, while in the interval $110 < Ta < 500$ it monotonically decreases (Fig. 5a). For $Ta \ge 500$ with a further increase of Taylor number (Fig. 5b), a growth in k_x is observed, that approximately follows:

$$k_x \approx 3.36 + 0.0154\sqrt{Ta - 500}\,. \qquad (30)$$

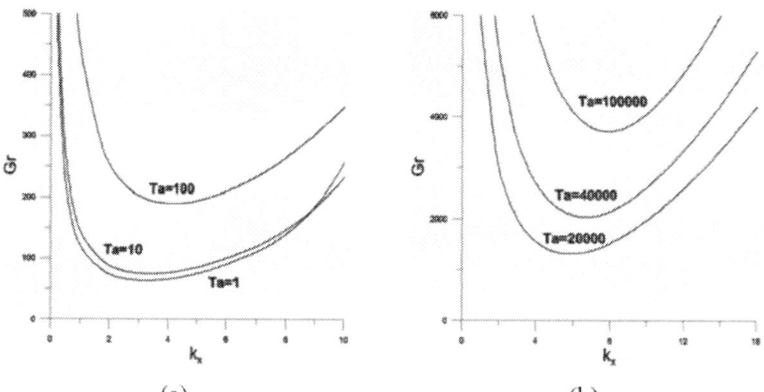

Figure 4. The neutral curves (a) for a small *Ta*, and (b) for a large *Ta*.

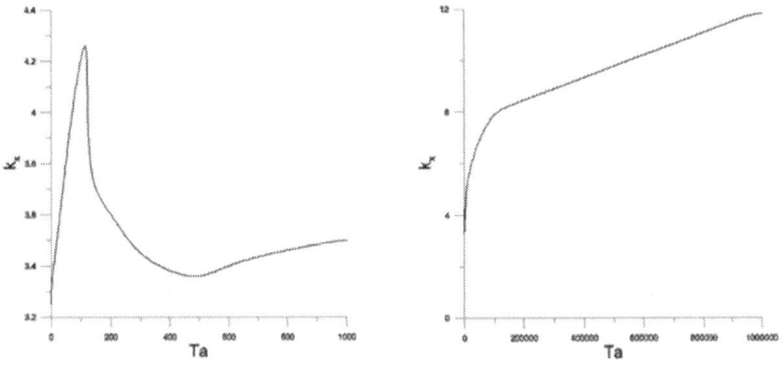

(a) (b)

Figure 5. The wave number k_x (corresponding to a critical Grashof number Gr_c) versus the Taylor number, (a) for a small, and (b) for a large *Ta*.

As for the critical value of Grashof number Gr_c which increases with the growth of Taylor number, and satisfactorily fits (see Fig. 6a).

$$Gr_k \approx 62.07 + 1.561 Ta^{0.675} \quad (31)$$

The numerical calculations show that, like in the absence of rotation, a case that was considered in (Gershuni G.Z., 1989), instability phenomenon preserves the oscillating character for all Taylor numbers (Fig. 6b). The period $\bar{\lambda} = 2\pi/\lambda_2$ grows with an increasing of Taylor number on the interval $0 \le Ta \le 2 \cdot 10^4$, and then, decreases for $2 \cdot 10^4 < Ta \le 10^6$ (see Fig. 6b). Maximum of a dimensionless phase velocity $c = 0{,}841$ is reached at $Ta = 5000$.

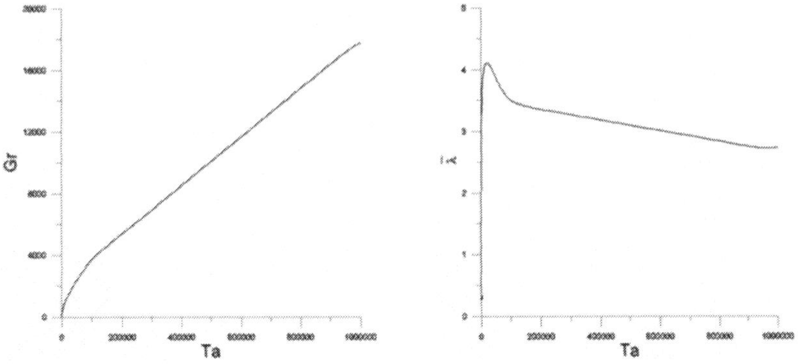

(a) (b)

Figure 6. Dependence (a) of the critical Grashof number (b) of the period $\bar{\lambda}$ on *Ta*.

3.2. The spatial disturbances in the form of rolls with axes parallel to x-direction

The equations for the spatial disturbances of the first type are obtained for $k_x = 0$. Reducing this other limiting case to the boundary problem for the system of linear one-dimensional partial differential equations, similarly to part 3.1., allow to solve the problem of linear stability. The disturbances of the three velocity components and temperature are here functions of time t and two spatial coordinates y

and z. Let us introduce the stream function $\Psi(t,y,z)$: $v=-\dfrac{\partial\Psi}{\partial z}$, $w=\dfrac{\partial\Psi}{\partial y}$ and vorticity : $\Phi=\dfrac{\partial v}{\partial z}-\dfrac{\partial w}{\partial y}=-\Delta\Psi$. All unknown functions F are introduced as $F=\left[F_1(t,z)+iF_2(t,z)\right]\exp(ik_y y)$. Substituting into the system (16)-(20) reduces to the following system of one-dimensional partial differential equations:

$$\frac{\partial\Phi_1}{\partial t}-k_y Gr\left[v_0(z)\Phi_2+v_0''(z)\Psi_2\right]+\sqrt{Ta}\frac{\partial u_1}{\partial z}=\frac{\partial^2\Phi_1}{\partial z^2}-k_y^2\Phi_1+k_y\theta_2, \tag{32}$$

$$\frac{\partial^2\Psi_1}{\partial z^2}-k_y^2\Psi_1+\Phi_1=0, \tag{33}$$

$$\frac{\partial\Phi_2}{\partial t}+k_y Gr\left[v_0(z)\Phi_1+v_0''(z)\Psi_1\right]+\sqrt{Ta}\frac{\partial u_2}{\partial z}=\frac{\partial^2\Phi_2}{\partial z^2}-k_y^2\Phi_2-k_y\theta_1, \tag{34}$$

$$\frac{\partial^2\Psi_2}{\partial z^2}-k_y^2\Psi_2+\Phi_2=0, \tag{35}$$

$$\frac{\partial u_1}{\partial t}-k_y Gr\left[v_0(z)u_2+u_0'(z)\Psi_2\right]+\sqrt{Ta}\frac{\partial\Psi_1}{\partial z}=\frac{\partial^2 u_1}{\partial z^2}-k_y^2u_1, \tag{36}$$

$$\frac{\partial u_2}{\partial t}+k_y Gr\left[v_0(z)u_1+u_0'(z)\Psi_1\right]+\sqrt{Ta}\frac{\partial\Psi_2}{\partial z}=\frac{\partial^2 u_2}{\partial z^2}-k_y^2u_2, \tag{37}$$

$$\frac{\partial\theta_1}{\partial t}-k_y Gr\left[v_0(z)\theta_2+\tau_0'(z)\Psi_2\right]+Gru_1=\frac{1}{\Pr}\left[\frac{\partial^2\theta_1}{\partial z^2}-k_y^2\theta_1\right], \tag{38}$$

$$\frac{\partial\theta_1}{\partial t}+k_y Gr\left[v_0(z)\theta_1+\tau_0'(z)\Psi_1\right]+Gru_1=\frac{1}{\Pr}\left[\frac{\partial^2\theta_2}{\partial z^2}-k_y^2\theta_2\right] \tag{39}$$

with the boundary conditions

$$\begin{aligned}&\Psi_{1,2}=\frac{\partial\Psi_{1,2}}{\partial z}=u_{1,2}=\theta_{1,2}=0 \text{ at z = -1,}\\ &\Psi_{1,2}=\Phi_{1,2}=\frac{\partial u_{1,2}}{\partial z}=\theta_{1,2}=0 \text{ at z = 1.}\end{aligned} \tag{40}$$

The stability is numerically investigated by the normal disturbance methodology, similar to that applied in part 4.1 [17–18]. The system (32)-(40) is here solved by finite difference method with an implicit scheme for numerical calculations too. The number of points is between 101 and 201. The Taylor number ranges in the interval $0\le Ta\le 10^6$. The Prandtl number $Pr=6.7$ (water).

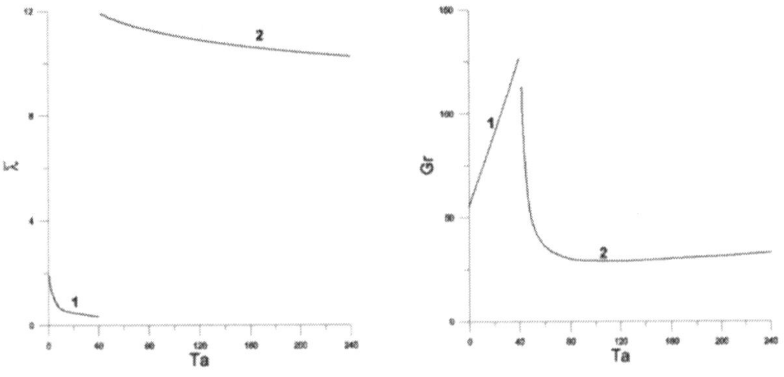

(a) (b)

Figure 7. (a) Period $\bar{\lambda}$ and (b) critical Grashof number for small *Ta*.

In the absence of rotation, i.e. *Ta=0,* the instability of advective flow in a horizontal layer of a liquid has a monotonous character, $\lambda_2 = 0$ [2]. At $Ta > 0$, it becomes an oscillatory one as we can see in Fig. 7a. For small values of Taylor number, $0 < Ta < 42$, rotation stabilizes the advective flow in a similar way to the case $k_y = 0$, considered in part 4.1. The zones below the neutral curves in Fig. 8a are the regions of stability, and above these curves the flow is instable. In this interval, the critical Grashof number grows quasi linearly, as this is well shown in the left part of curve 1 of Fig. 7b :

$$Gr_c \approx 55.27 + 1.772Ta\,. \tag{41}$$

The corresponding numerical values of the wave number k_y belong to the narrow interval [*3.4*, 3.*74*]. As for the period $\bar{\lambda}$, it is proportional to $\sqrt{Ta}$ in these cases (see curve 1 of Fig. 7a).

For *Ta=42,* to which correspond $k_y = 0.12$ and $\bar{\lambda} = 11.9$, the long-wave oscillatory instability starts to develop (see curves 2 of Figs. 7a, b). It sharply reduces the critical Grashof number Gr_c, and the advective steady flow becomes less stable even when the rotation is absent (see Fig. 7b). From $42 \le Ta \le 120$, Gr_k decreases. After the value $Ta = 120$, it starts again to increase with the growth of Taylor number; Stated otherwise, sufficiently high rotation is a stabilizing factor for the advective flow.

The evolution of the neutral curves for different values of the Taylor number is presented in Figs. 8. Until the value $Ta = 42$, only one instability region, which is situated above any neutral curve, is observed (see Figs. 8a). Above this value of Taylor number, contrary to the case $k_y = 0$ of part 4.1, a second zone of instability starts, which, with the growth of *Ta* increasingly enlarges. The example of Fig. 8b, drawn for $Ta = 70$, illustrates well this situation. With the further increase of *Ta*, the two neutral curves unify in a solely one with two minima. That may happen approximately for $100 < Ta < 250$. For $Ta \ge 250$, the neutral curves find again their initial tendency with one minimum (see Fig.8c, d).

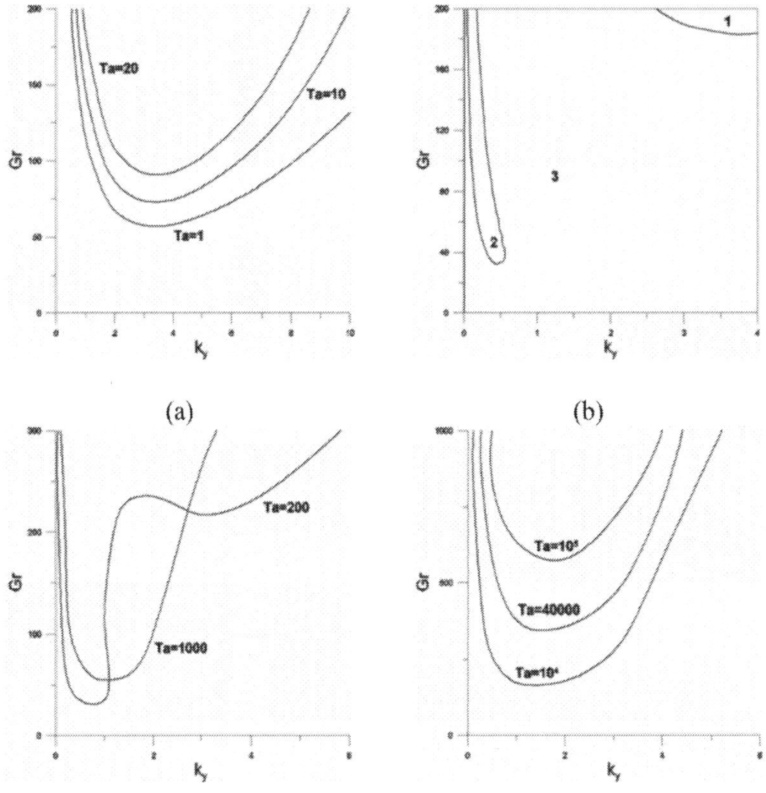

(a) (b) (c) (d)

Figure 8. The neutral curves, (a) for the small Taylor numbers, (b) for *Ta=70*, (c) in a transitional situation, (d) for high *Ta*.

In Figures 9 are presented the dependence of the critical Grashof number, Gr_c, and the corresponding wave number k_y, and period $\overline{\lambda}$ upon the Taylor number in the interval : $120 \le Ta \le 10^6$. As one can see from the figures the evolution of these functions approximately follows the forms:

$$Gr_k \approx 28.86 + 0.539(Ta - 120)^{0.6}, \tag{42}$$

$$k_y \approx 0.12 + 0.268(Ta - 42)^{0.162}, \tag{43}$$

and

$$\overline{\lambda} \propto Ta^{-\frac{1}{8}}. \tag{44}$$

Journal of Physics: Conference Series **216** (2010) 012005 doi:10.1088/1742-6596/216/1/012005

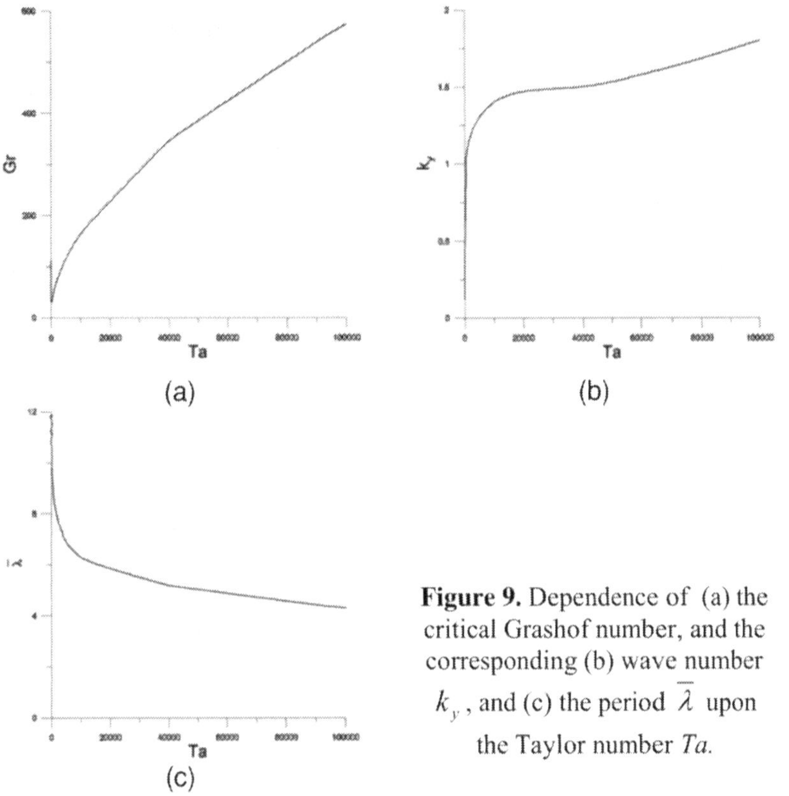

Figure 9. Dependence of (a) the critical Grashof number, and the corresponding (b) wave number k_y, and (c) the period $\overline{\lambda}$ upon the Taylor number Ta.

4. Conclusion

The advective flow that may develop in a horizontal layer of a rotating liquid with a free surface, bordered below by a horizontal solid plane subject to a constant temperature gradient on the two boundaries has been studied.

Starting from Navier-Stokes equations, an exact stationary solution that describes analytically this type of flow has been derived. The resulting calculations for the horizontal velocity components have shown that the structure of the advective flow is spiral and that boundary layers form at both the bottom and free surface. As for the temperature, only close to the solid plane that the boundary layer has been observed. The profiles of velocity components and temperature are depended on the Taylor number.

With an increase of Taylor number a decrease in thickness of both boundary layers have been shown. In the core of the fluid layer, when Ta increases the profile of the x– component of velocity becomes almost constant, while that of the y– one remains approximately linear. As for the temperature profile, with an increase of Taylor number the parabolic shape is closely achieved. The intensity of the movement has been shown much lower inner the core of the flow than outer.

Using small perturbation technique, linear one-dimensional systems of partial differential equations have been obtained. Numerical computations performed at fixed Prandtl number, Pr=6.7 (water) have shown that an oscillatory instability of the advective flow for any value of Taylor number can be developed. The study has revealed that the rotation is a stabilizing factor for the advective flow at spatial disturbances of the second type in form of rolls with axes perpendicular to x–direction. As for spatial disturbances of the first type in the form of rolls with axes parallel to x-direction, at low and high values of Ta, the rotation has also a stabilizing effect on the advective flow. At moderate Taylor numbers : $42 \leq Ta \leq 1000$, the inverted trend has been observed.

Helical Rayleigh instability is developed. One critical helical Rayleigh mode take place for spatial disturbances of the first type and two modes take place for spatial disturbances of the second type. Finally, as a main result of rotational effect on the stability we can retain that the first type of spatial disturbance is more dangerous than the second one.

5. References

[1] Birich R.V. 1966 *AMTP About thermocapillary convection in horizontal liquid layer* **3** 69.
[2] Gershuni G.Z., Zhukhovitsky E.M., Nepomnyashchy A.A. 1989 *Stability of convective flow.* (Nauka, Moscow)
[3] Gershuni G.Z., Laure P., Myznikov V.M., Roux B., Zhukhovitsky E.M. 1992 *Microgravity Q.* **2** 141
[4] Smith M.K., Davis S.H. 1983 *J. Fluid Mech.* **132** 119.
[5] Smith M.K., Davis S.H. 1983 *J. Fluid Mech.* **132** 145
[6] Hart J. 1983 . *J. Fluid Mech.* **132** 271.
[7] Ekman V.W. 1905 *Arhiv Mat., Astr., Phys.* **.2** 1
[8] Ekman V.W. 1923 *Arhiv Mat., Astr., Phys* . **17** 1
[9] Pedlosky J. 1979 *Geophysical Fluid Dynamics* (Springer - Verlag, New York).
[10] Faller A.J. 1963 *J. Fluid Mech.* **15.** 560.
[11] Aristov S.N., and Frick P.G. 1988 *Fluid Dynamic* **23** 522.
[12] Aristov S.N., and Schwarz K.G. 2006 *Vortices flows of the advective nature in a rotated layer* (Perm State University, Perm)
[13] Aristov S.N., and Shvarts K.G. *1990 Oceanology* **30**. 414
[14] Aristov S.N., and Schwarz K.G. 1991 *New two-dimensional model of large-scale oceanic circulation. Proc. of 2nd International Conference of Computer Modelling in Ocean Engineering'91* 49 (Balkema, Rotterdam)
[15] Schwarz K.G. 1998 *AIAA Progr. Ser.* **182**. 271
[16] Aristov S.N., and Shvartz K.G. 1999 *Fluid Dynamic* **34**.457.
[17] Schwarz K.G. 2001 *Computational Technology* **6**.702
[18] Schwarz K.G., and Tarunin E.L. 2001 *Computational Technology* **6**.108
[19] Schwarz K.G. 2004 *Instability of Advective Flow in Rotating Horizontal Layer of Liquid // Selected Papers of the International conference "Fluxes and Structures in Fluids". St. Petersburg, Russia, June 23-26, 2003* 164 (Moscow. IPM RAS)
[20] Gershuni G.Z., and Zhukhovitskii E.M. 1976. *Convective Stability of Incompressible Liquid.* Wiley, (Keter Press, Jerusalem)
[21] Szymczyk J.A. 1993 *ZAMM* **76(b)**. T539.

Journal of Physics: Conference Series **216** (2010) 012006 doi:10.1088/1742-6596/216/1/012006

Secondary instability of waves in annular two-phase flow with and without entrainment

S.V. Alekseenko[1,2], A.V. Cherdantsev[1,2, a)], S.M. Kharlamov[1], D.M. Markovich[1,2]

[1]Institute of Thermophysics, Siberian Branch of Russian Academy of Science, Lavrentiev ave., 1, Novosibirsk, 630090, Russian Federation.

[2]Novosibirsk State University, Pirogova str., 2, Novosibirsk, 630090, Russian Federation.

a)Electronic mail: cherdantsev@itp.nsc.ru

Abstract. Spatio-temporal evolution of disturbance waves and ripple waves in annular gas-liquid flow was studied using high-speed modification of laser-induced fluorescence technique. Both flow regimes with and without entrainment were studied. It was shown that all the ripples in regimes with entrainment arise due to instability of back slopes of the disturbance waves, and their further evolution is predefined by the point of appearance. In regimes without entrainment two types of waves were detected, where waves of one type arise due to instability of back slopes of waves of the other type. Thus, similarity of wavy structure of flow regimes with and without entrainment was established.

1. Introduction

Presence of intensive gas shear drastically changes wavy structure of liquid film in comparison to that of gravitationally falling films. The film surface under strong gas shear is covered by small scale ripple waves (also called "ripples"). According to generally accepted point of view [1], ripples are omnipresent at film surface, at any liquid Reynolds numbers. At higher liquid Reynolds numbers, the disturbance waves appear. The latter are characterized by high amplitude (several times greater than thickness of the residual liquid layer between waves); large longitudinal size (several centimeters); large transverse size (in channels with inner diameter less than 3 cm disturbance waves form full ring around the circumference of the channel); high velocity (several m/s).

The area of flow parameters where the investigators observed high-amplitude waves approximately coincides with the regime area where entrainment of liquid from the film surface occurs. The tiny liquid droplets are torn off the film surface and entrained into the core of gas stream. According to observations of Woodmansee & Hanratty [2], the droplets tearing-off happens due to scattering of ripples on crests of disturbance waves by the intensive gas shear. Woodmansee & Hanratty [2] proposed a model of appearance of ripples based on simple assumption of Kelvin-Helmholtz instability. An attempt to develop this approach was recently performed by Holowach et al. [3].

Nevertheless, no experimental investigation of ripples appearance process was performed. Such investigation requires the simultaneous measurements of local film thickness at large number of points along the longitudinal section of the channel with high spatial resolution and sampling frequency. The most efficient attempt of such measurements was made by Sekoguchi & Takeishi [4] using measuring system that consisted of large number of conductivity probes, separated by distance of 5 mm. Unfortunately, this was not sufficient to reach the necessary resolution.

At low liquid Reynolds numbers no entrainment occurs at any high gas velocities. It is commonly supposed that no disturbance waves appear at the film surface in absence of entrainment, and only ripples exist. Asali & Hanratty [5] tried to predict the ripples frequency in such regime, on the assumption that the frequency is equal to frequency of waves of maximum growth. The comparison has shown that experimental wavelength is approximately two times higher than wavelength predicted by theory.

The goal of present work is to investigate the appearance and further evolution of ripples in regimes with and without entrainment in details.

2. Experimental setup and measurement technique

Experiments were performed in vertical Plexiglas cylindrical channel with the inner diameter of 15 mm and the length of 1 meter, at distances 1-20 and 50-70 cm below the inlet. Liquid film was formed on channel wall using a slot distributor with the gap width of 0.5 mm, gas entered the channel through a co-axial tube of slightly smaller diameter.

Channel was made with square outer side to decrease optical distortions. Continuous green laser with the wavelength 532 nm and power 50 mW was used as the light source. Laser beam was converted into a vertical light sheet aimed at the region of measurements. Rhodamin-6G in concentration 30 mg/l was used as fluorescent matter. Fluorescent light was registered by the 10-bit CCD camera with linear (2048*1 pixels) matrix. Camera was aimed at the region of measurements with the focus on the nearest inner wall of the channel. Distance from camera to channel wall was 450 mm. Camera was equipped with low-pass orange filter. Axes of camera and laser were placed at a small angle to each other on horizontal plane. This allowed avoiding registration of light emitted by the film flowing on the remote wall of the channel. The sketch of experimental setup is given in Fig. 1.

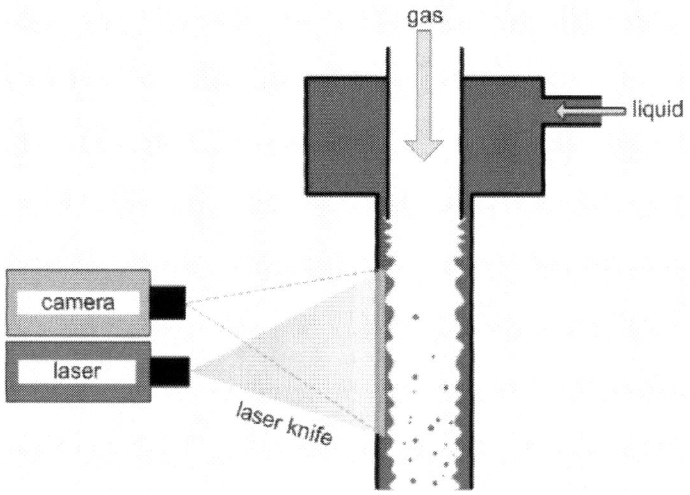

Figure 1. Experimental setup.

All experiments were conducted in the measurement area with the length of 20 cm with spatial resolution 0.1 mm. Width of measurements area was equal 0.1 mm, and the thickness of laser sheet was about 1 mm. Exposition time was 150 μs and registration frame rate - 2000 fps. Registered image brightness was converted into local film thickness using calibration curve obtained in a set of in situ calibration tests.

Flow regimes characterized by the presence of disturbance waves and liquid entrainment were investigated. Liquid Reynolds number Re (defined as ratio of volumetric liquid flow rate per unit

Journal of Physics: Conference Series **216** (2010) 012006 doi:10.1088/1742-6596/216/1/012006

perimeter of the tube to the kinematic viscosity of liquid) varied in the range of 140-350 for regimes with entrainment and 40-60 for regimes without entrainment; superficial gas velocity Vg varied in the range of 22-58 m/s (that corresponds to gas Reynolds number range 22000-58000). Distilled water was used as working liquid.

3. Experimental results

The picture of ripples generation in regimes with entrainment is shown in Figure 2. The sequence of instantaneous distributions of local film thickness is shown in reference system moving with the velocity of disturbance wave (3.5 m/s) – a large amplitude wave in the center of the frame. Ripples denoted by 1, 2 and 3 can be seen on the disturbance wave's surface. Ripples 2 and 3 have just appeared, and they are generated at the back slope of the disturbance wave. Our analysis of experimental data shows that all the ripples are generated only at the back slopes of disturbance waves, and, thus, only a part of film surface is unstable to ripples generation.

Returning to the Fig. 2, we can see that ripple 2 was generated closer to the crest of disturbance wave, and the ripple 3 – farther from the crest. In the following moments of time it can be seen that amplitude of ripple 2 will grow to rather high values and the ripple will travel faster than the parent disturbance wave. After reaching the front of the disturbance wave, it will disappear – either slowly or sharply – which can be seen on the example of ripple 1. This disappearance occurs due to scattering of the ripple by the gas shear, described in [2], contributing to the entrainment.

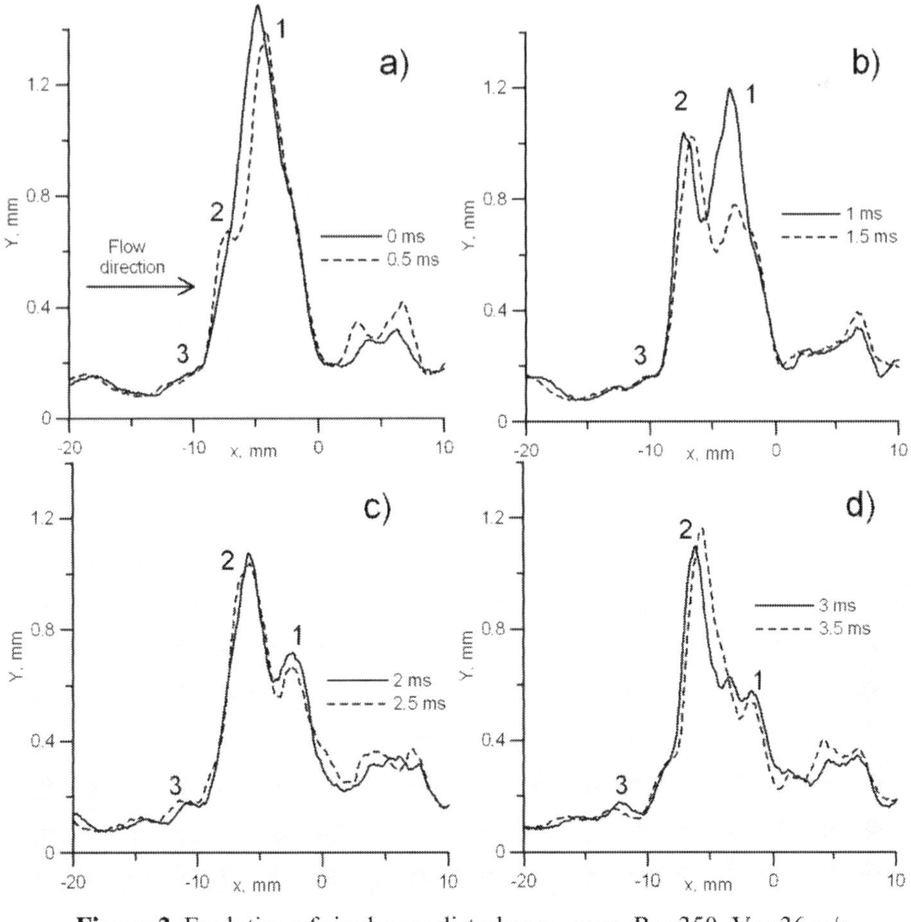

Figure 2. Evolution of ripples on disturbance wave. Re=350, Vg=36 m/s.

Ripple 3 appears farther from the crest of parent disturbance wave and lags behind it. Gradually it reaches the residual layer after the parent disturbance wave and travels along it with approximately constant speed until being overtaken by the following disturbance wave.

The similar sequence for the regimes without entrainment is shown in Figure 3. Despite that all the waves were expected to be ripples, we can mark out some waves that are faster, higher longer than the rest of the waves – similar to disturbance waves in entrainment regimes. Reference system in Fig. 3 is moving with velocity of such a fast wave in the center of the frame (1 m/s). It can be seen that smaller waves, denoted by 1 and 4, are generated at the back front of this fast wave. After inception, waves 1and 4 travel slower than the parent wave. Such smaller waves are promptly absorbed by the following fast wave (absorption can be seen for waves 2 and 3). So, in the regimes without entrainment, only a part of film surface – back slopes of the fast waves – is unstable to smaller waves generation.

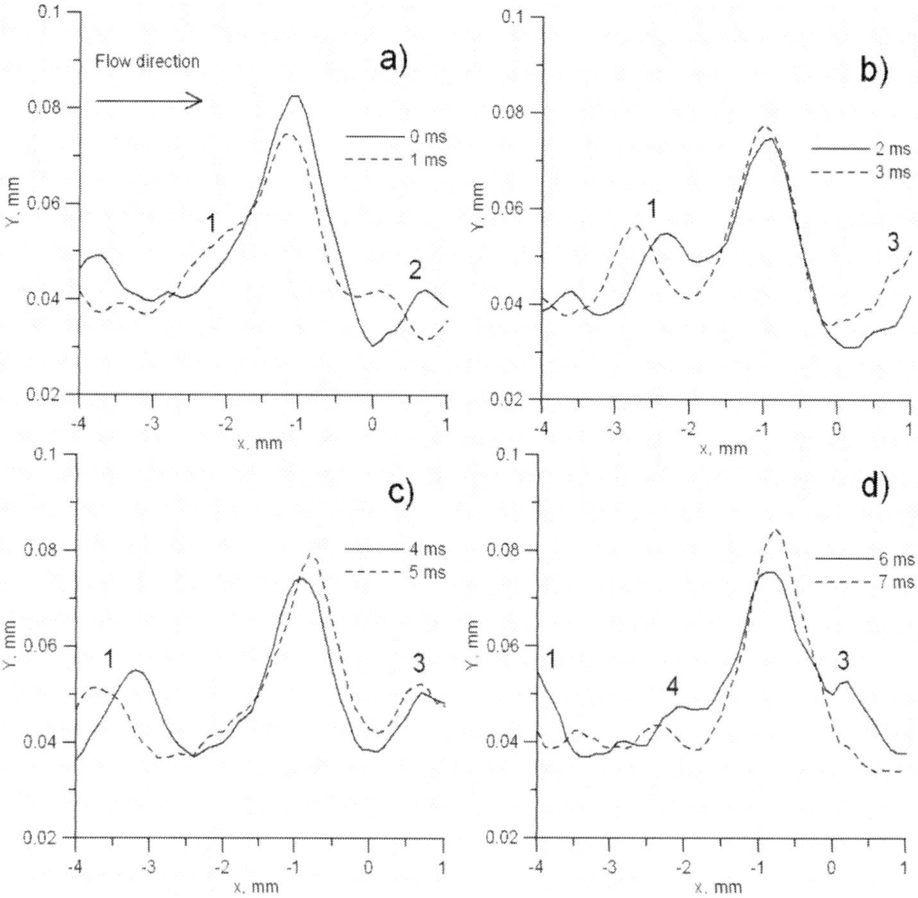

Figure 3. Generation of secondary waves by a primary wave in regime without entrainment. Re=40, Vg=36 m/s.

Thus, despite essential difference in amplitudes, wavelengths and velocities, the wavy structure in regimes with and without entrainment is rather similar. In both regimes two types of waves exist, that we will call "primary waves" and "secondary waves". The primary waves are characterized by high velocity and long lifetime, and the secondary waves are the short-living waves that arise only on the back slopes of primary waves. In regimes with entrainment primary waves are known as "disturbance waves", and the secondary waves are known as "ripples". In regimes without entrainment all the waves – primary and secondary ones – were previously called "ripples".

In regimes with entrainment fast and slow secondary waves exist: the former move faster than primary waves and disappear on their fronts, contributing to liquid entrainment; the latter move slower than primary ones and are promptly being absorbed by the following primary wave. In regimes without entrainment only slow secondary waves exist. This difference explains the transition to entrainment – no secondary waves can reach fronts of primary ones to be ruptured by the gas shear in regimes without entrainment.

To investigate further the processes of secondary waves generation, the statistical study of inception regions of secondary waves was performed. The coordinate of inception x_i was defined as the distance between point of inception and beginning of the primary wave. Since amplitude and length of primary waves may vary greatly at the same flow parameters, this distance was normalized by dividing it to the length λ of the primary wave. λ was defined as the distance between the points where slopes of primary wave cross the thickness of substrate. The latter was defined as the most probable value of film thickness. No less than 30 secondary waves of each sort were processed for each set of flow parameters. The relation x_i/λ is the dimensionless coordinate of waves inception, which will be dealt further.

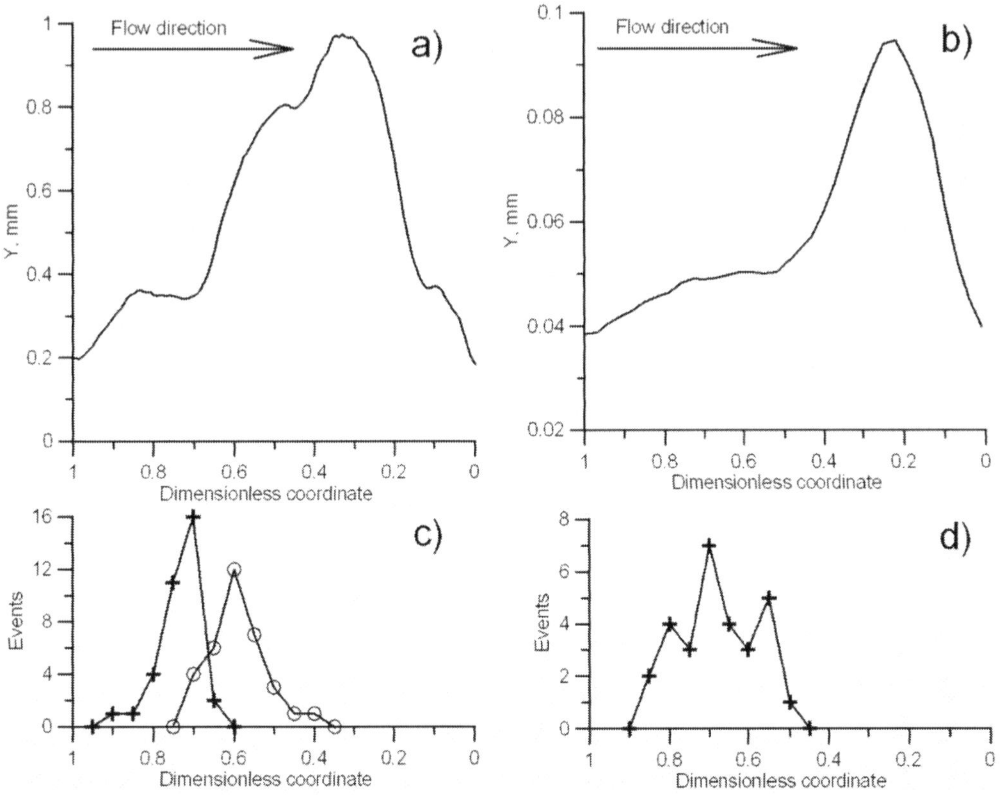

Figure 4. Example profiles of primary waves (a, b) and distributions of secondary waves inception coordinates (c, d). (a, c) – Re=350 (entrainment regime); (b, d) – Re=40 (without entrainment). Vg=36 m/s. Crosses denote slow secondary waves, circles denote fast secondary waves.

Figures 4 (a) and (b) show the examples of profiles of thickness of primary waves in regimes with and without entrainment, respectively. Point where the dimensionless coordinate is equal to zero corresponds to the beginning of the primary wave, point where the dimensionless coordinate equals to 1, corresponds to the end of the primary wave. Figures 4 (c) and (d) show the examples of distribution of inception coordinates of secondary waves in regimes with and without entrainment.

It can be seen from Figure 4 (c) that there exists some kind of separation point at the distance of 0.65 λ in regimes with entrainment. Fast secondary waves appear closer to primary wave's front, and slow secondary waves appear closer to primary wave's tail. So, the areas of secondary waves inception represent different adjoining parts of the primary wave's back slope.

From Figure 4 (d) it can be seen that no fast secondary waves exist in case of regimes without entrainment. The area of slow secondary waves generation is similar to the area of generation of all secondary waves in regimes with entrainment.

The relative areas of fast and slow secondary waves inception in case of entrainment do not change essentially with gas velocity growth (Fig. 5 (a)).

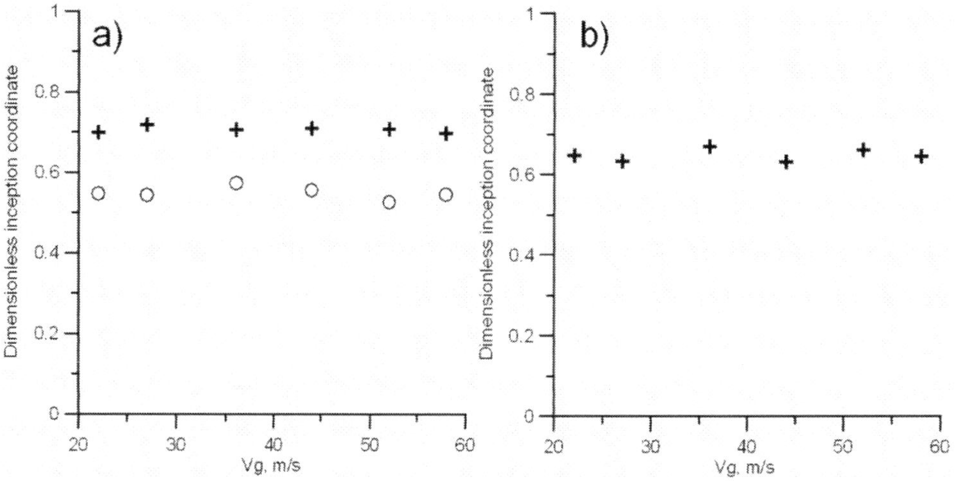

Figure 5. Dependence of the average dimensionless coordinate of secondary waves inception on the gas velocity. (a) – Re=350 (entrainment regime); (b) – Re=40 (without entrainment). Crosses denote slow secondary waves, circles denote fast secondary waves.

Figure 5 (b) shows corresponding data for flow regimes without entrainment. As it was noticed above, no fast secondary waves were observed for such regime; areas of slow secondary waves generation are rather close to that of entrainment regimes and their behavior with changing gas velocity is the same.

4. Discussion and conclusions

Several essentially new observations were made in present work. First, the similarity of wavy structure in regimes with and without entrainment was found: it was shown that in regimes without entrainment two types of waves exist, differing in length, amplitude, velocity, etc. Moreover, it was shown that the smaller waves are always generated by the larger waves. New terms "primary waves" for larger waves and "secondary waves" for smaller waves were introduced to simplify further investigation.

Second, it was shown that only a part of film surface in annular gas-liquid flow (the limited parts of back slopes of primary waves) is unstable to secondary waves generation in regimes with and without entrainment. These observations prejudice mechanism of ripples generation based on simple Kelvin-Helmholtz instability model, proposed by previous investigators.

Third, it was shown that in regimes without entrainment all the secondary waves move slower than parent primary waves, but in the regimes without entrainment some kind of bifurcation of the process of primary wave's back slope evolution takes place. Depending of the relative coordinate of initial disturbance (whether it is close or farther to the crest of parent primary wave than certain separation point), it may suffer two different scenarios of evolution, one of which leads to entrainment and the other doesn't.

Such separation can be explained under the assumption of existence of eddy motion inside the primary wave in regimes with entrainment. Such eddy for gravitationally falling films was predicted in many theoretical works (e.g., [6]). In case of eddy existence, the stagnation point appears at the back slope of the wave, separating the wave's back slope into two parts: one of these parts moves faster than the whole wave, and the other moves slower. In this case, external pressure in the vicinity of stagnation point can be regarded as the destabilizing force, leading to secondary waves inception.

According to simulation of [6], under certain values of wave's amplitude no eddy exists under the wave's hump and there is no stagnation point at the wave's back slope. Thus, no parts of wave move faster than the whole wave. This picture corresponds to our observations on the absence of fast secondary waves in regimes without entrainment. Moreover, the amplitude of primary waves in regimes without entrainment is essentially smaller than that of primary waves in entrainment regimes, and the absence of eddy wouldn't contradict to the model [6]. Of course, this is only one of possible explanations.

References

[1] Azzopardi B J 2006 Begell House, Connecticut
[2] Woodmansee D E and Hanratty T J 1969 Chem. Eng. Sci. 24 299
[3] Holowach et al 2002 Int. J. Heat Fluid Flow 23 807
[4] Sekoguchi K and Takeishi M 1989 Int. J. Mult. Flow 15 295
[5] Asali J C and Hanratty T J 1993 Int. J. Mult. Flow 19 229
[6] Moalem Maron D et al 1989 Int. Comm. Heat Mass Transfer 16 655

Journal of Physics: Conference Series **216** (2010) 012007 doi:10.1088/1742-6596/216/1/012007

Regular shock refraction in planar ideal MHD

P. Delmont[1,2] **& R. Keppens**[1,2,3,4]

[1] Centre for Plasma Astrophysics, K.U.Leuven, Belgium
[2] Leuven Mathematical Modeling and Computational Science Centre, K.U.Leuven, Belgium
[3] Astronomical Institute, Utrecht University, The Netherlands
[4] FOM institute for Plasma Physics Rijnhuizen, Nieuwegein, The Netherlands

E-mail: Peter.delmont@wis.kuleuven.be

Abstract. We study the classical problem of planar shock refraction at an oblique density discontinuity, separating two gases at rest, in planar ideal (magneto)hydrodynamics. In the hydrodynamical case, 3 signals arise and the interface becomes *Richtmyer-Meshkov* unstable due to vorticity deposition on the shocked contact. In the magnetohydrodynamical case, on the other hand, when the normal component of the magnetic field does not vanish, 5 signals will arise. The interface then typically remains stable, since the Rankine-Hugoniot jump conditions in ideal MHD do not allow for vorticity deposition on a contact discontinuity.

We present an exact Riemann solver based solution strategy to describe the initial self similar refraction phase. Using grid-adaptive MHD simulations, we show that after reflection from the top wall, the interface remains stable.

1. Introduction

The interaction of a shock wave with an inclined density discontinuity is a classical hydrodynamical (HD) shock tube problem. When the impinging shock refracts at the density discontinuity, 3 signals arise: a reflected signal, a transmitted signal and a shocked contact in between. Both the reflected and transmitted signal can be continuous (rarefaction fans) or discontinuous (shocks). Due to local Kelvin-Helmholtz instability, the shocked contact becomes unstable, and the Richtmyer-Meshkov instability (RMI) forms. In a recent paper, Delmont *et al.* [4] solved the initial self similar phase of the problem exactly, exploiting exact Riemann solver methods. This paper also predicts wave pattern transitions and critical angles for *regular shock refraction*, where regular means that all signals meet at 1 triple point. We also quantified the vorticity deposition on the shocked contact and investigated the effect of addition of a purely out-of-plane magnetic field. In this work, we extend that study to planar ideal magnetohydrodynamics (MHD).

For that planar case, Samtaney [14] showed by numerical grid-adaptive simulations that addition of a uniform magnetic field, aligned with the shock normal, suppresses the RMI. The setup of the MHD problem is shown in figure 1. The rectangular domain mimics a shock tube, thus the left and the right boundaries are modelled as open, while the upper and the lower boundaries are rigid walls. Completely to the left of the domain, there is a genuine right-moving shock, which moves with sonic Mach number M. It moves towards a contact discontinuity, which forms an inclination angle α with the shock normal and separates two gases at rest, with a density ratio η across it. The ratio of specific heats, γ, is considered a constant equation parameter, and the initially uniform applied magnetic field introduces a plasma-$\beta \equiv \frac{2p}{B^2}$ (in the

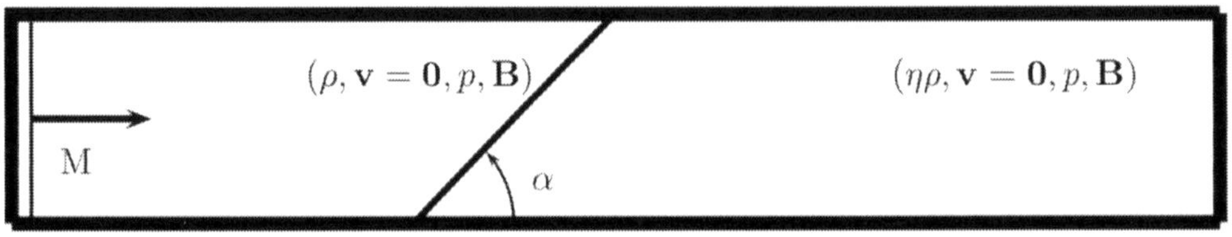

Figure 1. Initial configuration: a shock moves with shock speed M to an inclined density discontinuity. Both the upper and lower boundary are solid walls, while the left and the right boundaries are open.

pre-shock region). These 5 parameters define the problem uniquely. In the MHD case up to 5 signals arise: a fast (FR) and a slow (SR) reflected signal, a contact discontinuity (CD), and a slow (ST) and a fast (FT) transmitted signal, separating the 4 new formed states, as shown in figure 3.

Figure 2 shows simulation snapshots for the case where $(\alpha, \gamma, M, \eta) = (\frac{\pi}{4}, \frac{7}{5}, 2, 3)$, after reflection from the top wall. The upper snapshot shows the hydrodynamical case ($\beta^{-1} = 0$) and the lower snapshot shows an MHD case where $\beta = 2$. In the first case the RMI has formed on the interface, while it is clearly suppressed in the magnetohydrodynamical case. We refer to [4] for an exact solution of the HD case. In this paper we generalise the presented solution strategy to planar ideal MHD. Our results are in agreement with results from [17]. Our approach is inspired by [17]. However, our solution strategy for the RH jump conditions is essentially different and the iteration method used is more detailed.

We also compare our results with numerical grid adaptive simulations. These simulations are performed by the Adaptive Mesh Refinement code AMRVAC (see e.g. [11, 16]).

The suppression of the RMI in ideal MHD can be explained as a direct consequence of the Rankine-Hugoniot jump conditions across a shocked contact. It is well-known that contact discontinuities do not allow for a jump in tangential velocity, when the normal magnetic field component does not vanish (see e.g. [9]).

In section 2, we introduce the equations employed in the analytical solution strategy, namely the planar stationary MHD equations and the Rankine-Hugoniot conditions from their integral form. In section 3, we summarize some important features of MHD shocks. In section 4, we describe our solution strategy, which is based on the Riemann solver for ideal MHD presented by Torrilhon [15] and in section 5 we demonstrate the algorithm by solving a case presented first by Samtaney in [14]. In section 6 finally, we compare our results to AMRVAC simulations.

2. Governing equations

In order to describe the dynamical behaviour of ionized, quasi-neutral plasmas, we use the framework of ideal MHD. The initial configuration sketched in figure 1 will lead to a refraction pattern as sketched in figure 3. Regular refraction means that all signals meet at a single *quintuple point*. This quintuple point moves along the unshocked contact at speed $\mathbf{v}_{qp} = (M, M\tan\alpha)$. We will solve the problem in a comoving frame with the quintuple point. We can then assume that the solution is self similar and time independent: $\partial/\partial t = 0$. Therefore, we can exploit the stationary MHD equations, which written out in conservative form and for our planar problem, are given by

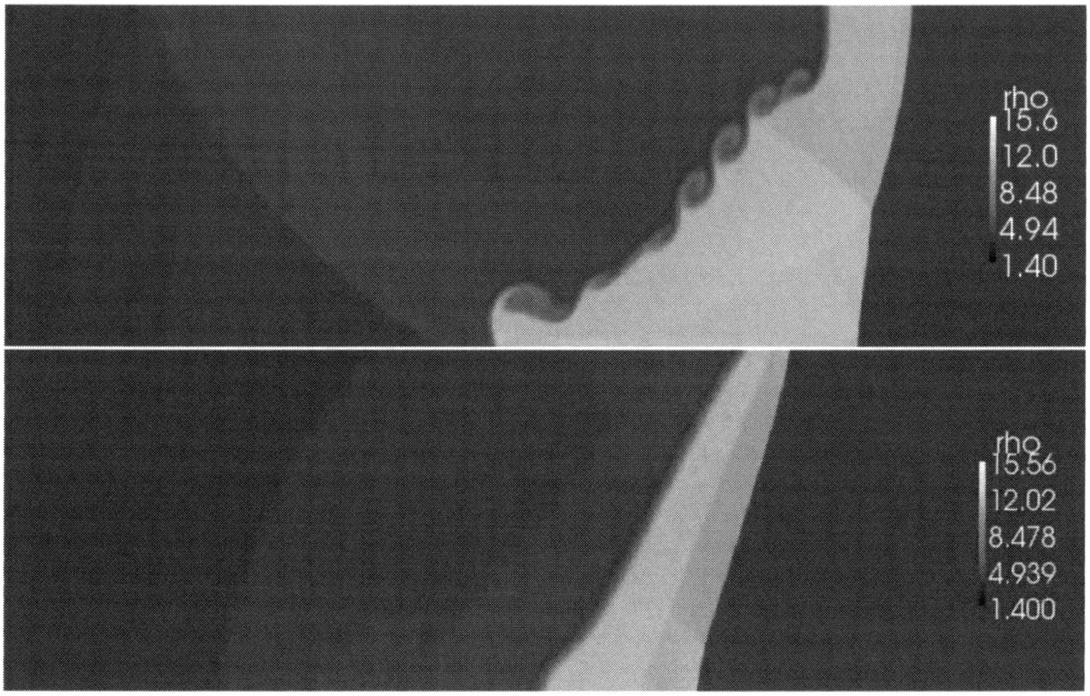

Figure 2. In both cases $(\alpha, \gamma, M, \eta) = (\frac{\pi}{4}, \frac{7}{5}, 2, 3)$. *Up:* $\beta^{-1} = 0$. When a hydrodynamical shock impinges on a contact discontinuity, the discontinuity becomes unstable due to local Kelvin-Helmholtz instability. *Down:* $\beta^{-1} = \frac{1}{2}$. When a normal magnetic field is applied, the Rankine-Hugoniot conditions do not allow for vorticity deposition on a contact discontinuity. Therefore, the interface remains stable and the RMI is suppressed.

$$\frac{\partial}{\partial x}\mathbf{F} + \frac{\partial}{\partial y}\mathbf{G} = \mathbf{0}, \tag{1}$$

where we introduced the flux terms for the orthogonal x and y directions given by

$$\mathbf{F} = \begin{pmatrix} \rho v_x \\ \rho v_x^2 + p - \frac{B_x^2}{2} + \frac{B_y^2}{2} \\ \rho v_x v_y - B_x B_y \\ v_x(\frac{\gamma}{\gamma-1}p + \rho\frac{v_x^2+v_y^2}{2} + B_y^2) - B_x B_y v_y \\ B_x \end{pmatrix} \tag{2}$$

and

$$\mathbf{G} = \begin{pmatrix} \rho v_y \\ \rho v_x v_y - B_x B_y \\ \rho v_y^2 + p + \frac{B_x^2}{2} - \frac{B_y^2}{2} \\ v_y(\frac{\gamma}{\gamma-1}p + \rho\frac{v_x^2+v_y^2}{2} + B_x^2) - B_x B_y v_x \\ B_y \end{pmatrix}. \tag{3}$$

This set of equations expresses conservation of mass density, momentum and energy. The conserved variables and the components of the flux terms are written in terms of the mass density ρ, the velocity components v_x and v_y, the thermal pressure p and the magnetic field components

B_x and B_y. Faraday's law ensures conservation of magnetic flux, which in the stationary case becomes $\nabla \times \mathbf{E} = 0$. The electric field E_z is therefore a global Riemann invariant:

$$E_z = v_y B_x - v_x B_y. \tag{4}$$

This allows us to eliminate $B_y = \frac{v_y B_x - E_z}{v_x}$ from the system, which is reduced to a 5×5-system. We hereby assume that $v_x(\mathbf{x}, t) < 0$ in the comoving frame with the quintuple point. In the rest frame this means that the velocity in the x-direction is bounded by the shock speed, during the refraction process.

We allow *weak solutions* of the system, which are solutions of the integral form of the MHD equations. The discontinuities occuring in the problem setup (both the impinging shock and the initial contact discontinuity), as well as those that may appear as FR, SR, ST or FT signals later on obey the *Rankine-Hugoniot conditions*. In appendix A we give the planar stationary Rankine-Hugoniot conditions and rewrite them as a 4×4-system in $[[B_t]], [[p]], [[\rho]]$ and $[[v_t]]$, where $[[Q]]$ refers to the jump in the quantity Q across the CD, and tangential vector components are referred to by a subscript t.

We now discuss the characteristics of the system (1), by writing it out in quasilinear form

$$\mathbf{u}_x + \mathbf{A}\mathbf{u}_y = \mathbf{0}, \tag{5}$$

where the matrix $\mathbf{A}$ is computed from the flux Jacobian matrices $\mathbf{F_u} = \frac{\partial \mathbf{F}}{\partial \mathbf{u}}$ and $\mathbf{G_u} = \frac{\partial \mathbf{G}}{\partial \mathbf{u}}$ as follows:

$$\mathbf{A} = \mathbf{F_u}^{-1} \cdot \mathbf{G_u}. \tag{6}$$

Here $\mathbf{u}$ can be any arbitrary vector for which $\mathbf{F_u}$ is invertible. We set $\mathbf{u} = (\rho, v_x, v_y, p, B_x)$. As argued in [4], the eigenvalues of $\mathbf{A}$ coincide with the tangent of the refraction angles $\phi_{FR}, \phi_{SR}, \phi_{CD}, \phi_{ST}$ and ϕ_{FT} (see figure 3). Since the system has 5 distinct eigenvalues, the shock indeed refracts in 5 distinct signals.

Computing the eigenvalues λ of $\mathbf{A}$, leads to the characteristic equation

$$det(\mathbf{A} - \lambda \mathbf{I}) = (v_x \lambda - v_y)\left(\Sigma_{i=0,4} t_i \lambda^i\right) = 0. \tag{7}$$

This is a quintic with 5 solutions. One eigenvalue $\lambda = \frac{v_y}{v_x}$ is linearly degenerate and corresponds to the CD. The 4 other eigenvalues are found as roots of a quartic, with coefficients given by

$$\begin{aligned}
t_4 &= v_x^4 + v_x^2(4c^2 - 2a^2) + 2c^2 a_x^2, &(8)\\
t_3 &= 4v_x v_y(c^2 + a^2 - v_x^2) - 8c^2 a_x a_y, &(9)\\
t_2 &= 6v_x^2 v_y^2 - 2a^2(v^2 + c^2) + 2v^2 c^2, &(10)\\
t_1 &= 4v_x v_y(c^2 + a^2 - v_y^2) - 8c^2 a_x a_y, &(11)\\
t_0 &= v_y^4 + v_y^2(4c^2 - 2a^2) + 2c^2 a_y^2. &(12)
\end{aligned}$$

In these expressions we denote the *sound speed* $c = \sqrt{\frac{\gamma p}{\rho}}$, the directional Alfvén speeds $a_x = \sqrt{\frac{B_x^2}{\rho}}$, $a_y = \sqrt{\frac{B_y^2}{\rho}}$ and the *total Alfvén speed* $a = \sqrt{a_x^2 + a_y^2}$.

We can rewrite the characteristic equation (7) as

$$v_\perp(v_\perp^2 - v_{f,\perp}^2)(v_\perp^2 - v_{s,\perp}^2) = 0, \tag{13}$$

where the (squared) normal speed is given by

$$v_\perp^2 = \frac{(v_x \lambda - v_y)^2}{1 + \lambda^2}, \tag{14}$$

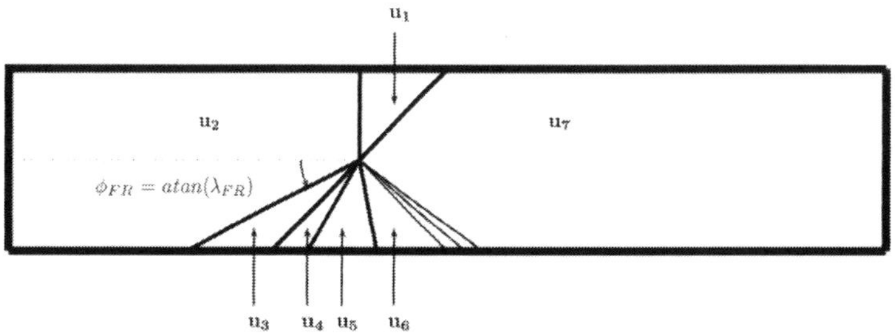

Figure 3. The wave pattern during interaction of the shock with the CD. The eigenvalues of the matrix **A** coincide with the tangent of the refraction angles. Since **A** has 5 distinct eigenvalues, 5 signals arise.

the (squared) fast speed by

$$v_{f,\perp}^2 = \frac{1}{2}\left(c^2 + a^2 + \sqrt{(c^2+a^2)^2 - 4c^2\frac{(B_x\lambda - B_y)^2}{\rho(1+\lambda^2)}}\right), \tag{15}$$

and the (squared) slow speed by

$$v_{s,\perp}^2 = \frac{1}{2}\left(c^2 + a^2 - \sqrt{(c^2+a^2)^2 - 4c^2\frac{(B_x\lambda - B_y)^2}{\rho(1+\lambda^2)}}\right). \tag{16}$$

This then clearly shows that the MHD system has the fast and the slow magnetosonic speeds in the direction perpendicular to the shock front as its basic characteristic speeds.

All the refracted magnetosonic signals can be expansion fans or shocks.

3. MHD shocks

MHD shocks are a topic of extensive research (see e.g. [?, 10]). Introducing the normal Alfvén speed $a_n \equiv \sqrt{\frac{B_n^2}{\rho}}$, which is an eigenvalue of the full three dimensional stationary MHD equations, and $a_t \equiv \sqrt{\frac{B_t^2}{\rho}}$ leads to $a = \sqrt{a_n^2 + a_t^2}$. The full set of eigenvalues of the full set of MHD equations is given by $\{\pm v_f, \pm a_n, \pm v_s, 0\}$, where the fast speed v_f and the slow speed v_s are defined as

$$v_f^2 \equiv \frac{1}{2}\left(c^2 + a^2 + \sqrt{(c^2+a^2)^2 - 4c^2a_n^2}\right); \tag{17}$$

$$v_s^2 \equiv \frac{1}{2}\left(c^2 + a^2 - \sqrt{(c^2+a^2)^2 - 4c^2a_n^2}\right). \tag{18}$$

Since $0 \leq v_s \leq a_n \leq v_f$, the full set of MHD equations is non-strictly hyperbolic. Those speeds define the up- and downstream states into 4 types (and 3 transition types plus 1 stationary type for completeness), which in order of increasing entropy $S \equiv \frac{p}{\rho^\gamma}$:

- (1) superfast: $v_f < |v_n|$;
- (1=2) fast: $|v_n| = v_f$;

- (2) subfast: $a_n < |v_n| < v_f$;
- (2=3) Alfvén: $a_n = v_n$;
- (3) superslow: $v_s < |v_n| < a_n$;
- (3=4) slow: $|v_n| = v_s$;
- (4) subslow: $v_s < |v_n| < a_n$;
- (∞) stationary: $v_x = 0$.

The second law of thermodynamics enforces $[[S]] > 0$ across a shock. If this inequality is satisfied, we call a shock *admissible*. When the upstream state is of type i and the downstream state is of type j, then we define the shock to be of type $i \rightarrow j$. The RH conditions (A.1) now allow for the following types of shocks:

- $(1 \rightarrow 2)$-shocks are called fast shocks. They satisfy $B_{t,2} > B_{t,1} > 0$ or $B_{t,2} < B_{t,1} < 0$;
- $(3 \rightarrow 4)$-shocks are called slow shocks. They satisfy $B_{t,1} > B_{t,2} > 0$ or $B_{t,1} < B_{t,2} < 0$;
- $(1 \rightarrow 2 = 3)$-shocks are called switch-on shocks, since they satisfy $B_{t,1} = 0$;
- $(2 = 3 \rightarrow 4)$-shocks are called switch-off shocks, since they satisfy $B_{t,2} = 0$;
- $(1 \rightarrow 3)$, $(1 \rightarrow 4)$, $(2 \rightarrow 3)$ and $(2 \rightarrow 4)$-shocks are called intermediate shocks. They satisfy $B_{t,2} \geq 0 \geq B_{t,1}$ or $B_{t,2} \leq 0 \leq B_{t,1}$;
- $(1 \rightarrow 4)$-shocks for which $B_{t,1} = B_{t,2} = 0$ also satisfy $v_{t,1} = v_{t,2} = 0$ and are essentially 1-dimensional. They are called hydrodynamical shocks;
- $(2 = 3 \rightarrow 2 = 3)$-discontinuities are called Alfvén discontinuities (or, alternatively, rotational discontinuities). They satisfy $B_{t,1} + B_{t,2} = 0$;
- $(\infty \rightarrow \infty)$-discontinuities can be contact discontinuities, where only ρ jumps across them, or tangential discontinuities, where B_n and v_n vanish.

The latter two cases are called linear discontinuities, since $[[v_n]]$ vanishes. The other cases are called MHD shocks. The set of planar MHD equations reduces the number of possibilities, since we now only have characteristic speeds given by $\{\pm v_f, \pm v_s, 0\}$, making the planar system strictly hyperbolic.

The existence of intermediate shocks is controversial since they are believed to be unstable under small perturbations. This topic is widely discussed in the literature (theoretically in e.g. [1, 6, 7, 12, 13], observationally in e.g. [3, 8] and numerically in e.g. [2, 5]). We allow for intermediate shocks in our solution strategy, since they form the central Alfvénic transition for MHD shocks [10], and as we show later, they naturally emerge in high resolution numerical simulations.

4. Riemann Solver based solution strategy

4.1. Dimensionless representation

We now present how we tackle the problem in a dimensionless manner. We choose a representation in which the initial shock speed s equals its sonic Mach number M. We determine the value of the primitive variables in the post-shock region by applying the stationary Rankine-Hugoniot conditions in the shock rest frame. Note $\mathbf{u}_i = (\rho_i, v_{x,i}, v_{y,i}, p_i, B_{x,i})$, where the index i refers to the value taken in the i−th region (figure 3). Note that we did not include $B_{y,i}$, since it is completely determined by equation (4).

We arbitrarily scale by setting $p_1 = 1$ and $\rho_1 = \gamma$. Now all velocity components are scaled with respect to the sound speed in this region between the impinging shock and the initial CD. Since this region is initially at rest, the *sonic* Mach number M of the shock equals its shock speed s. When the shock intersects the CD, the quintuple point follows the unshocked contact slip line. It does so at a fixed speed $\mathbf{v}_{qp} = (M, M tan\alpha)$, therefore we will solve the problem in the frame of the stationary quintuple point. We will look for self similar solutions in this frame, $\mathbf{u} = \mathbf{u}(\phi)$, where all signals are stationary. We now have that $\tilde{v}_x = v_x - M$ and $\tilde{v}_y = v_y - M\tan\alpha$, where $\tilde{v}$ refers to this new frame. From now on we will drop the tilde and only use this new frame. We now have $\mathbf{u}_1 = (\gamma, -M, -Mtan\alpha, 1, \sqrt{\frac{2}{\beta}})^t$ and $\mathbf{u}_7 = (\eta\gamma, -M, -M\tan\alpha, 1, \sqrt{\frac{2}{\beta}})^t$. The Rankine-Hugoniot relations now immediately give a unique planar solution for $\mathbf{u}_2$, namely

$$\mathbf{u_2} = \left(\frac{(\gamma^2+\gamma)M^2}{(\gamma-1)M^2+2}, -\frac{(\gamma-1)M^2+2}{(\gamma+1)M}, -Mtan\alpha, \frac{2\gamma M^2-\gamma+1}{\gamma+1}, \sqrt{\frac{2}{\beta}}\right)^t. \quad (19)$$

This then completely determines the initial condition, containing only states $\mathbf{u}_1, \mathbf{u}_2$ and $\mathbf{u}_7$ in terms of the dimensionless parameters α, β, γ, M and η.

4.2. Path variables

We know that $\mathbf{v}, \mathbf{B}$ and p are continuous across the CD (see appendix A). Since B_y is uniquely determined by the other 4 (independent) variables, we have 4 independent scalars which should vanish. Hence we can guess 4 path variables, 1 across each magnetoacoustic signal, and express the vanishing quantities in function of those path variables $\chi \equiv (\chi_{FR}, \chi_{SR}, \chi_{ST}, \chi_{FT})$. The fast magnetoacoustic signals are postulated to be shocks, while the slow magnetoacoustic signals can be both shocks and expansion fans.

In the case of a shock, fast signals are postulated to be fast shocks, while slow signals can be slow or intermediate shocks. We then select the correct solution to the Rankine-Hugoniot jump conditions as explained in Appendices B and C. In the case of a rarefaction wave, we numerically integrate the MHD equations as explained in Appendix D. In this manner we reduce the problem to solving

$$[[v_x, v_y, B_x, p]](\mathbf{u}_4(\mathbf{u}_3(\mathbf{u}_2, \chi_{FR}), \chi_{SR}), \mathbf{u}_5(\mathbf{u}_6(\mathbf{u}_7, \chi_{FT}), \chi_{ST})) = \mathbf{0}. \quad (20)$$

The function $\chi \mapsto [[v_x, v_y, B_x, p]]$ is not partially differentiable in all points of its domain. The price to pay is that we will postulate the wave configuration.

As said before, every magnetoacoustic signal is controlled by one path variable, such that $\mathbf{u}_u(\mathbf{u}_k, \chi_{signal})$, where *signal* is the signal separating $\mathbf{u}_u$ from $\mathbf{u}_k$. The path variables should have a one-to-one relationship with the refraction angles ϕ_{signal}. The actually used path variables are given by

$$\chi_{FR} = \tan(\phi_{FR}), \quad (21)$$

$$\chi_{FT} = \tan(\phi_{FT}), \quad (22)$$

$$\chi_{SR} = \begin{cases} \tanh^{-1}\left(2\left(\frac{\tan\phi_{SR}-\tan\phi_{a,R}}{\tan\phi_{cr,R}-\tan\phi_{a,R}}\right)-1\right), & \forall\phi_{SR} \in [\phi_{cr,R}, \phi_{a,R}] \\ \tanh^{-1}\left(2\left(\frac{\tan\phi_{SR}-\tan\phi_{sl,R}}{\tan\phi_{a,R}-\tan\phi_{sl,R}}\right)-1\right), & \forall\phi_{SR} \in [\phi_{a,R}, \phi_{sl,R}[\\ \tanh^{-1}\left(2\left(\frac{\tan\phi_{SR}-\tan\phi_{st,R}}{\tan\phi_{sl,R}-\tan\phi_{st,R}}\right)-1\right), & \forall\phi_{SR} \in [\phi_{sl,R}, \phi_{st,R}[\end{cases} \quad (23)$$

$$\chi_{ST} = \begin{cases} \tanh^{-1}\left(2\left(\frac{\tan\phi_{ST}-\tan\phi_{a,T}}{\tan\phi_{cr,T}-\tan\phi_{a,T}}\right)-1\right), & \forall\phi_{SR} \in [\phi_{a,T}, \phi_{cr,T}] \\ \tanh^{-1}\left(2\left(\frac{\tan\phi_{ST}-\tan\phi_{sl,T}}{\tan\phi_{a,T}-\tan\phi_{sl,T}}\right)-1\right), & \forall\phi_{SR} \in [\phi_{sl,T}, \phi_{a,T}[\\ \tanh^{-1}\left(2\left(\frac{\tan\phi_{ST}-\tan\phi_{st,T}}{\tan\phi_{sl,T}-\tan\phi_{st,T}}\right)-1\right), & \forall\phi_{SR} \in [\phi_{st,T}, \phi_{sl,T}[\end{cases} \quad (24)$$

(25)

where we introduced the Alfvénic angles, the critical angles, the slow angles and the stationary angles. The Alfvénic angles are those for which the upstream state is exactly Alfvénic and these are defined by

$$\phi_{a,R} = atan\left(\frac{B_{3,y}-\sqrt{\rho_3}v_{3,y}}{B_{3,x}-\sqrt{\rho_3}v_{3,x}}\right), \quad (26)$$

$$\phi_{a,T} = atan\left(\frac{B_{6,y} + \sqrt{\rho_6}v_{6,y}}{B_{6,x} + \sqrt{\rho_6}v_{6,x}}\right), \tag{27}$$

and the critical angles at which the transition from 1 to 3 real solutions to the RH conditions across the slow signal takes place, by

$$\phi_{cr,R} \equiv \max\{\phi < \phi_{a,R}|(N^2 - D^3)(\phi) = 0\}, \tag{28}$$
$$\phi_{cr,T} \equiv \min\{\phi > \phi_{a,T}|(N^2 - D^3)(\phi) = 0\}. \tag{29}$$

Finally, the slow angles $\phi_{sl,R}$ and $\phi_{sl,T}$, are those for which the upstream state is exactly slow (and are found numerically) and the stationary angles are defined by

$$\phi_{st,R} \equiv atan\frac{v_{y,3}}{v_{x,3}}, \tag{30}$$
$$\phi_{st,T} \equiv atan\frac{v_{y,6}}{v_{x,6}}. \tag{31}$$

The functions N and D are defined in appendix C, and these critical angles are dependent on the location of the fast signal.

Once the wave configuration is determined, χ determines the solution to the problem. A secant method iteration solves equation (20) and $\phi_{SC} = atan\frac{v_{y,4}}{v_{x,4}}$ closes the procedure.

Once the wave configuration is postulated, we know if the slow signal angles are bigger or smaller than the Alfvénic angles. The initial values of the path variables controlling the fast signals are taken such that the upstream states $\mathbf{u}_2$ and $\mathbf{u}_7$ are exactly fast. For the slow signals, initially we put $\chi_{SR} = \chi_{ST} = 0$. In the case of an intermediate shock, this means that our starting angles ϕ_{SR} and ϕ_{ST} respectively satisfy $2\phi_{SR} = \phi_{a,R} + \phi_{cr,R}$ and $2\phi_{ST} = \phi_{a,T} + \phi_{cr,T}$. The procedure also ensures that in every iteration step $\phi_{a,T} < \phi_{ST} < \phi_{cr,T}$ and $\phi_{cr,R} < \phi_{SR} < \phi_{a,R}$. Similarly, in the case of a slow shock, the initial guesses ϕ_{SR} and ϕ_{ST} respectively satisfy $2\phi_{SR} = \phi_{sl,R} + \phi_{a,R}$ and $2\phi_{ST} = \phi_{sl,T} + \phi_{a,T}$. Finally, when the slow signal is located at $\phi_{SR} > \phi_{sl,R}$ there is no shock solution possible which satisfies the entropy condition, and in the case the signal is a slow rarefaction fan. The initial iteration angles ϕ_{SR} and ϕ_{ST} respectively satisfy $2\phi_{SR} = \phi_{st,R} + \phi_{sl,R}$ and $2\phi_{ST} = \phi_{st,T} + \phi_{sl,T}$. We hereby postulated that the slow signals are expansion fans when $v^2_{n,3/6}(\phi_{SR/ST}) < v^2_{sl,3/6}$. In this case, the expansion fan is located between $\phi_{SR/ST}$ and $\phi_{sl,R/T}$. This criterion is equivalent to the criterion that a rarefaction fan will only occur at a given position if no shock solutions satisfying the entropy condition are possible, and is in this sense a generalisation of the criterion exploited in [4].

4.3. Tackling the signals

The solution to the stationary Rankine-Hugoniot conditions is given in appendix B. One problem is that the uniqueness of its solution is not guaranteed. We express the tangential component of the magnetic field in the downstream state, $B_{t,u}$, as the root of a cubic, which coefficients are expressed in terms of the known upstream state. The unknown state is then expressed in terms of $B_{t,u}$.

It can be shown that if there is a unique solution to the RH conditions, it is a fast or a slow shock. If on the other hand there are 3 solutions, they are all of a different shock type. Thus when the shock types of all magnetoacoustic signals are known, the Rankine-Hugoniot conditions can be solved uniquely. We solve the cubic analytically in Appendix C, noting that roots can be real or complex. In principle this completes the solution algorithm for the RH conditions of a genuine shock. However, the complications are that

- We need to select the appropriate root from the (up to) 3 mathematical possibilities, at each magnetosonic signal;

- The analytical expressions contain a square root and a cube root in the complex plane. These expressions are discontinuous in the negative real numbers.

In appendix C, we also provide additional technical information on how permutation of the root indices can make the analytical expressions for the roots (C.8) continuous and differentiable, and hence allows for a secant iteration method applied on equation (20). The only remaining discontinuities are reached, when the refraction angles cross the critical angles $\phi_{cr,R}$ or $\phi_{cr,T}$, where the transition from a unique solution to 3 solutions for the RH conditions takes place. Also $[[p]]$ is discontinuous across the Alfvénic angles. This leads to the restriction that we cannot cross those critical angles in subsequent iteration steps. This is taken care of by our choice of path variables.

In Appendix D finally, we describe which relations hold across expansion fans and how the numerical integration through the fan is performed.

5. Demonstration of result

We study the case where $(\alpha, \beta, \gamma, \eta, M) = \left(\frac{\pi}{4}, 2, \frac{7}{5}, 3, 2\right)$, a case which was studied in detail before by Samtaney [14] and Wheatley *et al.* [17]. For this case we have that

$$\mathbf{u}_1 = (1.4000, -2.0000, -2.0000, 1.0000, 1.0000), \quad (32)$$
$$\mathbf{u}_2 = (3.7333, -0.7500, -2.0000, 4.5000, 1.0000), \quad (33)$$
$$\mathbf{u}_7 = (4.2000, -2.0000, -2.0000, 1.0000, 1.0000). \quad (34)$$

The initial guesses for the iteration procedure are $\chi_{FR}, \chi_{SR}, \chi_{ST}, \chi_{FT} = (0.5314, 0, 0, 3.7306)$. The iteration converges to the exact solution for the angles $\phi_{FR}, \phi_{SR}, \phi_{ST}, \phi_{FT} = (0.405694, 0.917019, 1.19283, 1.27678)$. The corresponding cubics whose roots need to be properly selected are shown in figure 4.

Discussing the solution of the refraction pattern in the order of integration, we encounter:

- the fast reflected (FR) signal, located at $\phi_{FR} = 0.405694$. The cubic across the interface is then evaluated and found to be $0.0346\hat{B}_t^3 - 0.5184\hat{B}_t^2 + 7.2999\hat{B}_t + 3.9011 = 0$, which has, as predicted, only one real solution $\hat{B}_t = -0.5149$, such that

$$\mathbf{u}_3 = (4.424, -0.834, -1.774, 5.710, 1.159),$$

  and $B_{y,3} = 0.0685$. Note that FR is a fast shock;
- the slow reflected (SR) signal, located at $\phi_{SR} = 0.917019$. The cubic relation connected to this interface has 3 real solutions: $\hat{B}_t \in \{0.13075\ 10^{-3}, 0.34361, -0.02656\}$. Further physical argumentation is needed to select the correct root. The first possibility, $\hat{B}_t = 0.13075\ 10^{-3}$, leads to negative pressure. The second option, $\hat{B}_t = 0.34361$, corresponds to an intermediate shock, and yields

$$\mathbf{u}_{4,intermediate} = (4.395, -1.290, 2.374, 5.656, 0.198),$$

  together with $B_{y,4} = 1.186$. The entropy condition $\frac{p_4}{p_3} < \left(\frac{\rho_4}{\rho_3}\right)^\gamma$ is satisfied. The last possible solution, $\hat{B}_t = -0.02656$ corresponds to a slow shock. In this case

$$\mathbf{u}_{4,slow} = (4.596, -1.048, -2.028, 6.025, 0.736),$$

  and $B_y = 0.484$, which also satisfies the entropy condition $\frac{p_4}{p_3} < \left(\frac{\rho_4}{\rho_3}\right)^\gamma$;

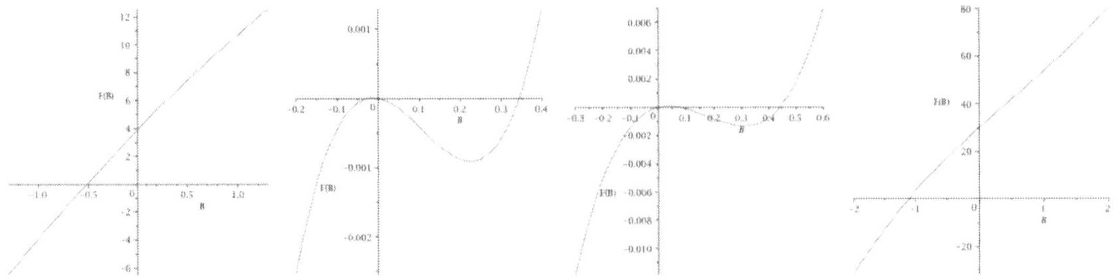

Figure 4. The cubic for respectively FR, SR, ST, FT.

- the fast transmitted (FT) signal, located at $\phi_{FT} = 1.27678$. The cubic relation across this fast interface again has a unique real solution $\hat{B}_t = -1.10819$. This solution corresponds to a fast shock and leads to

$$\mathbf{u}_6 = (11.932, -1.213, -2.384, 5.275, 1.237),$$

  and $B_{y,6} = 0.783$;
- the slow transmitted (ST) signal, located at $\phi_{ST} = 1.19283$. The cubic relation controlling this interface has 3 different real solutions $\hat{B}_t \in \{0.44121, -0.70666\ 10^{-3}, 0.07837\}$. The first possibility, $\hat{B}_t = 0.44121$ can be eliminated since it does not lead to a solution which satisfies the entropy condition. The second possibility, $\hat{B}_t = -0.70666\ 10^{-3}$ would lead to negative pressure. The remaining possibility, $\hat{B}_t = 0.07837$ corresponds to an intermediate $(2 \rightarrow 4)$-shock from $\mathbf{u}_6$ to $\mathbf{u}_5$, and yields

$$\mathbf{u}_5 = (13.092, -1.048, -2.028, 6.025, 0.736),$$

  with $B_{y,5} = -0.484$, which satisfies the entropy condition.

We now notice that the correct reflected slow signal is the slow shock and $\mathbf{u}_{SR} = \mathbf{u}_{4,slow}$, since the shocked contact must satisfy the matching conditions $(A.7)$.

6. Suppression of the Richtmyer-Meshkov Instability

Since v_n vanishes along a contact discontinuity, the Rankine-Hugoniot jump conditions simplify as

$$[[v_t, p, B_t]] = \mathbf{0}.$$

Therefore, they do not allow for vorticity deposition on the contact, and the interface must remain stable. Figure (5) shows snapshots of a numerical simulation of the parameter regime $(\alpha, \beta, \gamma, M, \eta) = \left(\frac{\pi}{4}, 2, \frac{7}{5}, 2, 3\right)$ performed by the Adaptive Mesh Refinement code AMRVAC. The first frame shows the density, ρ, during the shock refraction. Note that the CD remains stable, since the jump in tangential velocity, $[[v_t]]$, vanishes. The second frame shows the tangential magnetic field component, B_t. The density is discontinuous in every new-formed signal, while the CD is not visible in the B_t frame. SR is a slow shock, therefore B_t does not change sign across this signal, while ST on the other hand is an intermediate shock and B_t changes sign across it.

7. Conclusion and Future Work

We developed an exact Riemann solver based solution for regular shock refraction in planar ideal MHD. We are able to reproduce results from the literature and results by numerical simulations performed by AMRVAC. We are working on a generalisation allowing for arbitrary (2.5 dimensional) impinging shocks.

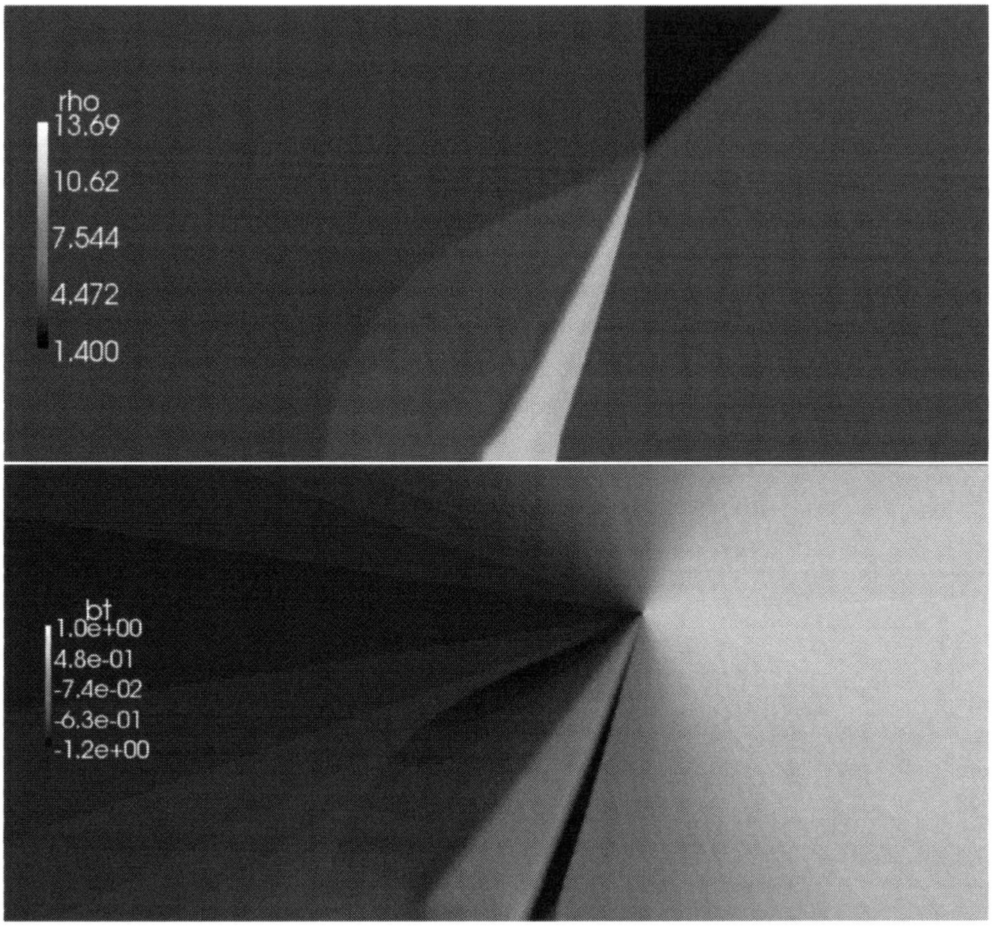

Figure 5. *Up:* Density plot for $(\alpha, \beta, \gamma, \eta, M) = \left(\frac{\pi}{4}, 2, \frac{7}{5}, 3, 2\right)$. Five signals arise and their location is found by the iteration procedure. Iterating one more time yields $\mathbf{u}(x)$ and solves the problem. Note that the CD remains stable. *Down:* A plot of the transverse magnetic field, B_t. Note that the transverse magnetic field is continuous across the CD: $[[B_t]] = 0$. Also note that B_t changes sign across ST. Indeed, ST is an intermediate shock.

Appendix A. Stationary planar Rankine-Hugoniot conditions

We allow *weak solutions* of the system, which are solutions of the integral form of the MHD equations, that may contain discontinuities. The shock occuring in the problem setup, as well as those that may appear as FR, SR, ST or FT signals later on obey the *Rankine-Hugoniot conditions.* In the case where the shock speed $s = 0$, the Rankine-Hugoniot conditions follow from equation (1). When considering a thin continuous transition layer in between the two regions, with thickness δ, solutions for the integral form of equation (1) should satisfy

$\lim_{\delta\to 0}\int_1^2(\frac{\partial}{\partial x}\mathbf{F}+\frac{\partial}{\partial y}\mathbf{G})dl=0$. For vanishing thickness of the transition layer, this yields

$$\left[\left[\begin{array}{c} \rho v_n \\ \rho v_n^2+p-\frac{B_n^2}{2}+\frac{B_t^2}{2} \\ \rho v_n v_t - B_n B_t \\ v_n(\frac{\gamma}{\gamma-1}p+\rho\frac{v_n^2+v_t^2}{2}+B_n^2)-B_nB_tv_t \\ v_nB_t - v_tB_n \\ B_n \end{array}\right]\right]=\mathbf{0}, \tag{A.1}$$

where the index n refers to the direction normal to the shock front and the index t refers to the direction tangential to the shock front (see e.g. [9]). From (A.1) we know that $m\equiv\rho v_n$ and B_n are constant across a shock. Inserting these constants in the other equations of (A.1) yields

$$m^2[[\frac{1}{\rho}]]+[[p]]+\frac{1}{2}[[B_t^2]] = 0, \tag{A.2}$$

$$m[[v_t]]-B_n[[B_t]] = 0, \tag{A.3}$$

$$m[[\frac{B_t}{\rho}]]-B_n[[v_t]] = 0, \tag{A.4}$$

$$\frac{\gamma}{\gamma-1}[[\frac{p}{\rho}]]+\frac{m^2}{2}[[\frac{1}{\rho^2}]]+[[\frac{B_t^2}{\rho}]]-\frac{B_n^2}{2m^2}[[B_t^2]] = 0, \tag{A.5}$$

where we used equation (A.4) to eliminate v_t from $[[v_n(\frac{\gamma}{\gamma-1}p+\rho\frac{v_n^2+v_t^2}{2}+B_n^2)-B_nB_tv_t]]=0$ to arrive at

$$\frac{\gamma}{\gamma-1}m[[\frac{p}{\rho}]]+\frac{m^3}{2}[[\frac{1}{\rho^2}]]+m[[\frac{B_t^2}{\rho}]]-\frac{B_n^2}{2m}[[B_t^2]]=0, \tag{A.6}$$

which is equivalent to equation (A.5), under the assumption that $m\neq 0$, which is true for all magnetoacoustic signals.

Since the signal located at $\frac{dy}{dx}=\frac{v_y}{v_x}$ is a contact discontinuity, it clearly obeys $m=0$. This simplifies (A.2-A.4) drastically:

$$[[(v_t,B_t,p)^t]]=\mathbf{0}. \tag{A.7}$$

Hence, in terms of the primitive variables, this means that v_x, v_y, p and B_x, and thus also B_y, remain constant across a contact discontinuity.

Appendix B. Relations across a shock

Suppose we know the primitive variables $\mathbf{u}_k$ at one side of a stationary shock, and denote the unknown primitive variables at the other side of the shock $\mathbf{u}_u$.

We will discuss the consequences of the RH condition relating up- and downstream states, for the case where $B_{t,k}\neq 0$. After elimination of $[[v_t]]$ from (A.4), we arrive at the following 3×3-system (see also Torrilhon [15]):

$$\hat{p}-1+C(\frac{1}{\hat{\rho}}-1)+\frac{1}{2}(\hat{B}_t^2-A^2) = 0, \tag{B.1}$$

$$C(\frac{\hat{B}_t}{\hat{\rho}}-A)-B^2(\hat{B}_t-A) = 0, \tag{B.2}$$

$$\frac{1}{\gamma-1}(\frac{\hat{p}}{\hat{\rho}}-1)+\frac{1}{2}(\frac{1}{\hat{\rho}}-1)(\hat{p}+1)+\frac{1}{4}(\frac{1}{\hat{\rho}}-1)(\hat{B}_t-A)^2 = 0, \tag{B.3}$$

where we introduced the dimensionless quantities connecting the up-and downstream values given by $\hat{\rho} = \frac{\rho_u}{\rho_k}$, $\hat{p} = \frac{p_u}{p_k}$, $\hat{v}_t = \frac{v_{t,u}}{c_k}$ and $\hat{B}_t = \frac{B_{t,u}}{\sqrt{p_k}}$ and the dimensionless parameters, quantifying the known values in state $\mathbf{u}_k$ by $A \equiv \frac{B_{t,k}}{\sqrt{p_k}}$, $B \equiv \frac{B_{n,k}}{\sqrt{p_k}}$ and $C = \frac{\rho_k v_{n,k}^2}{p_k}$. These parameters allow for simple criteria for fast, slow or intermediate shocks: fast shocks are characterised by $\frac{\hat{B}_t}{A} > 1$, slow shocks are characterised by $0 < \frac{\hat{B}_t}{A} < 1$ and intermediate shocks satisfy $\frac{\hat{B}_t}{A} < 0$, since the known state is the upstream state and the unknown state is the downstream state.

Note that super-Alfvénic flow in the direction normal to the shock front in the state $\mathbf{u}_k$ now implies

$$C > B^2 \Leftrightarrow v_{n,k}^2 > a_{n,k}^2. \tag{B.4}$$

The solution to equations (B.1-B.3) is given by

$$\hat{\rho} = \frac{C\hat{B}_t}{AC + B^2(\hat{B}_t - A)}, \tag{B.5}$$

$$\hat{p} = \frac{2C}{\gamma+1} - \frac{\gamma-1}{\gamma+1}\left(\frac{(A-\hat{B}_t)^2}{2} + 1\right) - \frac{\hat{B}_t C(A+\hat{B}_t)}{(\gamma+1)(C-B^2)}, \tag{B.6}$$

where $\hat{B}_t$ must satisfy a cubic relation

$$\Sigma_{i=0,3}\tau_i \hat{B}_t^i = 0, \tag{B.7}$$

with its coefficients given by

$$\tau_3 = B^2, \tag{B.8}$$

$$\tau_2 = \Big((\gamma-1)(B^2-C) + C\Big)A, \tag{B.9}$$

$$\tau_1 = ((\gamma+1)B^2 - (\gamma-1)C - (2+A^2)\gamma)(B^2-C), \tag{B.10}$$

$$\tau_0 = -(\gamma+1)A(B^2-C)^2. \tag{B.11}$$

Once the solution $\hat{B}_t$ from equation (B.7) is determined, and used to calculate $\hat{p}$ and $\hat{\rho}$ from (B.5)-(B.6), we find the upstream state from $\rho_u = \hat{\rho}\rho_k$, $p_u = \hat{p}p_k$ and $B_{t,u} = \hat{B}_t\sqrt{p_k}$. Finally equation (A.1) delivers $v_{n,u}$ and $B_{n,u}$, while from (A.4), we know $v_{t,u}$:

$$v_{n,u} = \frac{m}{\hat{\rho}\rho_k}, \tag{B.12}$$

$$v_{t,u} = v_{t,k} + \frac{B_{n,k}}{\rho_k v_{n,k}}(B_{t,n} - B_{t,u}), \tag{B.13}$$

$$B_{n,u} = B_{n,k}. \tag{B.14}$$

Finally, note that the regular solution discussed above is only valid when $B^2 \neq C$, i.e. when the normal velocity in the known region does not equal the Alfvén velocity, or equivalently when $B_{t,k}$ does not vanish. In this special case the system (B.1-B.3) has four mathematical solutions possible, one of which is the trivial solution, one of which is a rotational shock (which is the limit case of solution (B.5-B.6)) and finally the following non-trivial switch-off shock:

$$\hat{B}_t = 0, \tag{B.15}$$

$$\hat{\rho} = \frac{2(\gamma+1)C}{\gamma(A^2+2C+2) \pm \sqrt{(\gamma(A^2+2))^2 + 4C(C+A^2-2\gamma)}}, \tag{B.16}$$

$$\hat{p} = \frac{2+2C+A^2 \mp \sqrt{(\gamma(A^2+2))^2 + 4C(C+A^2-2\gamma)}}{2(\gamma+1)}. \tag{B.17}$$

Appendix C. Solving the cubic analytically

The cubic (B.7) is solved analytically by the following procedure. Defining the real quantities

$$D = 4(\tau_2^2 - 3\tau_3\tau_1), \quad \text{(C.1)}$$
$$N = 4(2\tau_2^3 - 9\tau_3\tau_2\tau_1 + 27\tau_3^2\tau_0), \quad \text{(C.2)}$$

one notes that D is the discriminant of the derivative function of the cubic. When $D > 0$, the inequality $N^2 - D^3 < 0$ gives limiting values for τ_0, for which the 2 extrema of the cubic have opposite signs. Since $N^2 - D^3 < 0$ implies $D > 0$, the criterium on the coefficient of the cubic to have 3 different real solutions is

$$N^2 - D^3 < 0. \quad \text{(C.3)}$$

From these real-valued quantities, we define the following, possibly complex, quantities:

$$H = \left(\sqrt{N^2 - D^3} - N\right)^{\frac{1}{3}}, \quad \text{(C.4)}$$
$$J = \frac{D}{H}. \quad \text{(C.5)}$$

In terms of these introduced quantities, the 3 complex solutions of the cubic equation (B.7) are given by

$$\hat{B}_{t,0} = \frac{J + H - 2\tau_2}{6\tau_3}, \quad \text{(C.6)}$$
$$\hat{B}_{t,1} = -\frac{J + H + 4\tau_2 - \sqrt{3}(H - J)i}{12\tau_3}, \quad \text{(C.7)}$$
$$\hat{B}_{t,2} = -\frac{J + H + 4\tau_2 + \sqrt{3}(H - J)i}{12\tau_3}. \quad \text{(C.8)}$$

The evaluation of expressions (C.6 - C.8) requires the evaluation of $\sqrt{N^2 - D^3}$ and $\left(\sqrt{N^2 - D^3} - N\right)^{\frac{1}{3}}$. $\sqrt{N^2 - D^3}$ is discontinuous when $N^2 - D^3$ is a negative (and thus real) number and $\left(\sqrt{N^2 - D^3} - N\right)^{\frac{1}{3}}$ is discontinuous when $\sqrt{N^2 - D^3} - N$ is a negative (and thus real) number. Since $N^2 - D^3$ is an (even) polynomial of degree 18 in M, it renders a mathematical classification rather difficult.

Before we explain how to permute the indices of the roots (C.6-C.8), let us first rewrite these expressions as

$$\hat{B}_\iota = \hat{H}_\iota + \hat{J}_\iota, \quad \text{(C.9)}$$

where $\hat{H}_\iota = H_\iota - \frac{\tau_2}{6\tau_3}$, $\hat{J}_\iota = J_\iota - \frac{\tau_2}{6\tau_3}$, $H_\iota = \frac{\left(-\frac{1}{2}+\frac{\sqrt{3}}{2}i\right)^\iota H}{6\tau_3}$, $J_\iota = \frac{\left(-\frac{1}{2}+\frac{\sqrt{3}}{2}i\right)^{-\iota} J}{6\tau_3}$ and $\iota \in \{0, 1, 2\}$.

Let us make some small technical remarks.

- Note that $\left(\left(\frac{\sqrt{3}}{2}i\right)^\iota\right)^3 = 1$, for each natural value of ι;
- Note that $\tau_3 > 0$. Therefore, equation (C.9) is well-defined. Also, when we know that one of those expressions is real, then its sign equals the sign of its nominator;
- It turns out that dividing the complex plane into 6 sextants is useful for our analysis. We therefore define the sextants

$$S_j \equiv \{r(\cos\varphi + i\sin\varphi) | (j-1)\frac{\pi}{3} < \varphi < j\frac{\pi}{3}, r > 0\}, \quad \text{(C.10)}$$

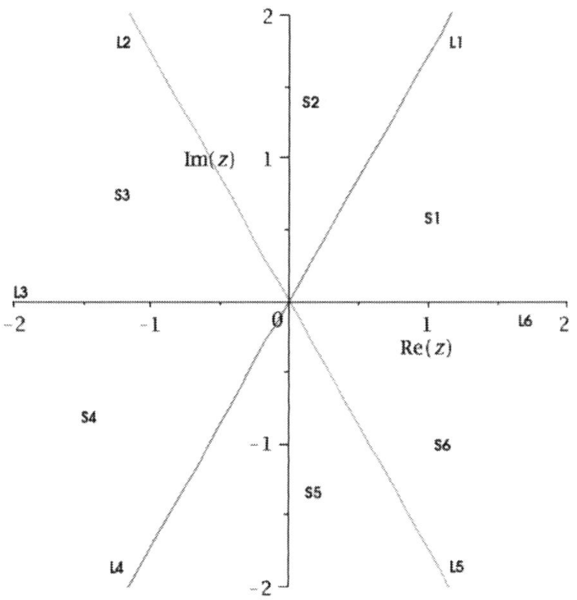

Figure C1. The sextants S_i and their separators, L_i.

and L_i, the lines separating those sextants, i.e.

$$L_j \equiv \{r(\cos\varphi + i\sin\varphi) | \varphi = j\frac{\pi}{3}, r > 0\}. \tag{C.11}$$

These regions are shown in figure (C1).

One can distinguish the following cases.

- Case I: $D > 0$, $N > 0$ and $N^2 - D^3 > 0$.
  Note that $H \in L_1$ and since $D > 0$, $J \in L_5$. Therefore, $H_1, J_1 \in L_3$, $J_2 \in L_1$ and $H_2 \in L_5$. It follows that $\hat{B}_{t,1}$ is a real number. Note that $\sqrt{N^2 - D^3} < N$, thus $2(N^2 - D^3) < 2N\sqrt{N^2 - D^3}$ or $\left(\sqrt{N^2 - D^3} - N\right)^2 < D^3$, thus $H < \sqrt{D}$ or $|H| < |J|$. Therefore, $\hat{B}_{t,2} \in S_1$ and $B_0 \in S_6$.

- Case II: $D > 0$, $N < 0$ and $N^2 - D^3 > 0$.
  Note that $H \in L_6$ and since $D > 0$ also $J \in L_6$. Therefore, $H_2, J_1 \in L_4$ and $H_1, J_2 \in L_2$. It follows that B_0 is a real number. Note that $\sqrt{N^2 - D^3} > N$, thus $2(N^2 - D^3) > 2N\sqrt{N^2 - D^3}$ or $\left(\sqrt{N^2 - D^3} - N\right)^2 > D^3$, thus $H > \sqrt{D}$ or $|H| > |J|$. Therefore, $\hat{B}_{t,1} \in S_3$ and $\hat{B}_{t,2} \in S_4$.

- Case III: $D > 0$, $N > 0$ and $N^2 - D^3 < 0$.
  Note that $H \in S_1$ and since $D > 0$, $J \in S_6$. Therefore, $H_1 \in S_3$, $H_2 \in S_5$, $J_1 \in S_4$ and $J_2 \in S_2$. Since $|\sqrt{N^2 - D^3} - N|^2 = D^3$, $|H| = \sqrt{D}$ and $|H| = |J|$. Hence all of the roots are real. After some algebra, we know that $\hat{B}_{t,0} > -\frac{\tau_2}{3\tau_3} + \frac{\sqrt{D}}{2}$, $\hat{B}_{t,1} < -\frac{\tau_2}{3\tau_3} - \frac{\sqrt{D}}{2}$ and $-\frac{\tau_2}{3\tau_3} - \frac{\sqrt{D}}{2} < \hat{B}_{t,2} < -\frac{\tau_2}{3\tau_3} - \frac{\sqrt{D}}{2}$.

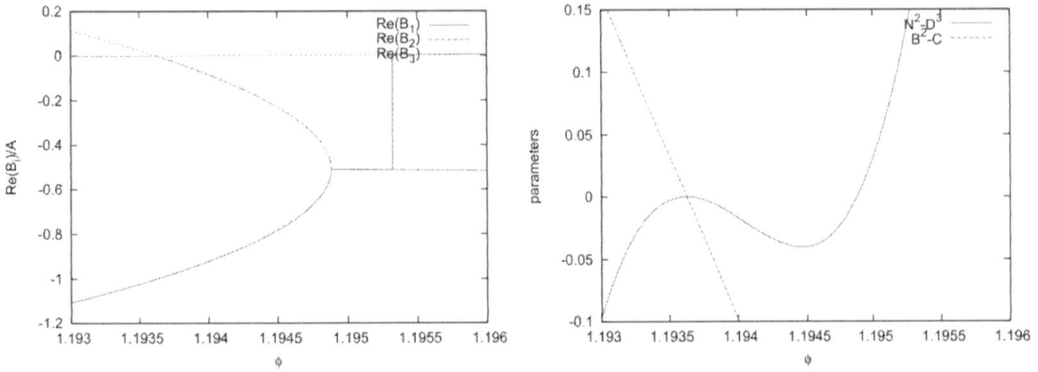

Figure C2. *Left:* the real parts of the roots of the cubic (B.7) for ST. *Right:* The graphs of $(B^2 - C)(\phi)$ and $(N^2 - D^3)(\phi)$. Note that the Alfvénic angle is a root of $B^2 - C$ and a double root of $N^2 - D^3$.

- Case IV: $D > 0$, $N < 0$ and $N^2 - D^3 < 0$.
  Note that $H \in S_6$ and since $D > 0$, $J \in S_1$. Therefore, $H_1 \in S_2$, $H_2 \in S_4$, $J_1 \in S_3$ and $J_2 \in S_5$. Since $|\sqrt{N^2 - D^3} - N|^2 = D^3$, $|H| = \sqrt{D}$ and $|H| = |J|$. Hence all the roots are real. We deduce that $\hat{B}_{t,0} > -\frac{\tau_2}{3\tau_3} + \frac{\sqrt{D}}{2}$, $\hat{B}_{t,2} < -\frac{\tau_2}{3\tau_3} - \frac{\sqrt{D}}{2}$ and $-\frac{\tau_2}{3\tau_3} - \frac{\sqrt{D}}{2} < \hat{B}_{t,1} < -\frac{\tau_2}{3\tau_3} - \frac{\sqrt{D}}{2}$.

- Case V: $D < 0 \Rightarrow N^2 - D^3 > 0$.
  Note that $H \in L_6$ and thus $J \in L_3$. Hence $\hat{B}_{t,0}$ is a real number. Since D is the discriminant of the derivative of the cubic, we know that the cubic only has one real root. Note that $J_1 \in L_1$, $H_1 \in L_2$, $H_2 \in L_4$ and $J_2 \in L_5$, which tells us that $\mathcal{I}m\hat{B}_{t,1} > 0$ and $\mathcal{I}m\hat{B}_{t,2} < 0$.

The Rankine-Hugoniot conditions indeed allow for a unique real solution if and only if $N^2 - D^3 > 0$.

When N, D or $N^2 - D^3$ change sign, we might need to permute the indices. We will not describe all the possible permutations in detail, but instead will illustrate this in one example. Suppose that in 2 subsequent iteration steps, $D > 0$, but in the first iteration step $N > 0$, while in the following step $N < 0$. Comparing case 1 and case 2, we come to the conclusion that we need to permute the indices of $\hat{B}_{t,0}$ and $\hat{B}_{t,1}$.

Let us illustrate these findings by a simplified example. Take the initial guess $\phi_{FT} = 1.27678$, which is the exact solution, and let the guess for ϕ_{ST} vary. In figure C2 we show the real part of the unpermuted roots of the cubic (B.7), together with $N^2 - D^3$ and $B^2 - C$ across the corresponding signal.

Note that the Alfvénic angle $\phi_{a,T} = 1.1936$, since at this position $B^2 - C$ vanishes. Therefore, when $\phi_{ST} < \phi_{a,T}$, the slow transmitted signal is a slow shock. On the other hand, when $\phi_{ST} > \phi_{a,T}$, the signal is an intermediate shock. Also note that $\phi_{a,T}$ is a double root of $N^2 - D^3$. Since $B^2 - C$ changes sign, the roots $\hat{B}_{t,1}$ and $\hat{B}_{t,2}$ are permuted.

The critical angle $\phi_{cr,T} = 1.1948$ is actually the smallest root of $N^2 - D^3$, bigger than $\phi_{a,T}$. For $\phi > \phi_{cr,T}$, the cubic (B.7) has only one real solution and three real solutions for $\phi < \phi_{Cr,T}$.

N changes sign at $\phi = 1.1953$, therefore, $\hat{B}_{t,0}$ and $\hat{B}_{t,1}$ are permuted here, since $D > 0$. The exact solution is located at $\phi = 1.19283$, and the root we select is $\hat{B}_{t,1}$.

Appendix D. Integration across rarefaction waves

When the slow signal is located at a position where no shock solution is possible which satisfies the entropy condition, we postulate the solution to be a rarefaction fan. In this case, we rewrite the stationary MHD equations (1) in cylindrical form, by local decomposition in normal and tangential components. We additionally assume self similarity, i.e. $\frac{\partial}{\partial t} = 0$, where the index t again refers to the tangential direction. Since the entropy S can be shown to be invariant in expansion fans (see e.g. [5]), we replace the energy equation by the isentropic equation. Doing so, we obtain

$$\begin{bmatrix} v_n & \rho & 0 & 0 & 0 \\ 0 & \rho v_n^4 - (v_t B_n - E_z)^2 & v_n B_n (v_t B_n - E_z) & v_n^3 & 0 \\ 0 & -B_n(v_t B_n - E_z) & -\rho v_n^3 + v_n B_n^2 & 0 & 0 \\ \gamma p & 0 & 0 & \rho & 0 \\ 0 & 0 & 0 & 0 & v_n \end{bmatrix} \begin{bmatrix} \rho' \\ v_n' \\ v_t' \\ p' \\ B_n' \end{bmatrix} = \begin{bmatrix} -\rho v_t \\ v_t (v_t B_n - E_z)^2 \\ 2 v_n^2 B_n^2 - \rho v_n^4 + v_t B_n (v_t B_n - E_z) \\ 0 \\ -v_t B_n + E_z \end{bmatrix} \tag{D.1}$$

where the prime $'$ denotes partial derivation in the normal direction. Also note that B_t is eliminated from the system since it is completely determined by the other variables. Only when (D.1) is singular it has a solution, in other words

$$v_n \left(v_n^4 - \left(\frac{B_n^2}{\rho} + \frac{(v_t B_n - E_z)^2}{\rho v_n} + \frac{\gamma p}{\rho} \right) v_n^2 + \frac{\gamma p}{\rho} \frac{B_n}{\rho} \right) = 0, \tag{D.2}$$

$$\Updownarrow$$

$$v_\perp (v_\perp^2 - v_{f,\perp}^2)(v_\perp^2 - v_{s,\perp}^2) = 0, \tag{D.3}$$

which is equivalent to characteristic equation (7). This relation should hold inside of the expansion fans, since $\mathbf{u}$ is partially differentiable there. The solution to (D.1) is given by

$$\rho'(\phi) = 2\rho \frac{v_n v_t (2v_n^2 - a^2 - c^2) + c^2 \tilde{a_n} \tilde{a_t}}{(4+2\gamma) v_n^4 - ((2\gamma+1)a^2 + (3+\gamma)c^2) v_n^2 + \gamma a_n^2 c^2}, \tag{D.4}$$

$$p'(\phi) = 2\gamma p \frac{v_n v_t (2v_n^2 - a^2 - c^2) + c^2 \tilde{a_n} \tilde{a_t}}{(4+2\gamma) v_n^4 - ((2\gamma+1)a^2 + (3+\gamma)c^2) v_n^2 + \gamma a_n^2 c^2}, \tag{D.5}$$

$$B_n'(\phi) = -B_t, \tag{D.6}$$

$$v_n'(\phi) = \frac{v_t (v_n^2 ((2\gamma - 1)a^2 + (\gamma+1)c^2 - 2\gamma v_n^2) - \gamma a_n^2 c^2) + 2 v_n c^2 \tilde{a_n} \tilde{a_t}}{(4+2\gamma) v_n^4 - ((2\gamma+1)a^2 + (3+\gamma)c^2) v_n^2 + \gamma a_n^2 c^2}, \tag{D.7}$$

$$v_t'(\phi) = \frac{\Sigma_{i=0}^{6} \tau_i v_n^i}{((4+2\gamma) v_n^4 - ((2\gamma+1)a^2 + (3+\gamma)c^2) v_n^2 + \gamma a_n^2 c^2) v_n \tilde{a_t}}. \tag{D.8}$$

Here we introduced the symbols $\tilde{a_n} \equiv \frac{B_n}{\sqrt{2p}}$ and $\tilde{a_t} \equiv \frac{B_t}{\sqrt{2p}}$, and the coefficients τ_i are given by

$$\tau_0 = -\gamma a_n^4 c^3, \tag{D.9}$$

$$\tau_1 = 0, \tag{D.10}$$

$$\tau_2 = ((\gamma - 2)\tilde{a_n}\tilde{a_t} + (\gamma+3)c^2 + (2\gamma+1)a^2)\tilde{a_n} c^2, \tag{D.11}$$

$$\tau_3 = 2(a^2 + c^2)\tilde{a_n} v_t, \tag{D.12}$$

$$\tau_4 = -(2\gamma+1)\tilde{a_t}a^2 - ((\gamma+3)\tilde{a_t} + (2\gamma+4)\tilde{a_n})c^2, \quad (D.13)$$
$$\tau_5 = -4\tilde{a_n}v_t, \quad (D.14)$$
$$\tau_6 = (2\gamma+4)\tilde{a_t}. \quad (D.15)$$

These expressions are valid for slow rarefaction fans. The integration has to be performed, starting from $\phi_{sl,R/T}$ till $\phi_{SR/ST}$.

References

[1] Barmin, A. A., Kulikovskiy, A. G. & Pogorelov, N. V. 1996 Shock-capturing approach and nonevolutionary solutions in magnetohydrodynamics. *J. Comp. Phys.* **126**, 77–90.
[2] Brio, M. & Wu, C. C. 1988 An upwind differencing scheme for the equations of ideal magnetohydrodynamics. *J. Comp. Phys.* **75**, 400–422.
[3] Chao, J. K., Lyu, L. H., Wu, B.H., Lazarus, A. J., Chang, T.S. & Lepping, R.P. 1993 Observation of an intermediate shock in interplanetary space. *J. Geophys. Res.* **98**, 17443–17450.
[4] Delmont, P., Keppens, R. & van der Holst, B. 2009 An exact Riemann-solver-based solution for regular shock refraction. *J. Fluid Mech.* **627**, 33–53.
[5] De Sterck, H., Low, B. C. & Poedts, S. 1998 Complex magnetohydrodynamic bow shock topology in field-aligned low-β flow around a perfectly conducting cylinder. *Phys. of Plasmas* **11**, 4015–4027.
[6] Falle, S. A. E. G. & Komissarov, S. S. 1997 On the existence of intermediate shocks. *Mon. Not. R. Astron. Soc.* **123**, 265–277.
[7] Falle, S. A. E. G. & Komissarov, S. S. 2001 On the inadmissibility of non-evolutionary shocks. *J. Plasma Phys.* **65**, 29–58.
[8] Feng, H. & Wang, J. M. 2008 Observations of a 2 → 3 Type Interplanetary Intermediate Shock. *Solar Phys.* **247**, 195–201.
[9] Goedbloed, H. & Poedts, S. 2004 *Principles of Magnetohydrodynamics With Applications to Laboratory and Astrophysical Plasmas*. Cambridge University Press.
[10] Goedbloed, J.P. 2008 *Time reversal duality of magnetohydrodynamical shocks. Phys. Plasmas*
[11] Keppens, R., Nool, M., Tóth, G. & Goedbloed, H. 2003 Adaptive Mesh Refinement for conservative systems: multi-dimensional efficiency evaluation. *J. Comp. Phys.* **153**, 317–339.
[12] Myong, R. S. & Roe, P. L. 1997 Shock waves and rarefaction waves in magnetohydrodynamics. Part 1. A model system. *J. Plasma. Phys.* **58**, 485.
[13] Myong, R. S. & Roe, P. L. 1997 Shock waves and rarefaction waves in magnetohydrodynamics. Part 2. The MHD system. *J. Plasma. Phys.* **58**, 521.
[14] Samtaney, R. 2003 Suppression of the Richtmyer-Meshkov instability in the presence of a magnetic field. *Phys. Fluids* **15**, L53–L56.
[15] Torrilhon, M. 2002 Exact Solver and Uniqueness Conditions for Riemann Problems of Ideal Magnetohydrodynamics, *Technical Report 2002-06, SAM, ETH Zurich.*
[16] van der Holst, B. & Keppens, R. 2007 Hybrid block-AMR in cartesian and curvilinear coordinates: MHD applications. *J. Comp. Phys.* **26**, 925–946.
[17] Wheatley, V., Pullin, D. I. & Samtaney, R. 2005 Suppression of the Richtmyer-Meshkov instability in the presence of a magnetic field. *J. Fluid Mech.* **552**, 179–217.

Journal of Physics: Conference Series **216** (2010) 012008 doi:10.1088/1742-6596/216/1/012008

Clustering and phase separation in dense shear granular flow

Evgeniy Khain

Department of Physics, Oakland University, Rochester, Michigan 48309

E-mail: khain@oakland.edu

Abstract. We investigate various regimes of steady dense Couette flow of inelastically colliding hard disks in the absence of gravity. The two governing parameters in this two-dimensional system are the inelasticity of particle collisions and the average density of particles. The simplest steady state is the uniform shear flow (USF), where the temperature and the density profiles are homogeneous over the system and the velocity of the flow changes linearly between the two moving walls. The USF becomes unstable when the inelasticity of particle collisions exceeds a certain threshold, which depends on the average density of particles. Then the USF gives a way to a "plug flow" regime, where a solid-like cluster coexists with one or two fluid layers. These regimes are investigated using equations of granular hydrodynamics with constitutive relations that interpolate between low and high densities. The results are tested in event-driven molecular dynamics (MD) simulations, and a good agreement is observed.

1. Introduction

Granular materials (materials composed of many macroscopic particles) are ubiquitous in nature and industry. We can find them in the kitchen (sugar, salt) and on the beach (sand), in geophysics (gravel) and astrophysics (interstellar dust). The industrial applications are very wide: from coal mining to food and pharmaceutical industry. Although the founders of the field of granular mechanics were famous physicists like Coulomb and Faraday, for the most part of the 20th century granular mechanics was regarded as a branch of engineering: civil, mechanical, or chemical, depending on the particular application. This era in the development of the field saw a considerable progress in understanding the phenomenology of granular materials. In the last two decades, this field has taken the center stage of condensed matter physics and statistical physics, with the realization that granular materials pose fundamental problems in the behavior of disordered systems, with strong links to subjects such as glasses, chaos and pattern formation, critical phenomena and other areas.

Granular materials are of a great scientific interest [1]-[4]. They exhibit a lot of fascinating phenomena like avalanches [5] and "Brazil Nut" effect [6], surface waves [7] and oscillons [8]. In the middle of the 1980-s, a hydrodynamic description of granular flow was developed by Haff [9] and Jenkins and Richman [10]. Working along the lines of Enskog theory [11, 12], they started from the "microscopic" dynamics of inelastic hard sphere model and kinetic theory of gases and arrived at a granular hydrodynamics in the form of the continuity, momentum and energy equations. However, since the discovery of the clustering instability [13], the validity of these hydrodynamic equations (and, actually, of any continuum theory) became questionable [2]. Granular clustering results from energy losses in inelastic collisions. It was first observed in a freely "cooling" granular gas in a pioneering work by Goldhirsch and Zanetti [13]. It was found that at large inelasticities, the size of clusters can be of the order of few particle diameters. It is clear that no coarse grained theory is able to describe effects on the particle scale. Quoting from the Review of Modern Physics by Kadanoff (1999): this "might invalidate any hydrodynamic-style theory of the granular material" and "we cannot be at all sure that hydrodynamic equations describe the behavior of granular materials in any general way" [2].

Nevertheless, granular hydrodynamics is expected to be valid in the limit of *nearly elastic collisions* we will consider throughout the paper. In this limit, the mean free path of the particles is usually much less than any length scale (and the inverse collision rate much less than any time scale) described hydrodynamically. Many different types of macroscopic granular flow has been investigated in past years both in experiments and in molecular dynamics (MD) simulations or theoretically, using the equations of granular hydrodynamics (see Ref. [14] for reviews). One can mention experimental works on Poiseuille flow [15] and Rayleigh-Taylor flow [16]. A lot of work was done on Couette flows [17].

A variety of symmetry-breaking instabilities in fluidized granular media were investigated employing granular hydrodynamics and molecular dynamics simulations. These include phase-separation instability [18] - [23], which is analogous, in many respects, to the spinodal decomposition of the gas-liquid transition of the van der Waals model [23], oscillatory instability [24], thermal granular convection [25], which resembles the Rayleigh-Benard convection, a famous example of pattern formation outside of equilibrium. Exciting clustering instabilities were investigated in a system consisting of array of two (or more) identical compartments, connected by slits in the walls between neighboring compartments [26].

Due to inelastic collisions between particles, one needs to continuously drive the system either by vibration or by shear in order to maintain the particles motion. As a result, a driven granular system can achieve a steady state, where an energy pumped into the system is dissipated by inelastic collisions. Therefore, driven granular media is intrinsically far from equilibrium [4, 27]. In this paper, we consider dense shear granular flow, which has recently attracted significant attention [17]. Different regimes of dense flows were investigated experimentally, theoretically, and using molecular dynamics (MD) simulations. The most widely studied shear flow regime in experiments is the regime of slow dense granular flow. Here particles in the bulk are not fluidized. Therefore, this regime can not be described by granular hydrodynamics. Moreover, even the model of inelastic hard spheres is not valid in this case, since particles collisions are not instantaneous, and multiples particle contacts become very important. In this study, we will focus on rapid (but dense) granular flow, where the granular media is fluidized, and the concept of granular temperature (introduced in granular hydrodynamics) is meaningful.

The most challenging and intriguing flow regime is the "plug flow" [28], where a solid-like cluster of particles coexists with (one or two) fluid-like layers [29] - [32]. Here the solid-like cluster is at rest or moves as a whole with some constant velocity. Figure 1 shows an example of such solid-fluid coexistence, as observed in molecular dynamics simulations. In this work, we investigate different regimes of a dense fluidized granular flow and theoretically describe the "plug flow" regime using both molecular dynamics (MD) simulations [33] and Navier-Stokes granular hydrodynamics [14].

2. The model of fluidized granular media

Let us consider the simplest model of fluidized granular matter [27]. We address here an idealized model that captures only one basic feature of macroscopic grains: inelastic collisions. The grains are assumed to be inelastic hard spheres (or disks in 2D) that interact with each other via a hard-sphere potential. During a collision between two particles, i and j, momentum is conserved, and some energy lost. Let us denote the component of the velocity perpendicular to the line connecting the centers of the two disks by $v_{i\perp}$ and the parallel component by $v_{i\parallel}$. Let us denote

Journal of Physics: Conference Series **216** (2010) 012008 doi:10.1088/1742-6596/216/1/012008

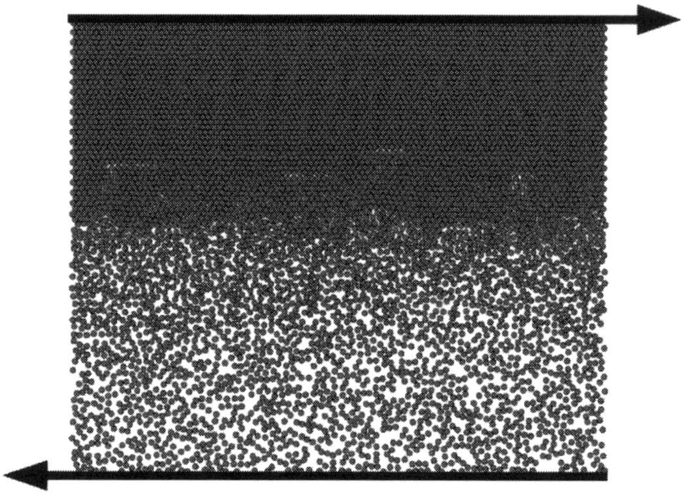

Figure 1. Snapshot of the two-phase flow for $r = 0.992$, $N = 6480$, and $< n >= 0.7244$ ($< \nu >= 0.657$) [32]. A large stripe-like cluster near the wall moves as a whole, showing solid-like behavior. The corresponding density, temperature, and velocity profiles are shown in Fig. 4. The granulate is fluidized everywhere including inside the cluster: the density is smaller than that of close packing, and the granular temperature is larger than zero. However, since $n_\eta < n$, the horizontal flow velocity does not vary across the cluster.

the velocities after the collisions by primes. The energy loss in binary collisions is characterized by the coefficient of normal restitution $0 < r \le 1$. This single parameter distinguishes our idealized granular fluid, $r < 1$, from the idealized "normal" fluid, $r = 1$. By momentum conservation $v'_{i\perp} = v_{i\perp}$ and $v'_{j\perp} = v_{j\perp}$. The perpendicular velocities after the collision are given by

$$\begin{pmatrix} v'_{i\|} \\ v'_{j\|} \end{pmatrix} = \frac{1}{2} \begin{pmatrix} 1-r & 1+r \\ 1+r & 1-r \end{pmatrix} \begin{pmatrix} v_{i\|} \\ v_{j\|} \end{pmatrix} \tag{1}$$

One can easily derive from these relations that the energy loss in each collision is proportional to $1 - r^2$:

$$\Delta E = \frac{1-r^2}{4} \left(v_{i\perp} - v_{j\perp}\right)^2 . \tag{2}$$

We assume the limit of nearly elastic collisions throughout this study, $1 - r^2 << 1$.

Let us consider the dynamics of a system of N inelastically colliding hard disks of unit mass and diameter d driven by two opposite walls at no gravity. The "upper" wall is located at $y = H/2$, the "lower" wall is located at $y = -H/2$, see Fig. 1. The two walls move in the x-direction with opposite velocities, u_0 and $-u_0$ respectively. Typically, the initial conditions are the following: particles have random velocities and are distributed homogeneously over the system. Boundary conditions in the x direction are periodic. Boundary conditions in the y direction can be chosen in different ways. One can choose the moving walls to be thermal, which would correspond to constant temperature boundary conditions in hydrodynamic description. Another option is to ensure that there is no heat flux through the walls. This can be implemented in the following way [32]: upon a collision with one of the driving walls, the normal particle velocity switches sign, while the tangential particle velocity is completely "forgotten". The new tangential velocity component is taken from Maxwell-Boltzmann distribution. The mean velocity of this distribution equals the wall velocity, and the variance corresponds to the instant temperature of the layer next to the wall.

3. Granular hydrodynamics and constitutive relations

The Navier-Stokes granular hydrodynamics operates with coarse grained quantities: number density of grains, $n(\mathbf{r}, t)$, granular temperature $T(\mathbf{r}, t)$, and mean flow velocity $v(\mathbf{r}, t)$ [9, 10]:

$$\begin{aligned} &\frac{dn}{dt} + n\,\nabla\cdot\mathbf{v} = 0, \\ &n\,\frac{d\mathbf{v}}{dt} = \nabla\cdot\mathbf{P}, \\ &n\,\frac{dT}{dt} = -\nabla\cdot\mathbf{Q} + \mathbf{P} : \nabla\mathbf{v} - \Gamma \end{aligned} \qquad (3)$$

Here $\mathbf{P}$ is the stress tensor, $\mathbf{Q}$ is the heat flux, and Γ is inelastic heat losses. The stress tensor $\mathbf{P}$ is given by $\mathbf{P} = [-p(n,T) + \mu(n,T)\,\mathrm{tr}\,(\mathbf{D}\,)]\,\mathbf{I} + 2\eta(n,T)\,\hat{\mathbf{D}}$, where $\mathbf{D} = (1/2)\left[\nabla v + (\nabla v)^T\right]$ is the rate of deformation tensor, $\hat{\mathbf{D}} = \mathbf{D} - \frac{1}{2}\,\mathrm{tr}\,(\mathbf{D}\,)\,\mathbf{I}$ is the deviatoric part of $\mathbf{D}$, and $\mathbf{I}$ is the identity tensor, $\eta(n,T)$ and $\mu(n,T)$ are the shear and bulk viscosities. The heat flux $\mathbf{Q}$ is given by $\mathbf{Q} = -\kappa(n,T)\,\nabla T$, where $\kappa(n,T)$ is the coefficient of thermal conductivity. In the limit of nearly elastic collisions, $1 - r^2 \ll 1$, an additional term in the expression for the heat flux, which is proportional to the *density* gradient [34], can be neglected. We also neglect inelastic corrections to transport coefficients [34] and replace $1-r^2$ factor in the inelastic heat losses term by $2(1-r)$.

For dilute and moderately dense flows, the Enskog-like shear viscosity, $\eta_E(n,T)$, the thermal conductivity $\kappa_E(n,T)$, the inelastic heat losses $\Gamma_E(n,T)$, and the equation of state p_E are given by

$$\begin{aligned} \eta_E &= \frac{4\nu\,T^{1/2}\,G_E}{\pi^{3/2}\,d}\left[1 + \frac{\pi}{8}\left(1 + \frac{1}{G_E}\right)^2\right], \\ \kappa_E &= \frac{8\nu\,T^{1/2}\,G_E}{\pi^{3/2}\,d}\left[1 + \frac{9\pi}{16}\left(1 + \frac{2}{3\,G_E}\right)^2\right], \\ \Gamma_E &= \frac{8(1-r)n\,T^{3/2}\,G_E}{\pi^{1/2}\,d}, \\ p_E &= n\,T(1 + 2\,G_E). \end{aligned} \qquad (4)$$

Here $G_E = \nu(1 - 7\nu/16)/(1-\nu)^2$, $\nu = n\,(\pi\,d^2/4)$ is the solid fraction, and index E stays for "Enskog".

In order to describe dense flow regimes, one needs to construct constitutive relations which are valid up to very high densities. Several attempts of extending granular hydrodynamics of inelastic hard sphere fluids to high densities have recently been performed [35, 36, 37]. Grossman *et al.* [35] investigated a two-dimensional system of inelastic hard disks driven by a thermal wall. The authors pioneered a practical approach, suggesting formulas for equation of state, granular heat conductivity and inelastic energy loss rate which interpolated between the dilute limit and the close vicinity of the hexagonal close packing, $n_{max} = 2/(\sqrt{3}d^2)$, where free-volume arguments are available [35, 38]. Meerson *et al.* [36] have taken the same approach, describing the "levitating cluster" in a similar system in the presence of gravity. In these two works only static steady states (with no mean velocity) were considered [35, 36]. The next step was done by Bocquet and coauthors [37], who investigated granular shear flow with the density approaching the random close packing density. In this case, the mean flow velocity is not zero, so the coefficient of shear viscosity needs to be specified in addition to other constitutive relations. The authors [37] proposed that the shear viscosity coefficient diverges at random close packing density, but with a larger exponent: $\eta \sim (n_{rcp} - n)^{-\beta}$, $\beta > 1$. Khain and Meerson [29] have recently suggested a different approach, incorporating a recent finding [39] that the shear viscosity coefficient diverges like $(n_\eta - n)^{-1}$ at a density $n_\eta < n_{max}$, where $n_{max} = 2/(\sqrt{3}d^2)$ is the density of crystalline close packing.

Therefore, while inelastic heat losses, thermal conductivity, and pressure diverge at the close packing density n_{max}, the shear viscosity diverges at a lower density [39], which may result in coexistence of fluid and solid phases [29, 30, 32]. In a recent work [30], the interpolation of constitutive relations between the dilute and dense limits was proposed (see also Refs. [39, 40]) and tested in a series of MD simulations; a viscosity divergence was also taken into account. The modified constitutive relations are:

$$\begin{aligned} p &= nT(1+2G), \\ \Gamma &= \frac{8(1-r)nT^{3/2}G}{\pi^{1/2}d}, \\ \kappa &= \kappa_E\left(1+0.1\nu-10\nu^{10}+\frac{0.11}{\nu_{max}-\nu}-\frac{0.11}{\nu_{max}}\right), \\ \eta &= \eta_E\left[1+a_\eta(\nu/\nu_\eta)^3/(\nu_\eta-\nu)\right], \end{aligned} \tag{5}$$

where $G = G_E + \theta\left[\nu_{max}/(\nu_{max} - \nu) - G_E\right]$. Here, θ is the interpolation function, given by $\theta = [1 + \exp((\nu_c - \nu)/m_0)]^{-1}$ with $\nu_c = 0.70$, $m_0 = 0.0111$, $\nu_\eta = 0.71$, $a_\eta = 0.037$, and $\nu_{max} = \pi/(2\sqrt{3})$ [39, 40]. In the next section we employ the resulting hydrodynamic equations to description of the phase separation between the fluid-like and solid-like regions in dense shear granular flow.

4. The plug flow

The plug in dense shear granular flow was first observed in MD simulations [28]. The basic physics of this regime relies on the fact that the shear viscosity of hard-disks-fluid diverges not at the density of crystalline close packing $n_{max} = 2/(\sqrt{3}\,d^2)$ (where pressure, inelastic heat losses and thermal conductivity diverge), but at a smaller density, $n_\eta < n_{max}$. Consider the simplest steady Couette flow: uniform shear flow. Here the density (along with temperature and velocity gradient) is constant over the system. It is well known that the uniform shear flow can become unstable [41], which leads to density inhomogeneities in the system: some regions become more dense, other become more dilute. If the mean density in the system is high, than the density in denser regions can become larger than the density of viscosity divergence: $n > n_\eta$. This region then starts behaving like a solid cluster: the velocity can not vary across the cluster since $n_\eta < n$. Interestingly, the density and temperature do vary (since $n < n_{max}$), and the particles are fluidized. Other regions behave like fluids, since $n < n_\eta$ there. This explains the fluid-solid coexistence, shown in Fig. 1. Figure 1 presents the results of MD simulations of $N = 6480$ inelastic hard disks. The initial state was homogeneous, but after a transient the system arrived at the state with broken symmetry with the solid plug next to the "upper" wall at $y = H/2$, moving as a whole with the velocity, which approximately equals the velocity of the wall. Obviously, the state with the cluster next to the "lower" wall at $y = -H/2$ is equally probable. To make sure that the system reached the steady state, we analyzed the time dependence of $y-$component of center of mass of the system. Figure 2 shows another example of steady plug-flow (the snapshot is taken at time $t = 1.4 \times 10^6$), while Figure 3 shows the dynamics of y_{cm}. After a transient, y_{cm} reaches a plato and does not change anymore. Note that the characteristic time of transient dynamics significantly depends on initial conditions. If the initial conditions are symmetric, the system needs to wait for a long time for a large fluctuation in order to decide what the final steady state should be and whether the solid plug should be near the upper or lower wall. One can make the transient time much smaller, starting with asymmetric conditions, see Fig. 3. These asymmetric conditions can be achieved, for example, by adding a small vertical component to all particles velocities.

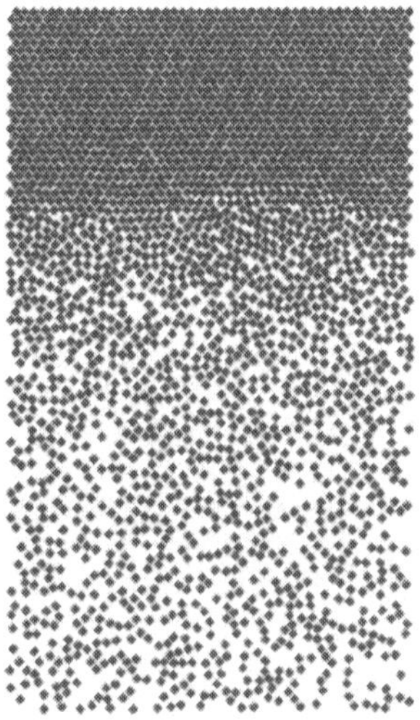

Figure 2. Snapshot of the two-phase flow for $r = 0.986$, $N = 2600$, and $< n >= 0.591$ ($< \nu >= 0.536$). A large stripe-like plug near the wall moves as a whole.

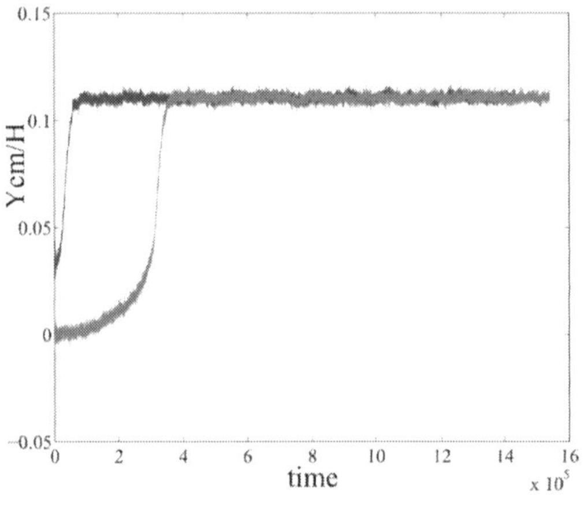

Figure 3. Y−component of center of mass of the system versus time. After a transient, Y_{cm} reaches a plato, which corresponds to the steady state flow. The two curves correspond to symmetric initial conditions (Y_{cm} starts from zero) and asymmetric initial conditions (Y_{cm} starts from some nonzero value). The transient dynamics is much faster if the initial conditions are asymmetric (see text). The parameters are the same as in Fig. 2.

Interestingly, the region inside the plug in Fig. 2 is crystallized; the particles form an ordered layer structure. The same ordered structure was observed in other dense granular clusters [36, 42]. An important difference between the steady states is that the states in Refs. [36, 42] are static, while the cluster shown in Fig. 2 moves as a block to the right with velocity, almost equal to the wall velocity. In both cases, the density inside the cluster is smaller than that of the maximal dense crystalline packing, n_{max}; the granulate is fluidized, the particles are not at rest, and the granular temperature is larger than zero.

Let us now describe hydrodynamically the possible steady states ($\partial/\partial t = 0$) of Eqs. (3), assuming flow in x−direction, zero vertical velocity and $\partial/\partial x = 0$. The coordinate y will be measured in units of the system "height" H (the distance between the two moving walls), the horizontal velocity u will be measured in units of the wall velocity u_0, the temperature T in units of u_0^2, the density n in units of n_{max}, and the pressure P in units of $n_{max}\,u_0^2$. Eqs. (3) can be written now in the following form

$$\frac{d}{dy}\left(f_2 T^{1/2}\frac{du}{dy}\right) = 0,$$
$$\frac{d}{dy}\left(f_1 T^{1/2}\frac{dT}{dy}\right) + \frac{f_2 T^{1/2}}{4}\left(\frac{du}{dy}\right)^2 - R f_3 T^{3/2} = 0\,,$$
$$f_4 T = \text{const.} \tag{6}$$

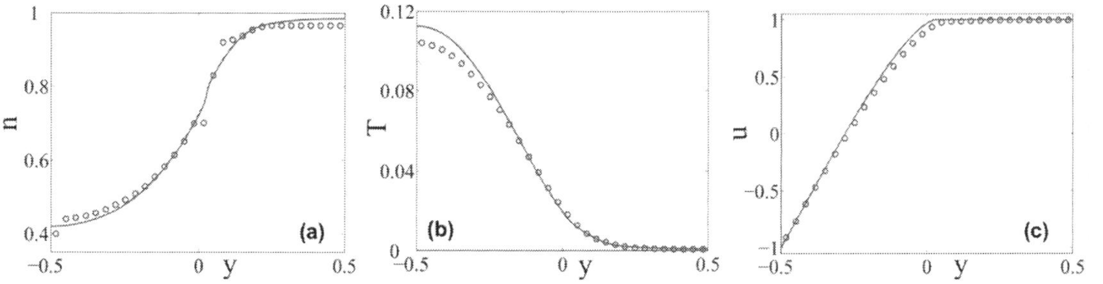

Figure 4. The density (a), temperature (b), and velocity (c) profiles of the two-phase flow for $R = 259.07$ and $< n >= 0.7244$ ($< \nu >= 0.657$) from Ref. [32]. The results of MD simulations are shown by circles; solid lines denote the theoretical curves obtained from numerical solution of Eqs. (6). The corresponding snapshot of the system is shown in Fig. 1.

Here $R = (16/\pi)(1-r)(H/d)^2$ is the hydrodynamic heat loss parameter, and the functions f_i are the density-dependent parts of the constitutive relations [Eqs. (5)]:

$$\begin{aligned} f_1 &= \kappa(\sqrt{\pi}d/2)T^{-1/2}, \\ f_2 &= \eta(2d\sqrt{\pi})T^{-1/2}, \\ f_3 &= \nu G, \\ f_4 &= n(1+2G). \end{aligned} \tag{7}$$

The total number of particles is conserved $\int_{-1/2}^{1/2} n(y)\,dy =< \nu > /\nu_{max}$ and no-flux and no-slip boundary conditions are assumed: $dT/dy(y = -1/2) = dT/dy(y = 1/2) = 0$, $u(y = -1/2) = -1$, $u(y = 1/2) = 1$.

Figure 4 shows the profiles of density (the left panel), temperature (the middle panel), and horizontal velocity (the right panel). The hydrodynamic profiles (solid lines) are computed from Eqs. (6). Details of the numerical procedure are given in Ref. [32]. MD simulations support the hydrodynamic predictions, Fig. 4 (circles). The corresponding snapshot of the system is shown in Fig. 1.

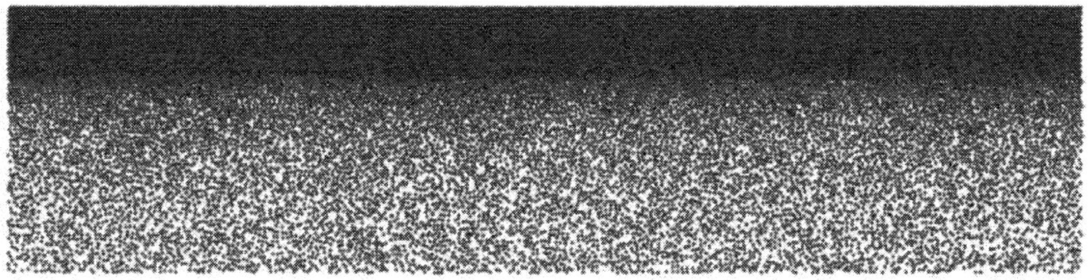

Figure 5. Snapshot of the two-phase flow for the same parameters as in Fig. 2, but with much higher aspect ratio: $L_x/H = 4.62$, the number of particles is $N = 19500$. A slight signature of a two-dimensional wavy structure can be seen.

The structures shown in Figs. 1 and 2 are essentially one-dimensional: the density, temperature, and velocity profiles depend only on the "vertical" coordinate y and do not depend on the coordinate x, along the flow direction. What happens in case of a wider system? Here, the layered structures may give way to wavy patterns [43]. A snapshot shown in Fig. 5 corresponds to the same parameters as in Fig. 2, except for the much larger aspect ratio L_x/H. Probably, there is some signature of two-dimensional structure in Fig. 5, but a more detailed analysis, both theoretical and using MD simulations is needed. If the moving boundaries are taken to be thermal walls, the wavy patterns occur in the middle of the system, like in Ref. [43]. For moderate densities, some of these structures were explained theoretically [44], while for higher densities the viscosity divergence should be taken into account.

For the fixed aspect ratio of the system, the steady flow regimes are governed by two parameters: the hydrodynamic heat loss parameter $R = (16/\pi)(1-r)(H/d)^2$ and the average area fraction $< \nu >$. In different regions of the phase diagram $(R, < \nu >)$, either uniform shear flow or the plug-flow can be found in MD simulations. Hydrodynamic analysis shows additional nonlinear steady flow regimes, which seem to be unstable [29, 30, 32]. Interestingly, in a certain region of the phase diagram both the uniform shear flow and the plug-flow are stable, and hysteretic behavior can be observed. The reader is referred to Ref. [32] for more details.

5. Summary and discussion

This study focuses on a dense fluidized granular flows in Couette geometry in the framework of the simplest model of fluidized granular media: inelastic hard sphere model. An intriguing plug-flow regime is investigated both theoretically by using Navier-Stokes granular hydrodynamics, and in event-driven molecular dynamics simulations. The plug is formed due to the nontrivial dependence of the coefficient of shear viscosity on density: the shear viscosity diverges at the density smaller than the density of crystalline close packing $n_\eta < n_{max}$. This explains the observed solid-fluid coexistence: the density in the solid-like plug is above n_η, while in the fluid-like regions it is below n_η. Interestingly, the particles inside the cluster form an ordered crystalline structure (though the density is smaller than n_{max}). Does this happen since the density of viscosity divergence n_η is very close to the density of the melting point? Or may be the crystallization is a unique feature of inelastically colliding hard disks, since it is observed in systems without any shear as well? Further studies are needed to address these questions.

6. References

[1] Jaeger H M, Nagel S R and Behringer R P 1996 *Rev. Mod. Phys.* **68** 1259
[2] Kadanoff L P 1999 *Rev. Mod. Phys.* **71**, 435
[3] de Gennes P G 1999 *Rev. Mod. Phys.* **71**, S374
[4] Aranson I S and Tsimring L S 2006 *Rev. Mod. Phys.* **78**, 641
[5] For example, Menta A and Barker G C 1994 *Rep. Prog. Phys.* **57** 383; Aranson I S and Tsimring L S 2001 *Phys. Rev. E* **64** 020301
[6] Kudrolli A 2004 *Rep. Prog. Phys.* **67** 209
[7] Pak H K and Behringer R P 1993 *Phys. Rev. Lett.* **71** 1832
[8] Umbanhowar P B, Melo F and Swinney H L 1996 *Nature* **382** 793
[9] Haff P K 1983 *J. Fluid Mech.* **134** 401
[10] Jenkins J T and Richman M W 1985 *Phys. Fluids* **28** 3485
[11] Enskog D and Sven K 1922 *Vetenskapsaked Handl.* **63** 4
[12] Chapman S and Cowling T G 1990 *The Mathematical Theory of Non-Uniform Gases* (Cambridge Univ. Press, Cambridge)
[13] Goldhirsch I and Zanetti G 1993 *Phys. Rev. Lett.* **70** 1619
[14] Campbell C S 1990 *Annu. Rev. Fluid Mech.* **22** 57; Goldhirsch I 2003 *Annu. Rev. Fluid Mech.* **35** 267
[15] Tsai J C, Losert W, Voth G A and Gollub J P 2002 *Phys. Rev. E* **65** 011306
[16] Duran J 2001 *Phys. Rev. Lett.* **87** 254301
[17] For example, Risso D and Cordero P 1997 *Phys. Rev. E* **56** 489; Sasvari M, Kertesz J and Wolf D E 2000 *Phys. Rev. E* **62** 3817; Volfson D, Tsimring L S and Aranson I S 2003 *Phys. Rev. Lett.* **90** 254301; Tsai J C, Voth G A and Gollub J P 2003 *Phys. Rev. Lett.* **91** 064301; Volfson D, Tsimring L S and Aranson I S 2004 *Phys. Rev. E* **69** 031302; Daniels K E and Behringer R P 2005 *Phys. Rev. Lett.* **94** 168001; Cordero P, Risso D and Soto R 2005 *Physica A* **356** 54; Saitoh K and Hayakawa H 2007 *Phys. Rev. E* **75** 021302
[18] Livne E, Meerson B and Sasorov P V 2002 *Phys. Rev. E* **65** 021302
[19] Brey J J, Ruiz-Montero M J, Moreno F and R. Garcia-Rojo 2002 *Phys. Rev. E* **65** 061302

[20] Khain E and Meerson B 2002 *Phys. Rev. E* **66** 021306
[21] Argentina M, Clerc M G and Soto R 2002 *Phys. Rev. Lett.* **89** 044301
[22] Livne E, Meerson B and Sasorov P V 2002 *Phys. Rev. E* **66** 050301(R)
[23] Khain E, Meerson B and Sasorov P V 2004 *Phys. Rev. E* **70** 051310
[24] Khain E and Meerson B 2004 *Europhys. Lett.* **65** 193
[25] Khain E and Meerson B 2003 *Phys. Rev. E* **67** 021306; Eshuis P, van der Weele K, van der Meer D, Bos R and Lohse D 2007 *Phys. Fluids* **19** 123301
[26] van der Meer D, van der Weele K, Reimann P and Lohse D 2007 *J. Stat. Mech.: Theory Exp.* P07021 and references therein
[27] *Granular Gases* edited by Pöschel T and Luding S 2001 (Springer, Berlin); *Granular Gas Dynamics* edited by Pöschel T and Brilliantov N 2003 (Springer, Berlin)
[28] Alam M and Luding S 2003 *Phys. Fluids* **15** 2298
[29] Khain E and Meerson B 2006 *Phys. Rev. E* **73** 061301
[30] Khain E 2007 *Phys. Rev. E* **75** 051310
[31] Alam M, Shukla P. and Luding S 2008 *J. Fluid Mech.* **615** 293
[32] Khain E 2009 *Europhys. Lett.* **87** 14001
[33] Rapaport D C 1995 *The Art of Molecular Dynamics Simulation* (Cambridge University Press, Cambridge)
[34] Brey J J, Moreno F and Dufty J W 1996 *Phys. Rev. E* **54** 445; Sela N and Goldhirsch I 1998 *J. Fluid Mech.* **361** 41; Brey J J, Dufty J W, Kim C S and Santos A 1998 *Phys. Rev. E* **58** 4638; Garzo V and Dufty J W 1999 *Phys. Rev. E* **59** 5895; Garzo V and Montanero J M 2002 *Physica A* **313** 336; Lutsko J F 2005 *Phys. Rev. E* **72** 021306
[35] Grossman E L, Zhou T and Ben-Naim E 1997 *Phys. Rev. E* **55** 4200
[36] Meerson B, Pöschel T and Bromberg Y 2003 *Phys. Rev. Lett.* **91** 024301
[37] Bocquet L, Losert W, Schalk D, Lubensky T C and Gollub J P 2002 *Phys. Rev. E* **65** 011307; Bocquet L, Errami J and Lubensky T C 2002 *Phys. Rev. Lett.* **89** 184301
[38] Salsburg Z W and Wood W W 1962 *J. Chem. Phys.* **37** 798 and references therein.
[39] Garcia-Rojo R, Luding S and Brey J J 2006 *Phys. Rev. E* **74** 061305
[40] Luding S 2001 *Phys. Rev. E* **63** 042201
[41] For example, Alam M and Nott P R 1998 *J. Fluid Mech* **377** 99; Alam M, Arakeri V H, Nott P R, Goddard J D and Herrmann H J *J. Fluid. Mech.* **523** 277
[42] Meerson B, Diez-Minguito M, Schwager T and Poschel T 2007 *Granular Matter* **10** 21
[43] Conway S L and Glasser B J 2004 *Phys. Fluids* **16** 509; Conway S L, Liu X and Glasser B J 2006 *Chem. Eng. Sc.* **61** 6404
[44] Alam M 2006 *J. Fluid Mech.* **553** 1

The Third International Symposium on Bifurcations and Instabilities in Fluid Dynamics IOP Publishing
Journal of Physics: Conference Series **216** (2010) 012009 doi:10.1088/1742-6596/216/1/012009

Bifurcation in a bubble translating and expanding in a sound field

Dinesh Khattar

Department of Mathematics, Kirori Mal College, University of Delhi, Delhi 110 007, India

E-mail: khattar_dinesh@yahoo.co.in

Abstract. We study the interaction of the motion of an expanding and translating bubble with a sound field, using the method of effective potential adopted by Landau and Lifshitz. We work under the assumption that the amplitude of the sound is small, while the frequency of the sound is large compared to the characteristic oscillation frequency of the bubble. We find that the equilibrium value of the mean radius R of the bubble exhibits a bifurcation as the amplitude of the high frequency sound field increases in a neighbourhood of the value zero.

1. Introduction

The importance of bubble dynamics in cavitation [2], underwater explosion [6], ultrasonic cleaning devices and medical application of ultrasonics [12], sonoluminescence [7] and other applications like processing of material [1] in a space laboratory is well known.

Rayleigh [14] studied the adiabatic expansion of a spherical gas bubble in an inviscid fluid. He derived the differential equation for growth of the radius $R(t)$ of the bubble. Plesset [13] modified this equation to include the viscosity of the outside liquid. The motion of a gas bubble, which translates with velocity U as it expands, has been considered by Chakraborty [4] and Chakraborty and Tuteja [5], assuming a spherical bubble shape. Detailed numerical results have been given in Ref. [5] when the pressure in the liquid far from the bubble is constant.

In the absence of the translational motion of the expanding bubble, its motion in a sound field has been discussed by Lauterborn [11] and Keller and Miksis [9]. In the present paper, we investigate the interaction of the motion of an expanding and translating bubble with a sound field. In our study we shall use a method applied by Landau and Lifshitz [10] in their study of stability of an inverted pendulum when the pivot oscillates vertically.

This method has been discussed by Blackburn, Smith and Grønbeck-Jensen [3]. In this method the concept of an "effective potential" is used. The oscillating amplitude of the pivot is taken as small and the period of oscillations is assumed to be small compared to a characteristic period of the pendulum. Using the same method, in our discussion of the bubble motion in a sound field, we work under the assumption that the amplitude of the sound field is small, while the frequency of this field is large compared to a characteristic frequency of the oscillation of the bubble. We find that the equilibrium value for the mean radius R of the bubble exhibits a bifurcation [8] as the sound field amplitude increases from a critical value which is zero or nearly so.

2. Formulation

We consider a spherical gas bubble of radius R at time t which is translating with velocity U in an inviscid liquid of density ρ and surface tension coefficient σ. If p_e is the pressure in the liquid at a large distance from the bubble, p_{g0}, U_0 and R_0 are the gas pressure, velocity of translation and radius initially, respectively, then the equations governing the motion of the bubble [4] are

$$R\frac{d^2R}{dt^2}+\frac{3}{2}\left(\frac{dR}{dt}\right)^2-\frac{U^2}{4}+\frac{1}{\rho}\left\{p_e-p_{g0}\left(\frac{R_0}{R}\right)^{3\gamma}+\frac{2\sigma}{R}\right\}=0 \tag{1}$$

and

$$UR^3=U_0R_0^3=k, \tag{2}$$

where k is constant. Using eq. (2), we eliminate U from (1) and obtain the equation

$$\frac{d^2}{dt^2}(R^{5/2})-\frac{5}{8}k^2R^{-11/2}+\frac{5}{2}\frac{R^{1/2}}{\rho}\left\{p_e-p_{g0}\left(\frac{R_0}{R}\right)^{3\gamma}+\frac{2\sigma}{R}\right\}=0 \tag{3}$$

Defining

$$r=R^{5/2} \tag{4}$$

and taking $\gamma=4/3$ for simplicity, we find that eq. (3) becomes

$$\frac{d^2r}{dt^2}-\frac{5}{8}k^2r^{-11/5}+\frac{5}{2}\frac{r^{1/5}}{\rho}\left\{p_e-p_{g0}\left(\frac{r_0}{r}\right)^{8/5}+2\sigma r^{-2/5}\right\}=0 \tag{5}$$

where $r_0=R_0^{5/2}$.

We consider the case when the pressure p_e in the liquid at a large distance from the bubble is modulated harmonically, due to a sound field, with frequency ω and a relative amplitude ε ($\ll 1$) around its static value p_{e0}, so that p_e is given by

$$P_e=p_{e0}(1+\varepsilon\cos\omega t). \tag{6}$$

We use p_{g0} and U_0 as characteristic pressure and speed. We also use r_0 and T_0 as scales of r and t, respectively, where $T_0=R_0/U_0$. We define the dimensionless quantities r', t', ω' and p_e' as

$$r'=r/r_0,\quad t'=t/T_0,\quad \omega'=\omega T_0,\quad p_e'=p_e/p_{g0}.$$

We may then write the equations (5) and (6) as

$$\frac{d^2r'}{dt'^2}=\frac{5}{8}r'^{-11/5}\left\{1-\frac{4p_{g0}}{\rho U_0{}^2}\left[p_e'r'^{12/5}+\frac{2\sigma}{p_{g0}R_0}r'^2-r'^{4/5}\right]\right\} \tag{7}$$

and

$$p_e'=p_{e0}(1+\varepsilon\cos(\omega't'))/p_{g0}. \tag{8}$$

Finally, equations (7) and (8) take the forms

$$\frac{d^2r}{dt^2}=\frac{5}{8}r^{-11/5}\left[1-\frac{4p_{g0}}{\rho U_0^2}\left\{p_er^{12/5}+\frac{2\sigma}{p_{g0}R_0}r^2-r^{4/5}\right\}\right], \tag{9}$$

and

$$p_e=\frac{p_{e0}}{p_{g0}}(1+\varepsilon\cos\omega t), \tag{10}$$

where the dashes have been omitted from r', t', ω' and p_e'. Following Landau and Lifshitz [10] we suppose that r in equation (9) can be split up into a fast component ξ and a slow component η so that

$$r=\xi+\eta. \tag{11}$$

We shall assume that $|\xi|$ is small, as ε is taken as small. In writing (11) we have assumed that

$$\omega \gg 1 \tag{12}$$

Equations (9) and (10), in view of (11), then yield

$$\frac{d^2\xi}{dt^2}+\left[\frac{1}{2}\frac{p_{g0}}{\rho U_0^2}\left\{\frac{p_{e0}}{p_{g0}}\eta^{-4/5}-\frac{2\sigma}{p_{g0}R_0}\eta^{-6/5}+7\eta^{-12/5}\right\}+\frac{11}{8}\eta^{-16/5}\right]\xi$$
$$= -\frac{5}{2}\varepsilon\frac{p_{g0}}{\rho U_0^2}\frac{p_{e0}}{p_{g0}}\eta^{1/5}\cos(\omega t) \tag{13}$$

and

$$\frac{d^2\eta}{dt^2} = \frac{5}{8}\eta^{-11/5}\left\{1-\frac{4p_{g0}}{\rho U_0^2}\left[\frac{p_{e0}}{p_{g0}}\eta^{12/5}+\frac{2\sigma}{p_{g0}R_0}\eta^2-\eta^{4/5}\right]-\frac{4p_{g0}p_{e0}}{5\rho U_0^2 p_{g0}}\eta^{7/5}\langle\varepsilon\xi\cos(\omega t)\rangle\right.$$
$$\left.+\left[\frac{360}{25}\frac{p_{g0}}{\rho U_0^2}\frac{p_{e0}}{p_{g0}}\eta^{2/5}-\frac{24}{25}\frac{p_{g0}}{\rho U_0^2}\frac{\sigma}{p_{g0}R_0}+\frac{168}{25}\frac{p_{g0}}{\rho U_0^2}\eta^{-6/5}+\frac{88}{25}\eta^{-2}\right]\langle\xi^2\rangle\right\} \tag{14}$$

where $\langle\ \rangle$ denotes the average over a period $\frac{2\pi}{\omega}$. Terms of order ε^3 and smaller have been neglected in writing (13) and (14). Equation (13) can be integrated to yield the fast component as

$$\xi = \frac{\frac{5\varepsilon}{2}\frac{p_{g0}}{\rho U_0^2}\frac{p_{e0}}{p_{g0}}\cos(\omega t)\eta^{17/5}}{\omega^2\eta^{16/5}-\frac{1}{2}\frac{p_{g0}}{\rho U_0^2}\left\{\frac{p_{e0}}{p_{g0}}\eta^{12/5}-\frac{2\sigma}{p_{g0}R_0}\eta^2+7\eta^{4/5}\right\}-\frac{11}{8}}. \tag{15}$$

In view of (15), the averages can be evaluated as

$$\langle\varepsilon\xi\cos\omega t\rangle = \frac{\frac{5}{4}\varepsilon^2\eta^{17/5}\frac{p_{g0}}{\rho U_0^2}\frac{p_{e0}}{p_{g0}}}{\left\{\omega^2\eta^{16/5}-\frac{1}{2}\frac{p_{g0}}{\rho U_0^2}\left[\frac{p_{e0}}{p_{g0}}\eta^{12/5}-\frac{2\sigma}{p_{g0}R_0}\eta^2+7\eta^{4/5}\right]-\frac{11}{8}\right\}} \tag{16}$$

$$\langle\xi^2\rangle = \frac{\frac{25}{8}\varepsilon^2\left(\frac{p_{g0}}{\rho U_0^2}\frac{p_{e0}}{p_{g0}}\right)^2\eta^{34/5}}{\left\{\omega^2\eta^{16/5}-\frac{1}{2}\frac{p_{g0}}{\rho U_0^2}\left[\frac{p_{e0}}{p_{g0}}\eta^{12/5}-\frac{2\sigma}{p_{g0}R_0}\eta^2+7\eta^{4/5}\right]-\frac{11}{8}\right\}^2} \tag{17}$$

We define the dimensionless quantities

$$U^* = \left(\frac{\rho U_0^2}{p_{g0}}\right)^{1/2}, P = \frac{p_{e0}}{p_{g0}}, S = \frac{2\sigma}{p_{g0}R_0}. \tag{18}$$

We note that U^*, P and S are proportional to translational velocity of the bubble, steady part of the pressure due to the sound field, surface tension coefficient of the liquid outside the bubble, respectively. Using (18) we can write, in view of (16) and (17), the equation (14) as

$$\frac{d^2\eta}{dt^2} = \frac{5}{8}\eta^{-11/5}(\Phi + E^2/\Omega + E^2\bar{P}/\Omega^2) \tag{19}$$

where

$$E = \varepsilon P U^{*^{-2}} \tag{20}$$

is the control parameter, and we have introduced the abbreviations

$$\Phi = 1 - 4U^{*^{-2}}(P\eta^{12/5} + S\eta^2 - \eta^{4/5}), \tag{21}$$

$$\Omega = -O\eta^{16/5} + \frac{U^{*^{-2}}}{2}(P\eta^{12/5} - S\eta^2 + 7\eta^{4/5}) + \frac{11}{8}, \tag{22}$$

$$\bar{P} = 45PU^{*^{-2}}\eta^{36/5} - \frac{3}{2}SU^{*^{-2}}\eta^{34/5} + 21U^{*^{-2}}\eta^{28/5} + 11\eta^{24/5} \tag{23}$$

and

$$O = \omega^2 \tag{24}$$

Following Landau and Lifshitz ([10]), we introduce the effective potential V and write equation (19) as

$$\frac{d^2\eta}{dt^2} = -\frac{dV}{d\eta}, \tag{25}$$

where

$$\frac{dV}{d\eta} = -\frac{5}{8}\eta^{-11/5}[\Phi + E^2/\Omega + E^2\bar{P}/\Omega^2]. \tag{26}$$

3. Bifurcation of the mean bubble radius

To discuss possible bifurcations of the bubble radius, we find the equilibrium points (E, η), in the $E - \eta$ plane, characterized by $dV/d\eta = 0$ (cf. (25)). From (26) we have, at an equilibrium point,

$$\Phi + E^2/\Omega + E^2\bar{P}/\Omega^2 = 0. \tag{27}$$

By back transformation from η to R, using (4), we identify the mean radius by

$$R = \eta^{2/5}. \tag{28}$$

The abbreviations (21)-(23) now become

$$\Phi = 1 - 4U^{*^{-2}}(PR^6 + SR^5 - R^2), \tag{29}$$

$$\Omega = -OR^8 + U^{*^{-2}}(PR^6 - SR^5 + 7R^2) + \frac{11}{8}, \tag{30}$$

$$\bar{P} = 45PU^{*^{-2}}R^{18} - \frac{3}{2}SU^{*^{-2}}R^{17} + 21U^{*^{-2}}R^{14} + 11R^{12} \tag{31}$$

On physical grounds, we shall always restrict ourselves to the considerations of positive and real roots R of (27), where $\Phi, \Omega, \bar{P}$ are now given by (29)-(31). We shall examine the change in the equilibrium value of R as the control parameter E increases from zero. A point of bifurcation or branch point $(E_{\text{critical}}, R^*)$ in $E-R$ plane, where R^* is a root of the equation (27) at $E = E_{\text{critical}}$, and where (cf.(20)) E_{critical} is small ($\varepsilon \ll 1$), is identified by a change of the number of real and positive roots R of (27) in a small neighbourhood of R^*, when E varies in a small neighbourhood of E_{critical}. As $E \to 0$ we assume that $E/\Omega \to 0$ (so that $\Phi \to 0$ (cf. (27)), and the equation (27) reduces (cf. (29)) to the equation

$$PR^6 + SR^5 - R^2 - \frac{1}{4U^{*-2}} = 0. \tag{32}$$

A real positive root $R = R_0$ of (32) gives an equilibrium point of (25) when $E = 0$, i.e. in the absence of a sound field. Since there is one change of sign in (32), in the coefficients of powers of R, there cannot be more than one positive real root of (32) by Decartes' rule of signs [15]. Since the left hand side of (32) is positive for large R and negative for small R, there is exactly one positive real root $R = R_0$ of (32), for a given set of values of P, S and U^*. When $E \neq 0$, but is small, the equation (27) shows that there is an equilibrium point (E, R), where R is a real positive root of

$$\Phi\Omega^2 + E^2\Omega + E^2\bar{P} = 0 \tag{33}$$

Before treating the case $E \neq 0$, we first examine if $E = 0$ and $R = R_0$ is a point of bifurcation, and for this purpose, we find out the positive real root $R = R_0$ of (32) for a given set of values of P, S and U^*. We then determine O such that for this value of O the equation

$$\Omega = 0 \tag{34}$$

is satisfied at $R = R_0$. As $E \to 0$, the equation (33) then reduces to

$$\Phi\Omega^2 = 0 \tag{35}$$

which has three repeated real positive roots $R = R_0$ (one from $\Phi = 0$, and two from $\Omega^2 = 0$), so that as E increases from the value 0, three branches of real solutions R of equation (33) may possibly emanate from the point $E = 0$, $R = R_0$ in the $E - R$ plane indicating that $E = 0$, $R = R_0$ is a bifurcation point. Different sets of values of P, S and U^* are chosen and corresponding values of R_0 and O are determined from (32) and (34). Only those sets of values of P, S and U^* are taken which give O large compared to 1, since (24) is true and (12) holds. Then E is increased gradually from the value $E = 0$, and real positive roots of R of equation (33) are determined. Only in a few exceptional cases we have found three branches of real positive roots R of (33), which emanate from the point $E = 0$, $R = R_0$ in the ER plane. In such an exceptional case we have a branch point or bifurcation point at $E = 0$, $R = R_0$. One such bifurcation point, and the branches of the real roots R of equation (33) emanating from this bifurcation point, constituting a bifurcation diagram [8], are shown in Fig. 1, where $S = 0.03, P = 0.11, U^* = 0.277$. The stability of the bubble at equilibrium, at different values of the mean radius R for different E, along different branches of the solution R are also indicated in the bifurcation diagram in Fig. 1. (There is stability of the bubble at an equilibrium point (E, R) if $\frac{-d^2V}{d\eta^2} < 0$, at the corresponding point (E, η), where $dV/d\eta$ is given by (26), otherwise there is instability at that point (E, R).) In most cases, for a given combination of values of S, P and U^*, the point $E = 0, R = R_0$ is found not to be a branch point or bifurcation point. One such case is shown in Fig. 2.

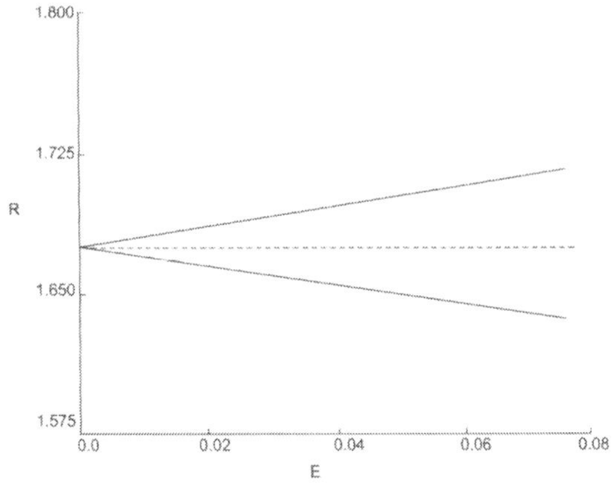

Figure 1. The bifurcation diagram when $S = 0.003$, $U^* = 0.277$, $P = 0.11$, $O = 2.296$, and critical E is 0.0. The variations of R, the real roots of (33), with E are shown. The stable and unstable cases are shown by continuous and broken lines, respectively

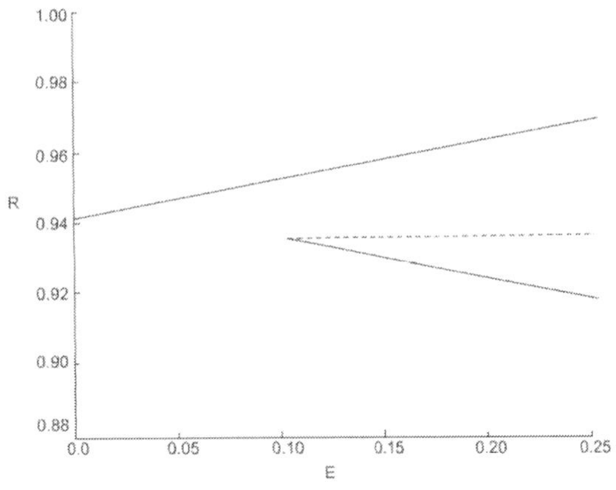

Figure 2. The bifurcation diagram when $S = 0.5$, $U^* = 1.0$, $P = 1.1$, $O = 7.764$, and critical E is 0.1028. The variations of R, the real roots of (33), with E, are shown. The stable and unstable cases are shown by continuous and broken lines, respectively

As E increases from the value $E = 0$, only one branch of the real solution for R is found to emanate from the point $E = 0, R = R_0$. Of the three repeated roots $R = R_0$, at $E = 0$ for the equilibrium equation (33), as E increases from $E = 0$ value, two roots become complex conjugate and the variation of positive imaginary part of one such root is shown in Fig. 3 for the combination $S = 0.5, P = 1.1, U^* = 1.0$. This imaginary part gradually increases with E, reaches a maximum and then decreases to become zero at a value $E = E_{\text{critical}}$. At value $E = E_{\text{critical}}, R = R^*$, a real quantity. When $E = E_{\text{critical}}$ in (33), we have three branches of real solutions for R of the equilibrium equation (33) of which one branch starts from $E = 0, R = R_0$ and other two branches originate from $E = E_{\text{crilical}}$ and $R = R^*$ in $E - R$ plane. The point $E = E_{\text{critical}}$ and $R = R$ in $E - R$ plane is now the point of bifurcation. The equilibrium value of R on one branch of the bifurcation diagram are stable, and those on other branch are unstable. Fig. 2 shows the bifurcation diagram for the combination of values $S = 0.5$, $P = 1.1$, $U^* = 1.0$ for which Fig. 3 is valid. Similarly Fig. 4 gives one more bifurcation diagram for a different set of values of S, P and U^*.

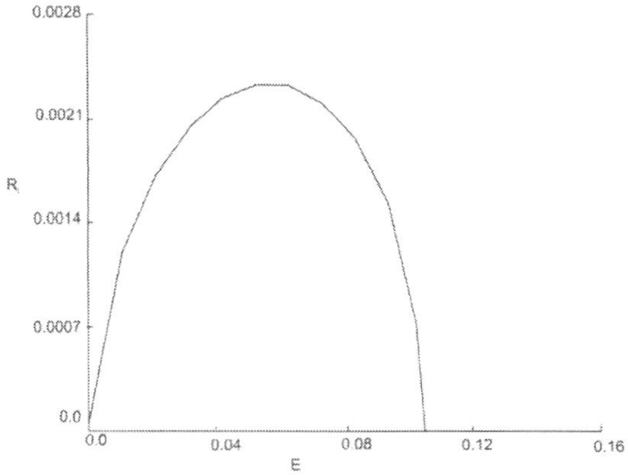

Figure 3. The variation of R_i, the imaginary part of the two complex conjugate roots R of (33), with E, when $S = 0.05$, $U^* = 1.0$, $P = 1.1$, $O = 7.764$ and critical E is 0.1028

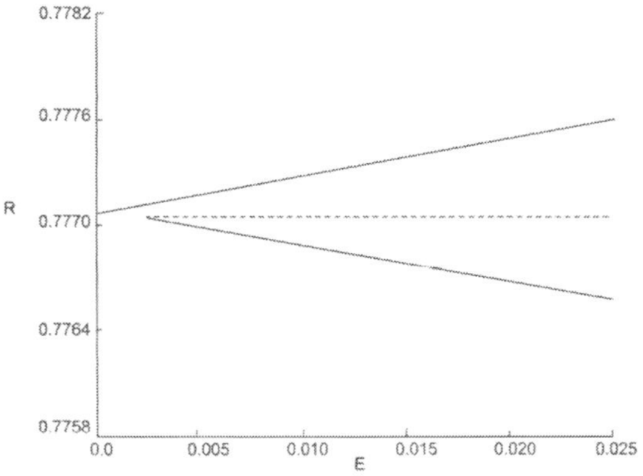

Figure 4. The bifurcation diagram when $S = 0.5$, $U^* = 0.5$, $P = 2.38$, $O = 79.69$ and critical E is 0.0021. The variations with E, of the real roots R of (33), are shown. The stable and unstable cases are shown by continuous and broken lines, respectively

In Fig. 5 we give the variation of O, where O is proportional to the square of frequency of sound field (cf. (24)) for which bifurcation is observed, with translating velocity U^* of the bubble, when $S = 0.5$, $P = 1.125$. We note that O decreases as velocity U^* of the translating bubble is increased. Similar results are observed for other values of the parameters S and P. In Fig. 6 we give the variation of E_{critical} as U^* changes but S and P kept fixed ($S = 0.5$, $P = 1.125$) Similar results are observed for other values of P and S.

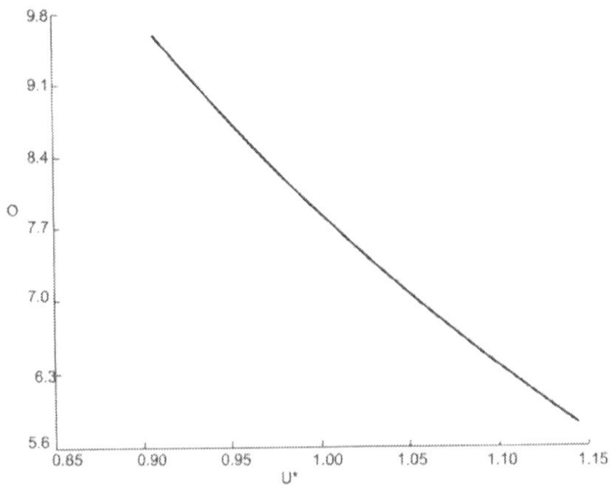

Figure 5. The variation of O, with U^*, for which bifurcation is observed, when $S = -0.5, P = 1.125$

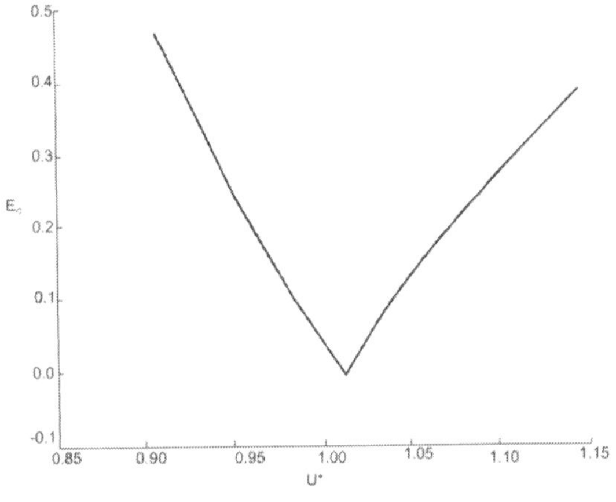

Figure 6. The variation of E_c, the critical value of E at which bifurcation occurs, with U^*, when $S = 0.5, P = 1.125$

4. Concluding remarks

In this paper, we have discussed the bifurcations in an expanding and translating bubble in a sound field when the outside fluid is incompressible and inviscid. We have examined the roots of an algebraic equation (33) of degree 22 in R for different values of S, P, U^* and E. A set of values of the parameters S, P and U^* are so chosen that the value of O, determined from (34), for which bifurcation is possible, is large compared to 1, and the value of E_{critical} is also small. Only in rare cases $E_{\text{critical}} = 0$ is found to be true.

For a given set of values of S and P, the value of O, for which bifurcation is possible, is found to decrease as U^* increases (Fig. 5).

For a given set of values of S and P, as U^* increases, E_{critical} is found to decrease first, to come to a minimum value, and then to increases again (Fig. 6.).

References

[1] Annamalai P, Shankar N, Cole R, Subramanian R S 1982. Bubble migration inside a liquid drop in a space laboratory. *Applied Scientific Research* **38** 179.
[2] Brikhoff G, Zarantonello E H 1975. *Jets, Waves and Cavities*, Academic Press Inc., New York.
[3] Blackburn J A, Smith H J T, Gøbeck-Jensen N 1992. Stability and Hopf bifurcations in an inverted pendulum. *Am. J. Phys.* **60** 903.
[4] Chakraborty B B 1990. Effect of a viscous fluid flow past a spherical gas bubble on the growth of its radius. *Proc. Ind. Acad. Sciences* **100** 185.
[5] Chakraborty B B, Tuteja G S 1993. Motion of an expanding gas bubble in a viscous liquid under gravity. *Phys. Fluids* **5**, 187.
[6] Cole R H 1948. *Underwater Explosions*, Princeton University Press, Princeton, New Jersey.
[7] Crum L A, Putterman Seth 1992. Sonoluminescence. *J. Acoust. Soc. Am.* **91** 517.
[8] Drazin P G 1992. *Nonlinear Systems*. Cambridge University Press, Cambridge.
[9] Keller J B, Miksis M 1980. Bubble oscillation of large amplitude. *J. Acoustic Soc. America* **68**, 628.
[10] Landau L D, Lifshitz E M 1976. *Mechanics*. Pergamon, New York, 9395.
[11] Lauterborn W 1976. Numerical investigations of nonlinear oscillations of gas bubbles in liquids. *J. Acoustic Soc. America* **59** 283.
[12] Lauterborn W, Hentschel W 1985. Cavitation bubble dynamics studied by high speed photography and holography: Part one. *Ultrasonics* **22** 216.
[13] Plesset M S 1949. The dynamics of cavitation bubbles. *J. Appl. Mech.* **16** 277.
[14] Rayleigh Lord 1917. On the pressure developed in a liquid during the collapse of a spherical void. *Phil. Mag.* (Ser. 6) **34** 94.
[15] Turnbull H W 1957. *Theory of Equations*, Oliver and Boyd, London.

Journal of Physics: Conference Series **216** (2010) 012010 doi:10.1088/1742-6596/216/1/012010

The Rayleigh-Benard instability in an enclosure having finite thickness walls

Geniy V Kuznetsov[1], **Mikhail A Sheremet**[2]
[1] Faculty of Thermal Power Engineering, Tomsk Polytechnic University, Tomsk, 634050, Russia
[2] Faculty of Mechanics and Mathematics, Tomsk State University, Tomsk, 634050, Russia

E-mail: Michael-sher@yandex.ru

Abstract. Natural convection in an enclosure having finite thickness heat-conducting walls at local heating at the bottom of the cavity has been numerically studied. Heat exchange with an environment due to convection and radiation has been considered on one of external sides of the decision region. The governing unsteady three-dimensional flow equations in the Boussinesq approximation for the gas cavity and heat conduction equation for the solid walls, written in dimensionless variables such as vector potential functions, the vorticity vector and the temperature, have been solved using finite difference method. Results have been obtained for a Prandl number of 0.7 and for a Grashof number ranging from 10^4 to 10^6.

Nomenclature

Bi	Biot number, $\left(=hL_x/k\right)$
Fo	Fourier number, $\left(=\alpha t_0/L_x^2\right)$
g	gravitational acceleration (m/s^2)
Gr	Grashof number, $\left(=g\beta\left(T_{hs}-T_0\right)L_x^3/\nu^2\right)$
h	heat-transfer coefficient (W/m^2 K)
H	vertical (along the *z*-axis) dimension of the solution region (m)
k	thermal conductivity (W/m K)
k_{ij}	thermal conductivity ratio, $\left(=k_i/k_j\right)$
L_x	horizontal (along the *x*-axis) dimension of the solution region (m)
L_y	horizontal (along the *y*-axis) dimension of the solution region (m)
Sk	Stark number, $\left(=\varepsilon\sigma L_x\left(T_{hs}-T_0\right)^3/k\right)$
Nu_{avg}	average Nusselt number
Pr	Prandtl number, $\left(=\nu/\alpha\right)$
Ra	Rayleigh number, $\left(=g\beta\left(T_{hs}-T_0\right)L_x^3/\nu\alpha\right)$
t	time (s)
t_0	time scale (s)

T	temperature (K)
T_0	initial temperature (K)
u, v, w	velocity components in x, y, z directions (m/s)
U, V, W	dimensionless velocity components in X, Y, Z directions
V_0	velocity scale, $\left(=\sqrt{g\beta(T_{hs}-T_0)L}\right)$ (m/s)
x, y, z	Cartesian coordinates (m)
X, Y, Z	dimensionless Cartesian coordinates

Greek symbols

α	thermal diffusivity (m^2/s)
β	thermal expansion coefficient (K^{-1})
ε	specific emissitivity factor
Θ	dimensionless temperature
Θ_0	dimensionless initial temperature
ν	kinematic viscosity (m^2/s)
σ	Stephen-Boltzman constant (W/m^2 K^4)
τ	dimensionless time
ψ_0	vector potential functions scale (m^2/s)
ψ_x, ψ_y, ψ_z	vector potential components in x, y, z directions (m^2/s)
Ψ_x, Ψ_y, Ψ_z	dimensionless vector potential components in X, Y, Z directions
ω_0	vorticity vector scale (1/s)
ω_x, ω_y, ω_z	vorticity vector components in x, y, z directions (1/s)
Ω_x, Ω_y, Ω_z	dimensionless vorticity vector components in X, Y, Z directions

Subscripts

e	environment
f	fluid
hs	heat source
i, j	number of the decision region element (Figure 1)
k	number of basis vector
s	solid

1. Introduction

The Rayleigh-Benard convection as one of kinds of a thermal instability of a fluid layer under the gravity is a flow pattern that attracts the greatest attention of researchers [1-10]. Recently heightened interest is linked to creation of the system for validation of new computing algorithms of mechanics of continua [5, 7, 10]. At the same time, effects of both finite heat-conducting walls of an enclosure and nonuniform heat exchange with an environment are not considered in [5, 6, 10]. As is well known [11] such effects can lead to essential changes of both velocity fields and temperature fields. Investigation of similar systems is important for microelectronics. Modern progress trends of the microelectronic technique, linked to reduction of overall dimensions, lead to increase in specific heat-flux density. The latter has the negative effect on operating efficiency of all system [12, 13].

The purpose of the present work is mathematical simulation of unsteady three-dimensional conjugate convective-conductive heat transfer in an enclosure with local heat source in conditions of convective-radiative heat exchange on one of the external sides of the decision region (Figure 1).

2. Mathematical formulation

We considered a boundary-value problem of an unsteady conjugate heat transfer in a region as shown in figure 1. The heat source located at the bottom of the gas cavity is kept at constant temperature during the whole process. The convective-radiative heat exchange with an environment is modeled on one of the external sides ($x = 0$). Other external sides are assumed to be adiabatic.

Thermophysical properties of solid material elements and gas are assumed to be temperature-independent, and the flow regime is laminar. The gas is supposed to be a Newtonian incompressible fluid satisfying the Boussinesq approximation. The gas motion and heat transfer in the internal volume are assumed three-dimensional, the heat exchange by an emission from the heat source and between the walls is assumed to be negligibly small as compared to the convective heat exchange, the gas is assumed to be absolutely transparent for thermal emission.

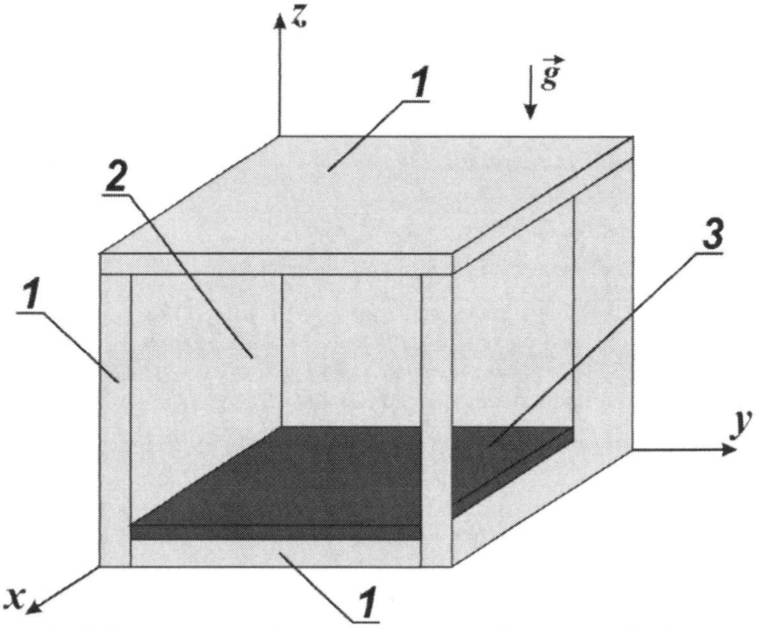

Figure 1. Schematic view of the problem: *1* – walls; *2* – gas; *3* – heat source

Heat transfer process in the considered area (Fig. 1) is governed by the system of unsteady three-dimensional convection equations in the Boussinesq approximation in the gas cavity [14–16]. The unsteady three-dimensional energy equation [17] with nonlinear boundary conditions is used for simulation of heat conduction in the solid walls.

The mathematical model is formulated in the dimensionless variables such as vector potential functions, vorticity vector and temperature [18–20].

The length of the gas cavity along x-axis is chosen as the scale distance. For the reduction to the dimensionless form of the equations system following correlations are used:

$$X = x/L_x,\ Y = y/L_x,\ Z = z/L_x,\ \tau = t/t_0,\ U = u/V_0,\ V = v/V_0,\ W = w/V_0,$$

$$\Theta = (T - T_0)/(T_{hs} - T_0),\ \Psi_x = \psi_x/\psi_0,\ \Psi_y = \psi_y/\psi_0,\ \Psi_z = \psi_z/\psi_0,\ \Omega_x = \omega_x/\omega_0,$$

$$\Omega_y = \omega_y/\omega_0,\ \Omega_z = \omega_z/\omega_0,\ \omega_0 = V_0/L_x,\ \psi_0 = V_0 L_x,\ V_0 = \sqrt{g_z \beta (T_{hs} - T_0) L_x}.$$

Based on the above-mentioned assumptions, the non-dimensional form of the governing equations for the fluid can be written as follows:

$$\frac{\partial \Omega_x}{\partial \tau}+U\frac{\partial \Omega_x}{\partial X}+V\frac{\partial \Omega_x}{\partial Y}+W\frac{\partial \Omega_x}{\partial Z}-\Omega_x\frac{\partial U}{\partial X}-\Omega_y\frac{\partial U}{\partial Y}-\Omega_z\frac{\partial U}{\partial Z}=\frac{1}{\sqrt{\mathrm{Gr}}}\left(\frac{\partial^2 \Omega_x}{\partial X^2}+\frac{\partial^2 \Omega_x}{\partial Y^2}+\frac{\partial^2 \Omega_x}{\partial Z^2}\right)+\frac{\partial \Theta}{\partial Y} \quad (1)$$

$$\frac{\partial \Omega_y}{\partial \tau}+U\frac{\partial \Omega_y}{\partial X}+V\frac{\partial \Omega_y}{\partial Y}+W\frac{\partial \Omega_y}{\partial Z}-\Omega_x\frac{\partial V}{\partial X}-\Omega_y\frac{\partial V}{\partial Y}-\Omega_z\frac{\partial V}{\partial Z}=\frac{1}{\sqrt{\mathrm{Gr}}}\left(\frac{\partial^2 \Omega_y}{\partial X^2}+\frac{\partial^2 \Omega_y}{\partial Y^2}+\frac{\partial^2 \Omega_y}{\partial Z^2}\right)-\frac{\partial \Theta}{\partial X} \quad (2)$$

$$\frac{\partial \Omega_z}{\partial \tau}+U\frac{\partial \Omega_z}{\partial X}+V\frac{\partial \Omega_z}{\partial Y}+W\frac{\partial \Omega_z}{\partial Z}-\Omega_x\frac{\partial W}{\partial X}-\Omega_y\frac{\partial W}{\partial Y}-\Omega_z\frac{\partial W}{\partial Z}=\frac{1}{\sqrt{\mathrm{Gr}}}\left(\frac{\partial^2 \Omega_z}{\partial X^2}+\frac{\partial^2 \Omega_z}{\partial Y^2}+\frac{\partial^2 \Omega_z}{\partial Z^2}\right) \quad (3)$$

$$\frac{\partial^2 \Psi_x}{\partial X^2}+\frac{\partial^2 \Psi_x}{\partial Y^2}+\frac{\partial^2 \Psi_x}{\partial Z^2}=-\Omega_x \quad (4)$$

$$\frac{\partial^2 \Psi_y}{\partial X^2}+\frac{\partial^2 \Psi_y}{\partial Y^2}+\frac{\partial^2 \Psi_y}{\partial Z^2}=-\Omega_y \quad (5)$$

$$\frac{\partial^2 \Psi_z}{\partial X^2}+\frac{\partial^2 \Psi_z}{\partial Y^2}+\frac{\partial^2 \Psi_z}{\partial Z^2}=-\Omega_z \quad (6)$$

$$\frac{\partial \Theta}{\partial \tau}+U\frac{\partial \Theta}{\partial X}+V\frac{\partial \Theta}{\partial Y}+W\frac{\partial \Theta}{\partial Z}=\frac{1}{\mathrm{Pr}\sqrt{Gr}}\left(\frac{\partial^2 \Theta}{\partial X^2}+\frac{\partial^2 \Theta}{\partial Y^2}+\frac{\partial^2 \Theta}{\partial Z^2}\right) \quad (7)$$

Energy equation for the solid walls

$$\frac{\partial \Theta_1}{\partial Fo_1}=\frac{\partial^2 \Theta_1}{\partial X^2}+\frac{\partial^2 \Theta_1}{\partial Y^2}+\frac{\partial^2 \Theta_1}{\partial Z^2} \quad (8)$$

Equations (1)–(8) are subjected to the following initial and boundary conditions.
Initial conditions are

$$\Psi_x(X,Y,Z,0)=0,\ \Psi_y(X,Y,Z,0)=0,\ \Psi_z(X,Y,Z,0)=0,$$
$$\Omega_x(X,Y,Z,0)=0,\ \Omega_y(X,Y,Z,0)=0,\ \Omega_z(X,Y,Z,0)=0,$$

$\Theta(X,Y,Z,0)=0$ except temperature for the heat source on which $\Theta=1$ during the whole process time.

Boundary conditions are:

- convective-radiative heat exchange with an environment is modeled at the wall $X=0$

$$\frac{\partial \Theta_1}{\partial X}=Bi_1\cdot\Theta_1-Bi_1\cdot\Theta^e+Sk_1\cdot\left[(\Theta_1+\zeta)^4-\left(\frac{T^e}{T_{hs}-T_0}\right)^4\right],\quad X=0;$$

- at the rest external walls for the equation (8) heat insulation conditions are set

$$\frac{\partial \Theta_1}{\partial X^i}=0,\qquad X^1\equiv X,\ X^2\equiv Y,\ X^3\equiv Z;$$

- at the solid-fluid interfaces parallel to plane XZ:

$$\Psi_x = \frac{\partial \Psi_y}{\partial Y} = \Psi_z = 0, \quad \begin{cases} \Theta_1 = \Theta_2, \\ \dfrac{\partial \Theta_1}{\partial Y} = k_{2,1} \dfrac{\partial \Theta_2}{\partial Y}; \end{cases}$$

- at the solid-fluid interfaces parallel to plane *XY*:

$$\Psi_x = \Psi_y = \frac{\partial \Psi_z}{\partial Z} = 0, \quad \begin{cases} \Theta_1 = \Theta_2, \\ \dfrac{\partial \Theta_1}{\partial Z} = k_{2,1} \dfrac{\partial \Theta_2}{\partial Z}; \end{cases}$$

- at the solid-fluid interfaces parallel to plane *YZ*:

$$\frac{\partial \Psi_x}{\partial X} = \Psi_y = \Psi_z = 0, \quad \begin{cases} \Theta_1 = \Theta_2, \\ \dfrac{\partial \Theta_1}{\partial X} = k_{2,1} \dfrac{\partial \Theta_2}{\partial X}. \end{cases}$$

Equations (1)–(8) with corresponding initial and boundary conditions have been solved by means of finite differences method [21–24]. Each time step was started from the temperature field computation both in the gas cavity and in solid walls [Eqs. (7) and (8)], then the Poisson equations for vector potential functions were solved [Eqs. (4)–(6)]. The boundary conditions for the vorticity vector components were determined thereafter and equations (1)–(3) were solved.

The locally one-dimensional scheme of Samarskii [21] was used to solve energy equations [Eqs. (7) and (8)] and equations for the vorticity vector components [Eqs. (1)–(3)] numerically. In this scheme, the solution of a three-dimensional scheme reduces to sequential solution of one-dimensional systems. In this case, the solution of one-dimensional system reduces to a sequential solution of systems of difference equations with tridiagonal matrices by the Thomas method. An implicit difference scheme was used. The evolutionary term represented a one-sided difference in time and had the first order of accuracy in time step. All the derivatives with respect to spatial coordinates were approximated with the second order of accuracy in the step along the coordinate.

2.1. Benchmark solutions

Accuracy of the program developed by the authors was checked by preparing the benchmark solutions both for non-conjugate and conjugate problems. In case of non-conjugate analysis, well-known benchmarks of 3D natural convection in a differently heated cubical enclosure with adiabatic side walls [25, 26] and with linear temperature law at side walls [27, 28]. These benchmark results are shown in Tables 1 and 2.

Table 1. Variations of average Nusselt number of heat wall with Rayleigh number (the first benchmark)

Ra	[25]	[26]	Present uniform grid	
			50×50×50	60×60×60
10^4	2.055	2.100	2.075	2.071
10^5	4.339	4.361	4.494	4.446
10^6	8.656	8.770	9.719	9.432

Table 2. Variations of average Nusselt number of heat wall with Rayleigh number (the second benchmark)

Ra	[27]	[28] uniform grid 95×95×95	Present uniform grid 50×50×50	Present uniform grid 60×60×60
10^4	1.520±0.015	1.497	1.517	1.513
10^5	3.097±0.028	3.106	3.229	3.194
10^6	6.383±0.070	6.681	7.254	7.027

For conjugate problem benchmark solution has been obtained by using results [29]. Table 3 shows the good comparison between the results.

Table 3. Variations of average Nusselt number with Grashof number and heat conductivity ratio

Gr	λ_s/λ_f	[11]	[29]	Present
	1	0.877	0.87	0.872
10^3	5	–	1.02	1.023
	10	–	1.04	1.046
	1	2.082	2.08	2.116
10^5	5	–	3.42	3.421
	10	–	3.72	3.781
	1	2.843	2.87	3.002
10^6	5	–	5.89	6.306
	10	–	6.81	6.935

3. Results and discussion

Numerical analysis of the boundary value problem (1)–(8) has been carried out at following dimensionless complexes such as $Gr=10^3-10^6$, $Pr=0.7$, $k_{2,1}=3.7\cdot10^{-2}, 3.7\cdot10^{-3}, 3.7\cdot10^{-4}$ describing the basic modes of conjugate convective heat transfer in enclosures. Dimensionless defining temperatures were $\Theta^e=-1$, $\Theta_{hs}=1$, $\Theta_0=0$.

3.1. Effect of the Grashof number

Streamlines and temperature fields at different values of the Grashof number at $\tau=60$, $X=0.6$ are presented in figure 2. The direction of gas motion in the gas cavity is indicated by arrows on streamlines.

Two convective cells are formed in the gas cavity at $X=0.6$ for $Gr=10^4$ (figure 2,*a*). The reason for the appearance of these cells is both the heat source located on the bottom of the gas cavity and diffusion of perturbations deep into the gas cavity from solid walls. The field of temperature is uniformly in the gas cavity that is linked to dominance of conductive heat transfer. Heating of solid walls from the heat source is observed. In the central part of the cavity $(Y=0.58)$ the thermal plume caused by influence of the heat source is shaped. The increase in Grashof number (figure 2,*b*) leads to change of both a configuration of streamlines and intensity of gas circulation in the cavity. Increase in flow velocities and displacement of a core of the convective cells in a vertical direction are observed. At the same time the temperature field is changed. The essential increase in temperature is observed in top layers of the gas cavity.

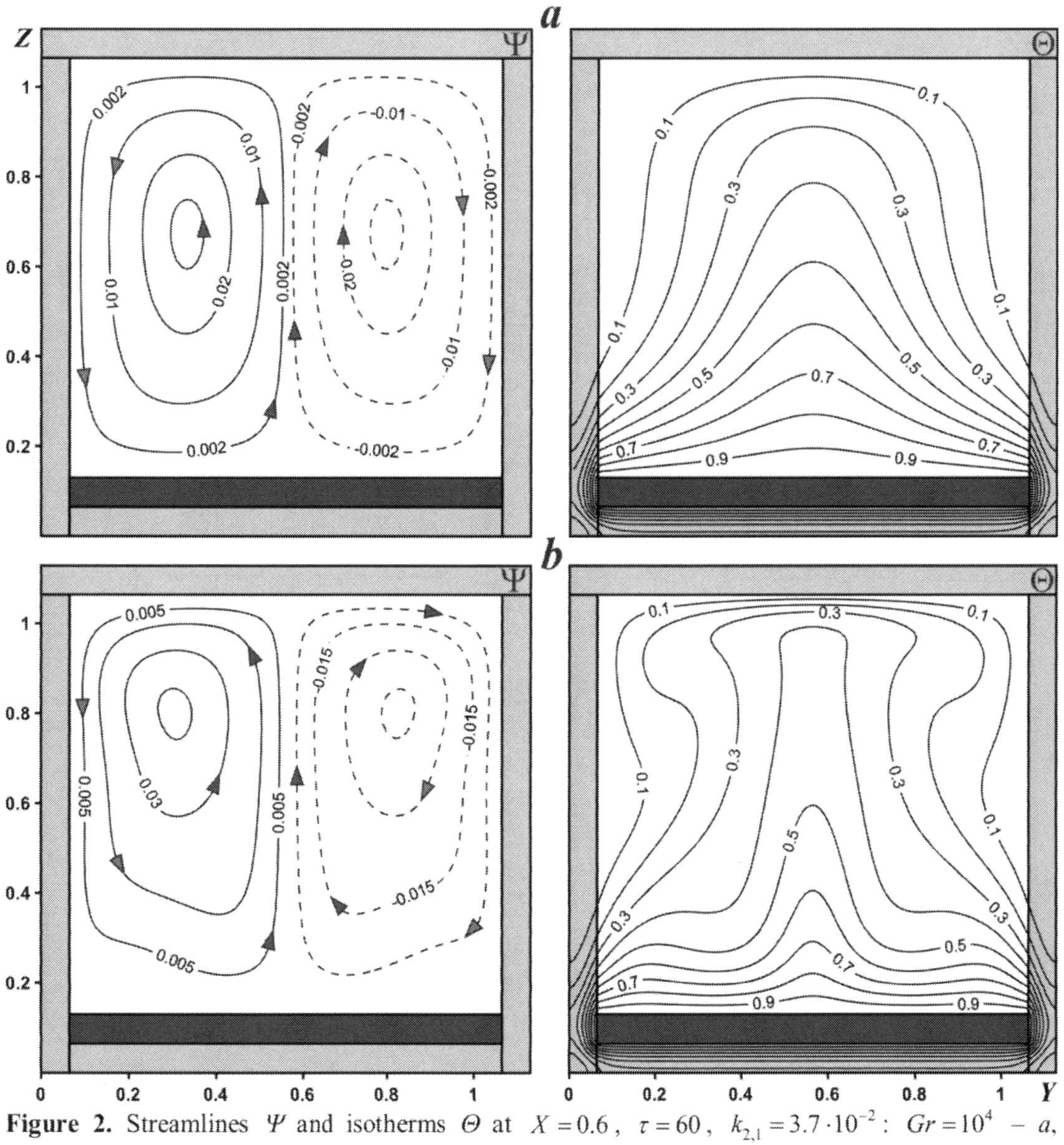

Figure 2. Streamlines Ψ and isotherms Θ at $X = 0.6$, $\tau = 60$, $k_{2,1} = 3.7 \cdot 10^{-2}$: $Gr = 10^4$ – *a*, $Gr = 10^5$ – *b*

Streamlines and temperature fields at $Z = 0.6$ are presented in figure 3. The Grashof number ranging from 10^4 to 10^5 is reflected on the increase in the circulation velocities of the convective cells. The temperature field is changed. The thermal plume is formed in the centre of the gas cavity at $Gr = 10^5$. There is the heating of the cavity from the thermal plume in a radial direction.

The temperature profiles at $X = 0.6, Z = 0.9$, $\tau = 60$ and at different values of the Grashof number are shown in figure 4.

Journal of Physics: Conference Series 216 (2010) 012010 doi:10.1088/1742-6596/216/1/012010

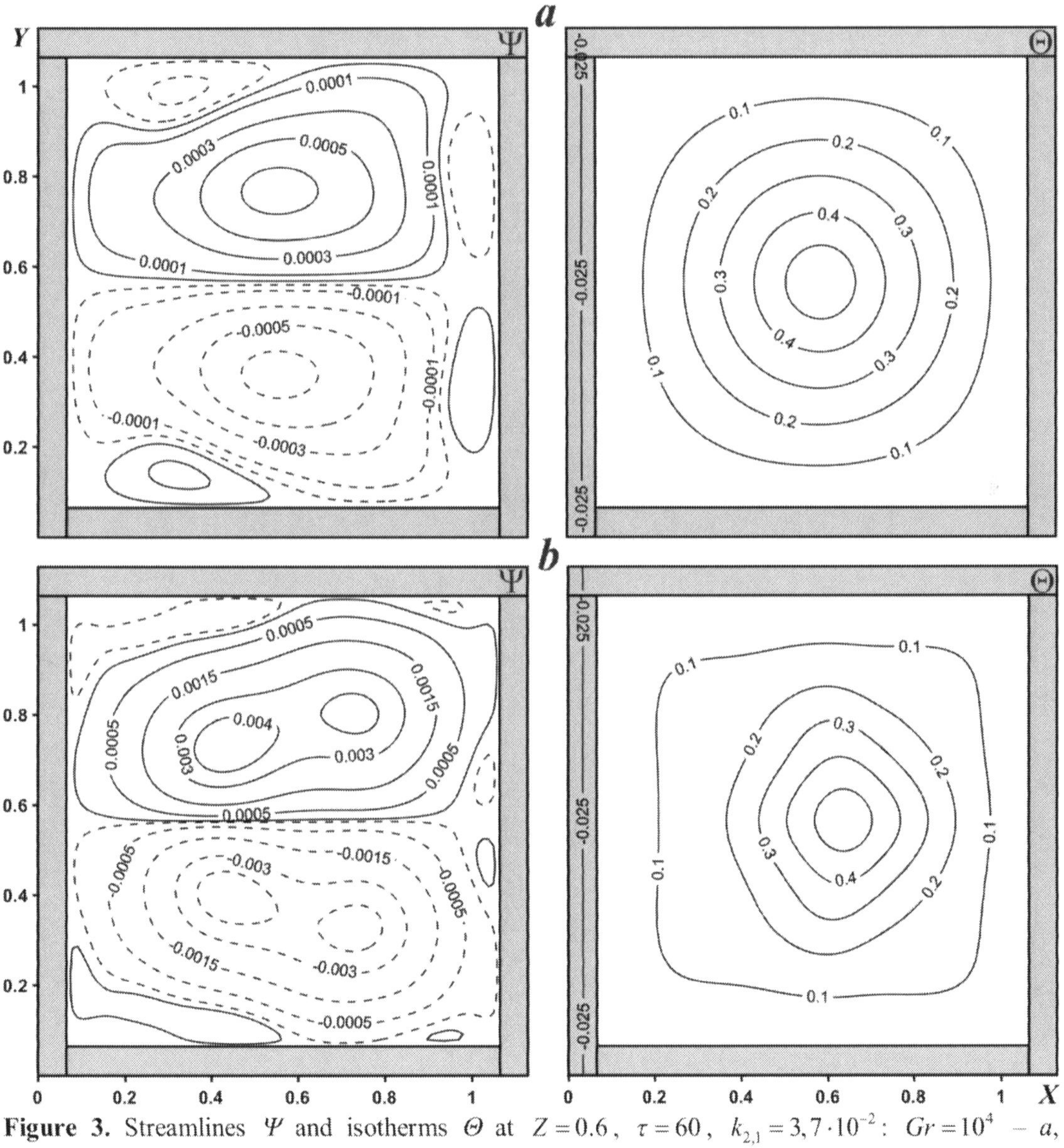

Figure 3. Streamlines Ψ and isotherms Θ at $Z=0.6$, $\tau=60$, $k_{2,1}=3,7\cdot10^{-2}$: $Gr=10^4$ – *a*, $Gr=10^5$ – *b*

Figure 4 shows the effect of the buoyancy force on the temperature field in the gas cavity. The increase in the Grashof number $10^3 \le Gr < 10^5$ leads to uniform increase in the temperature in the central part of the cavity where there is the thermal plume. The essential change of the temperature profile at $Gr=10^5$ and $Gr=5\cdot10^5$ is linked to the increase in the gas motion velocities. It should be noted that the increase in Grashof number leads to the diminution of the boundary layer thickness.

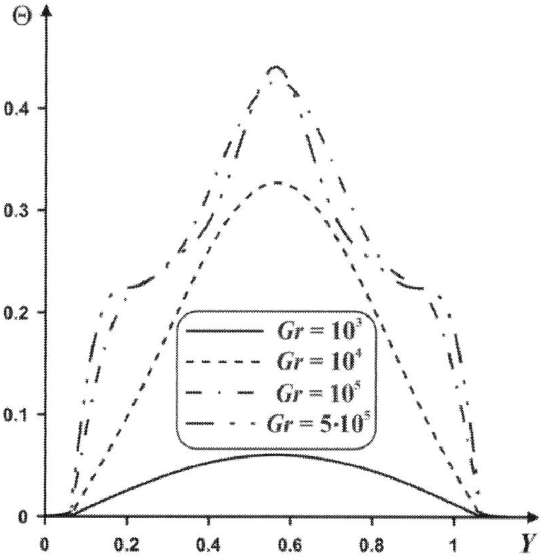

Figure 4. Temperature profiles at $\tau = 60$, $k_{2,1} = 3,7 \cdot 10^{-2}$, $X = 0.6$, $Z = 0.9$

The analysis of the Grashof number influence on the generalized heat transfer coefficient (the average Nusselt number $Nu_{avg} = \int\limits_{0.06}^{1.06}\int\limits_{0.06}^{1.06} \left|\frac{\partial \Theta}{\partial Z}\right|_{Z=0.13} dXdY$) on the heat source surface has been carried out (figure 5).

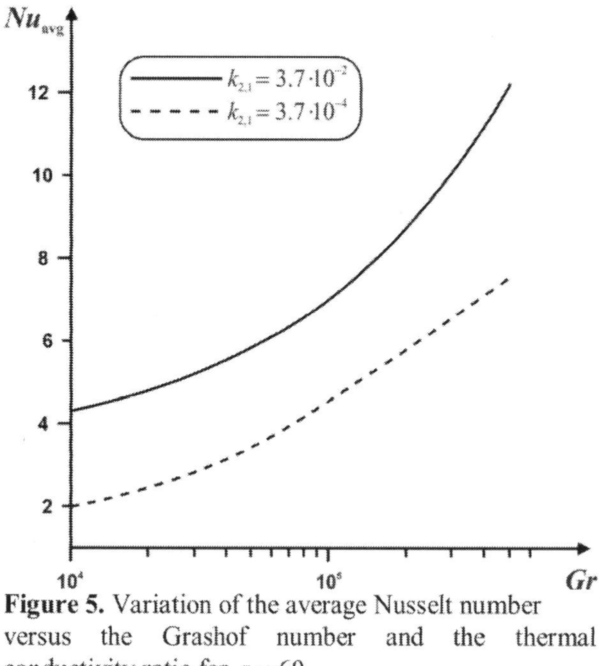

Figure 5. Variation of the average Nusselt number versus the Grashof number and the thermal conductivity ratio for $\tau = 60$

The presented graphic dependences of the average Nusselt number as a function of the Grashof number evidently show the typical increase in the heat transfer intensity on the heat source surface at the Grashof number ranging $10^4 \le Gr < 10^6$. The increase in a role of the buoyancy force in comparison with the internal friction force leads to the heat transfer intensification on the heater surface. The reason for this fact is both the increase in the motion velocities and more essential cooling of the descending gas flows. The latter leads to the significant heat sink from the heat source surface.

3.2. Effect of the transient factor

The transient factor in the conjugate heat transfer problems plays an essential role [20, 24] as it reflects not only dynamics of the velocity and temperature fields in the gas cavity, caused by formation, evolution and dissipation of the vortex structures from an initial quiescent state, but also it characterizes the thermal sluggishness of the solid walls in conditions of the environment influence. At the same time the advantage of such statement is definition of the temperature field at the solid-fluid interface on the basis of conservation laws without additional empirical data, for example, for the heat transfer coefficient. In turn the approach based on use of empirical heat transfer coefficients does not allow considering the transient factor as these coefficients are time functions.

The dynamics of the streamlines formation and temperature fields formation at $Y = 0.6$ for $Gr = 5 \cdot 10^4$ are shown in figure 6.

There are two convective cells of small intensity in the gas cavity at $\tau = 12$ (figure 6,*a*). This fact is explained by the initial stage of the flow evolution. The temperature distribution is already non-uniformly. There is an intensive heating of the bottom of the cavity.

The increase in the time up to $\tau = 36$ (figure 6,*b*) leads to the increase in the velocities of the gas circulation. The intensity of a motion in the left vortex and its sizes are greater in comparison with another vortex. This fact can be explained as the low temperature front has reached the gas cavity in a zone of the center of the left wall. The latter leads to the displacement of the thermal plume to the right wall. The further increase in the dimensionless time (figure 6,*c*) is reflected on dissipation of the right convective cell and leads to full displacement of the thermal plume (figure 7) that is caused by cooling of the left wall.

The temperature profiles at $Y = Z = 0.6$ for $Gr = 5 \cdot 10^4$ depending on time points are presented in figure 7. The increase in the dimensionless time leads to the displacement of the thermal plume to the right wall. There is reduction in the temperature at $X = 0$ and, accordingly, cooling of the solid wall such as $0 \le X \le 0.06$ as the ambient temperature is below initial temperature of the decision region.

The graphic dependences of the generalized heat transfer coefficient on the heat source surface versus dimensionless time and the Grashof number are presented in figure 8. Decrease in the average Nusselt number Nu_{avg} with the dimensionless time can be explained by diminution of the temperature gradient at a surface of the heat source.

3.3. Effect of the heat conductivity ratio

The thermal conductivity ratio characterizes the heat exchange conditions on the solid-fluid interface at analysis of the conjugate heat transfer. Accordingly, the variation of this parameter enables to define a range of change of the required characteristics.

The decrease in the thermal conductivity ratio (figures 9, 10) leads to both an intensification of the heat transfer in the solid walls and suppression of the convective motion in the gas cavity. The latter can be explained by intensive heat elimination into the solid walls. At the same time the average Nusselt number on the heat source surface decreases (figure 5).

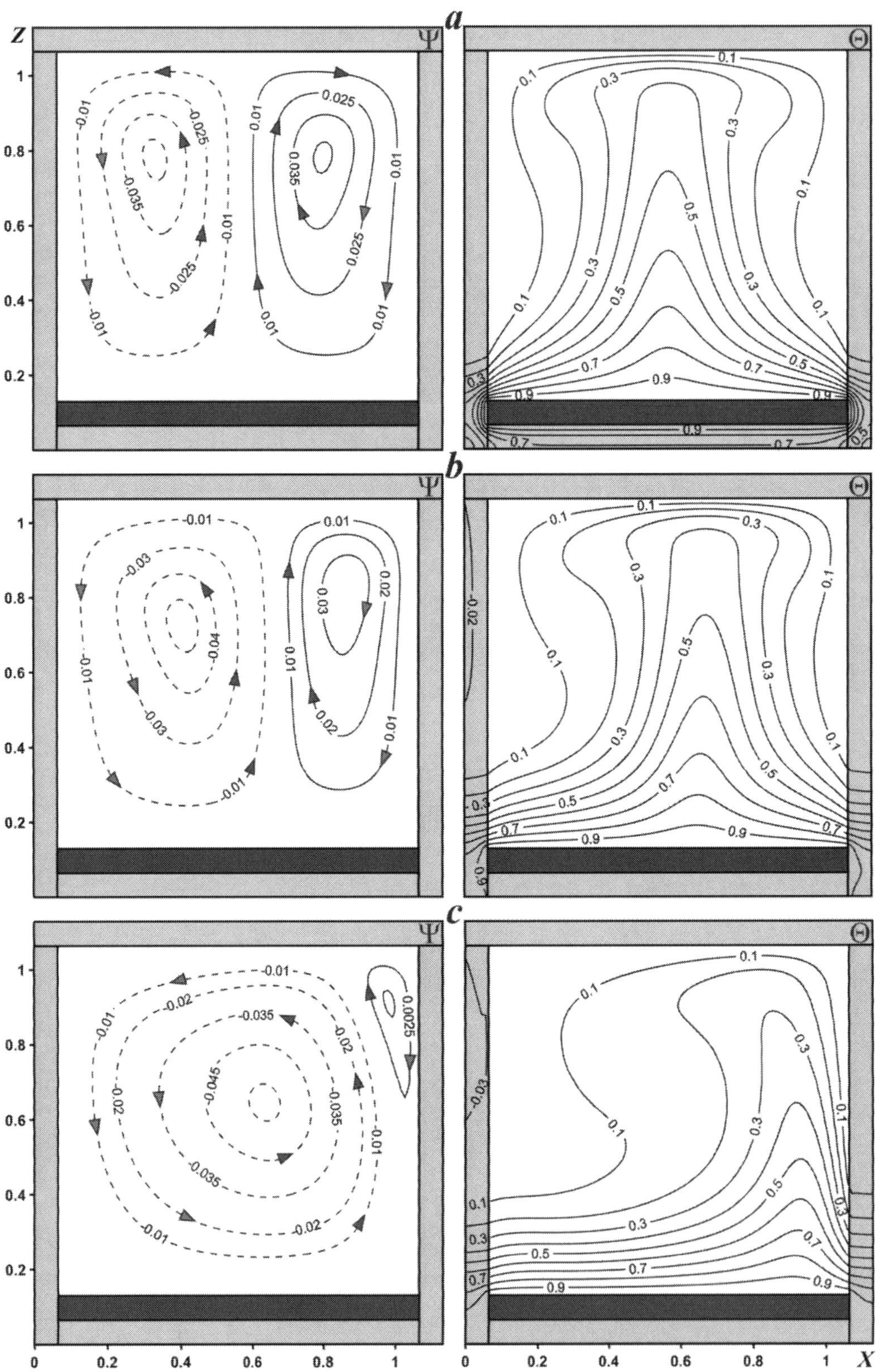

Figure 6. The dynamics of streamlines Ψ and temperature fields Θ at $Y = 0.6$ for $Gr = 5 \cdot 10^4$, $k_{2,1} = 3{,}7 \cdot 10^{-3}$: $\tau = 12$ – *a*, $\tau = 36$ – *b*, $\tau = 60$ – *c*

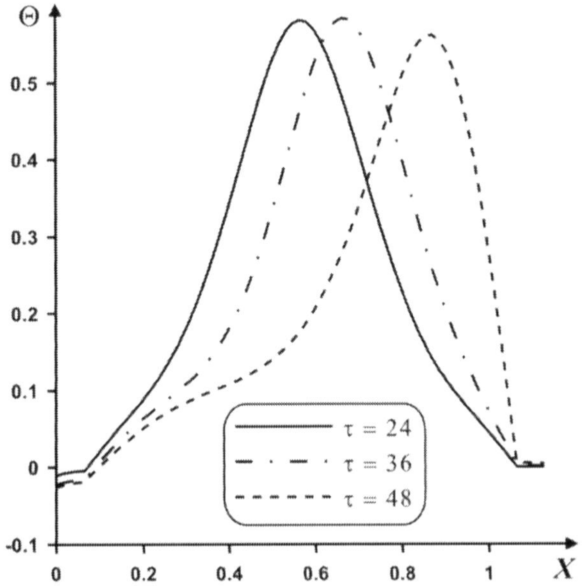

Figure 7. The temperature profiles at $Y = Z = 0.6$ for $Gr = 5 \cdot 10^4$, $k_{2,1} = 3,7 \cdot 10^{-3}$

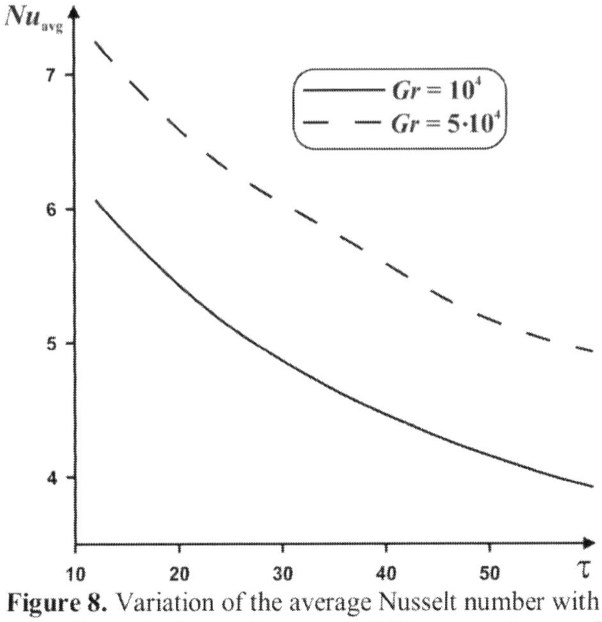

Figure 8. Variation of the average Nusselt number with the dimensionless time at different values of the Grashof number

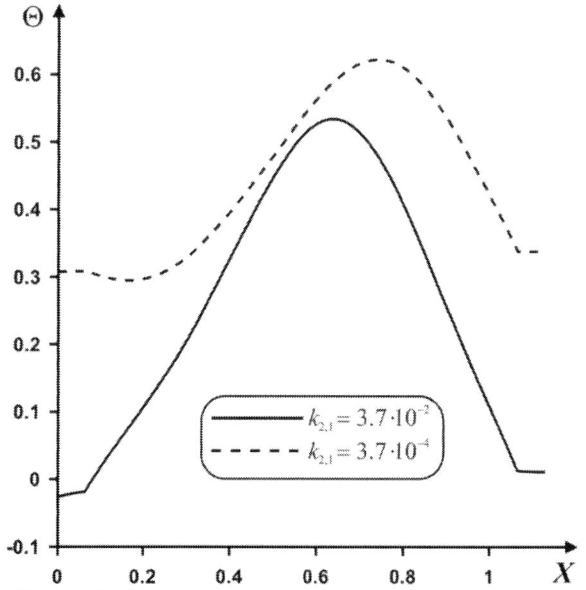

Figure 9. The temperature profiles at $Y = Z = 0.6$ for $Gr = 10^4$ depending on the heat conductivity ratio

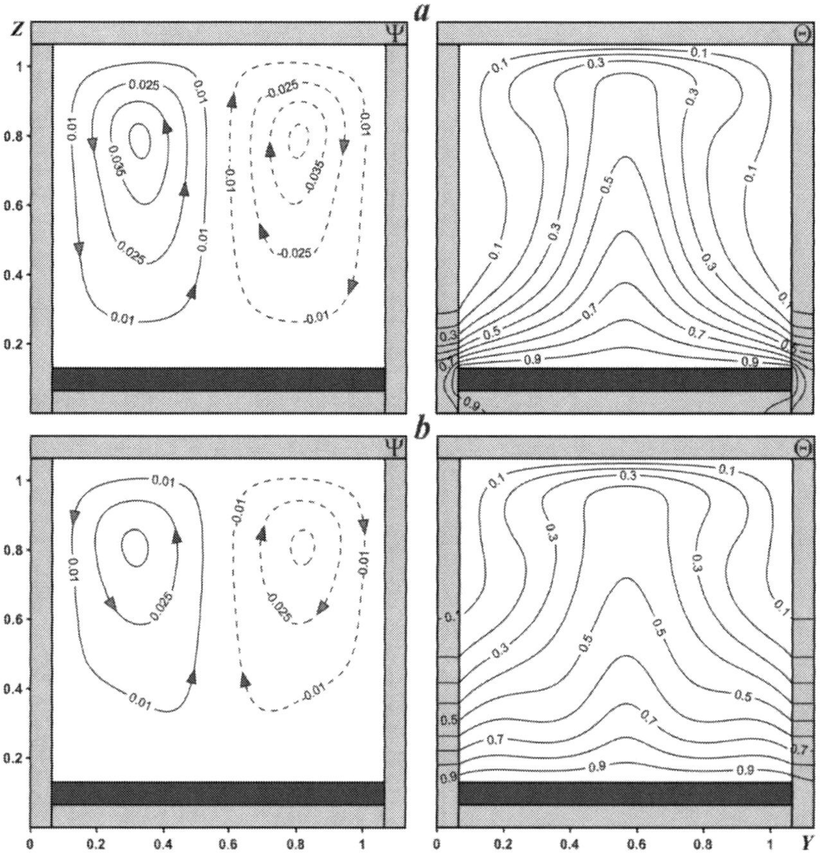

Figure 10. Streamlines Ψ and isotherms Θ at $X = 0.6$, $\tau = 24$, $Gr = 5 \cdot 10^4$: $k_{2,1} = 3.7 \cdot 10^{-3}$ – *a*, $k_{2,1} = 3.7 \cdot 10^{-4}$ – *b*

4. Conclusion

Mathematical simulation of unsteady conjugate heat transfer in an enclosure having finite thickness heat-conducting walls in the presence of heat source under the condition of convective-radiative heat exchange with an environment has been carried out. Distributions of streamlines and temperature fields in wide enough range of defining parameters $10^4 \le Gr < 10^6$, $Pr = 0.7$ have been obtained. The influence of the defining parameters such as the Grashof number, the transient factor and the heat conductivity ratio on formation of thermo-hydrodynamic modes has been analysed. The scopes of the nonlinear environmental effect, owing to conduction in the solid walls of the enclosure (Fig. 3, 6, 7) have been determined. It should be noted that the decrease in the heat conductivity ratio leads to reduction of the average Nusselt number on the heat source surface (Fig. 5).

Acknowledgements

This study has been supported by the Russian Foundation for Basic Research, through grant No. 08-08-00402-a.

References

[1] Getling A V 1999 *Rayleigh-Benard Convection: Structures and Dynamics* (Moscow: Editorial URSS)
[2] Wesfreid J, Pomeau Y, Dubois M, Normand C and Berge P 1978 *J. Physique* **39** 725
[3] Arter W and Newell A C 1988 *Phys. Fluids* **31** 2474
[4] Hernandez R and Frederick R L 1994 *Int. J. Heat Mass Transfer* **37** 411
[5] Pallares J, Cuesta I and Grau F X 2002 *Int. J. Heat Fluid Flow* **23** 346
[6] Natarajan E, Tanmay Basak and Roy S 2008 *Int. J. Heat Mass Transfer* **51** 747
[7] Valencia L, Pallares J, Cuesta I and Grau F X 2007 *Int. J. Heat Mass Transfer* **50** 3203
[8] Tanmay Basak, Roy S and Balakrishnan A R 2006 *Int. J. Heat Mass Transfer* **49** 4525
[9] Sathiyamoorthy M, Tanmay Basak, Roy S and Pop I 2007 *Int. J. Heat Mass Transfer* **50** 766
[10] Calcagni B, Marsili F and Paroncini M 2005 *Applied Thermal Engineering* **25** 2522
[11] Liaqat A and Baytas A C 2001 *Int. J. Heat Mass Transfer* **44** 3273
[12] Jaluria Y 2008 *Design and Optimization of Thermal Systems* (Boca Raton, FL: CRC Press)
[13] Sathe S and Sammakia B 1998 *ASME J. Heat Transfer* **120** 830
[14] Bejan A 2004 *Convection Heat Transfer* (New York: Wiley)
[15] Gebhart B, Jaluria Y, Mahajan R L and Sammakia B 1988 *Buoyancy – induced flows and transport* (New York: Hemisphere Publishing Corporation)
[16] Martynenko O G and Khramtsov P P 2005 *Free-convective heat transfer: with many photographs of flows and heat exchange* (Berlin: Springer)
[17] Luikov A V 1967 *Theory of Thermal Conductivity* (Moscow: Vysshaya shkola)
[18] Aziz K and Hellums J D 1967 *The physics of fluids* **10** 314
[19] Oosthuizen P H and Paul J T 2005 *Int. J. Heat Fluid Flow* **26** 587
[20] Kuznetsov G V and Sheremet M A 2009 *Int. J. Heat Mass Transfer* **52** 2215
[21] Samarskii A A 1977 *Theory of Difference Schemes* (Moscow: Nauka)
[22] Fletcher C A J 1988 *Computational Techniques for Fluid Dynamics* (Berlin: Springer)
[23] Roache P 1976 *Computational Fluid Dynamics* (Albuquerque: Hermosa)
[24] Kuznetsov G V and Sheremet M A 2006 *Thermophysics and Aeromechanics* **13** 565
[25] Bessonov O A, Brailovskay V A, Nikitin S A and Polezhaev V I 1997 Three- dimensional natural convection in a cubical enclosure: a benchmark numerical solution *Advances in Computational Heat Transfer: Proc. the First International Symposium (Cesme, Turkey, 26–30 May 1997)* ed G de Vahl Davis & E Leonardi (New York: Begell House Publishers) pp. 157–165
[26] Fusegi T, Hyin J M and Kuwahara K 1991 *Int. J. Heat Mass Transfer* **34** 1543
[27] Leong W H, Hollands K G T and Brunger A P 1999 *Int. J. Heat Mass Transfer* **42** 1979

[28] Ginkin V P and Ganina S M 2002 Method and program of three-dimensional convection by means of fine grid: *Proc. 3rd Russian Nat. Conf. on Heat Exchange (Moscow, Russia, 21-25 October,* Vol. 3*)* ed A I Leont'ev (Moscow: MEI) pp 116–119.
[29] Kaminski D A and Prakash C 1986 *Int. J. Heat Mass Transfer* **29** 1979.

Journal of Physics: Conference Series **216** (2010) 012011 doi:10.1088/1742-6596/216/1/012011

On Coupled Development of MHD Instabilities of Rayleigh-Taylor and Kelvin-Helmholtz Types in Nonuniform Gas-Plasmas Flows

A P Likhachev [1] and S A Medin

Joint Institute for High Temperatures of Russian Academy of Sciences, Izhorskaya 13 bld 2, Moscow 125412, Russia

E-mail: apl@ihed.ras.ru

Abstract. The simultaneous development of the MHD instabilities of Raylegh-Taylor and Kelvin-Helmholtz types at the interface between high-conducting plasmoid and surrounding non- or low-conducting gas is considered. The linear stage of the RTI development is studied analytically for incompressible and compressible fluids. The nonlinear stage of the individual development of the RTI and the coupled development of both instabilities has been investigated numerically. The time-dependent two-dimensional numerical model based on the solution of the Euler gasdynamic equations with body momentum and energy sources of MHD origin has been developed and used in calculations. A disturbance introducing in the background flow has been periodic with varied assignment type and wave length. Fundamental difference between the results of linear and nonlinear analysis has been revealed. In particular, the increment of the RTI development at nonlinear stage is one-two order of magnitude less than that predicted by linear theory and rather weakly depends on initial disturbance mode. In linear analysis the coupled development of the RTI and the KHI is determined by simple summing of the two effects in the expression of wave increment, whereas in nonlinear case the mutual influence of the instabilities leads to essential alterations in their development, main of which is the intensive "layer-by-layer" destruction of the plasmoid surface.

1. Introduction

The high-conducting plasmoid (i.e., localized plasma formation) intensively flowed over by non- or low-conducting compressible fluid is inherent in many MHD flows. Such flows may naturally arise (for instance, in astro- and magnetospheric physics [1, 2]) and can be realized in different plasmadynamic (PD) and magnetohydrodynamic (MHD) applications [3, 4]. The problem of hydrodynamic stability of plasmoids often defines their life time and is important for the evolution of such flows.

At the interface between plasmoid and ambient gas conditions exist for the coupled development of MHD instabilities of Rayleigh-Taylor and Kelvin-Helmholtz types (hereafter referred to as "RTI" and "KHI" respectively). The RTI is induced by the nonuniformity of the ponderomotive force (in general case, overall body force) component normal to the interface, the KHI is initiated by the nonuniformity of the tangential velocity component. The interaction of these instabilities may produce essential alterations in the character of the development of each of them and in the flow in the whole.

[1] To whom any correspondence should be addressed.

The peculiarities of the problem have not been adequately explored. Although the MHD RTI and to a smaller extend the KHI were considered before (see, for instance, review in monograph [5]), they were usually studied individually. The coupled development of instabilities and their interaction in nonlinear stage have been investigated rarely and in rather particular cases (for instance, [6]). The phenomena accompanying the instabilities development and occurring after the plasmoid disintegration were not studied. The results presented in this paper are the continuation of our work [7] in which the first attempt to consider the coupled development of these instabilities was taken. In the second paragraph of the paper the RTI linear stage is studied for incompressible and compressible fluids. The main goal of this analysis is to compare characteristic times of the instabilities development in linear and nonlinear formulations. Besides that a mode of maximum instability due to viscosity is estimated to obtain a lower limit of wave lengths such that the inviscid computational model used to investigate the nonlinear stage of instabilities development still remains valid.

The latter formulation is given in the third paragraph. The time-dependent two-dimensional model based on Euler gasdynamic equations is developed. The layered one-dimensional approximation of undisturbed flow structure is used. The velocity component tangential to the interface in the region upstream of the high-conducting layer is varied. Two types of artificial disturbance of the one-dimensional background flow are considered to provoke instabilities development. The results of the nonlinear analysis are discussed in the forth paragraph, and finally the conclusions of research are presented.

2. Linear analysis

Let us consider an isotropically conducting MHD flow containing a layer of elevated conductivity. It is supposed that the undisturbed flow velocity is uniform and parallel to the x-axis, i.e. $\mathbf{U}_0=u_0\,\mathbf{e}_x$, u_0=const. The pressure gradient ∇p_0 is balanced by ponderomotive force $\mathbf{f}_{pm}=\mathbf{j}_0\mathrm{x}\mathbf{B}$, where $\mathbf{B}=B\mathbf{e}_z$, $\mathbf{j}_0=j_0\mathbf{e}_y$, $\mathbf{E}=E\mathbf{e}_y$, $j_0=\sigma(\rho_0)(E-u_0B)$, B=const, E=const. Distributions of ρ_0, σ_0 are described below.

Let us conduct a linear analysis of stability of such a flow in relation to the disturbances occurring in the XZ-plane. In the analysis it is convenient to use the system of co-ordinates moving with velocity u_0. The variables can be presented in the form of a sum of the undisturbed value and a small disturbance:

$$\mathbf{U}=\delta\mathbf{U}=\delta u\mathbf{e}_x+\delta w\mathbf{e}_z,\quad p=p_0+\delta p,\quad \rho=\rho_0+\delta\rho,\quad \sigma=\sigma_0+\delta\sigma,\quad \mathbf{j}=\mathbf{j}_0+\delta\mathbf{j},\quad \delta\mathbf{j}=\delta j\mathbf{e}_y,\quad \delta j=E^*\delta\sigma-\sigma B\delta u,$$

where electric field strength in moving co-ordinates $E^*=E-u_0B$. Let the disturbances be plane waves propagating along magnetic field vector, wave vector $\mathbf{k}=k\mathbf{e}_z$. The disturbances are presented in the form $\delta f=\delta f_0(x)\exp(nt+ikz)$. The technique described by S. Chandrasekhar in monograph [8] has been used to obtain the dispersion relationships $n(k)$. For real values of n the disturbances are immobile waves with amplitudes changing exponentially in time. At real $n>0$ the solution is characterised by the unlimited increase of the disturbance amplitude, i.e. the instability case takes place.

2.1. Incompressible fluid

Let us consider the flow in inviscid incompressible approximation as described above. Although the density and electric conductivity in every fluid particle are constant they may differ in neighbouring particles. The undisturbed distribution of $\rho_0\,(x,z)$ is the following:

$$\rho_0(x,z)=\begin{cases}\rho_1 & at\;\; x<0,\; x>l\\ \rho_2 & at\;\; 0<x<l\end{cases},$$

where indices "1" and "2" refer to the low-conducting and high-conducting fluids, l is the layer thickness. The distributions of $\sigma_0(x,z)$ and $j_0(x,z)$ are characterized analogously. Omitting intermediate expressions the final dispersion relationship linking wave number and increment of instability development may be written as:

$$\left[\frac{k^2}{n^2}E^*B(\sigma_2-\sigma_1)\right]^2-(\rho_1{}^2\alpha_1{}^2+\rho_2{}^2\alpha_2{}^2)-2\rho_1\rho_2\alpha_1\alpha_2 cth(\alpha_2 l)=0, \qquad (1)$$

where $\alpha_1 = k\sqrt{1+\frac{\sigma_1 B^2}{\rho_1 n}}$, $\alpha_2 = k\sqrt{1+\frac{\sigma_2 B^2}{\rho_2 n}}$.

Let us consider several limiting cases:

a) thin layer: $l \to 0$.

Due to equation (1) one can obtain $n = 0$, i.e. the case of indifferent equilibrium is realized.

b) thick layer: $l \to \infty$.

In this case the undisturbed flow has two semi-infinite regions with different properties. Equation (1) takes the following form:

$$n^2\left(\rho_1\sqrt{1+\frac{\sigma_1 B^2}{\rho_1 n}}+\rho_2\sqrt{1+\frac{\sigma_2 B^2}{\rho_2 n}}\right) = -kE^* B(\sigma_2-\sigma_1) \tag{2}$$

c) fast modes: $\frac{\rho_1 n}{\sigma_1 B^2} \to \infty$, $\frac{\rho_2 n}{\sigma_2 B^2} \to \infty$

In this case $a_1 \to k$, $a_2 \to k$. This means physically that a disturbance of velocity has weak influence on disturbances of current density and ponderomotive force. Writing (1) in the following form:

$$n^2 = -k\frac{E^* B(\sigma_2-\sigma_1)}{\rho_1+\rho_2}\left(\frac{1-e^{-2kl}}{1-\left(\frac{\rho_1-\rho_2}{\rho_1+\rho_2}\right)^2 e^{-2kl}}\right)^{1/2} \tag{3}$$

one can see that the layered MHD flow may be unstable for both acceleration ($E^*>0$) and deceleration ($E^*<0$) regimes.

2.2. *Compressible fluid*

In the case of a gaseous medium it is necessary to consider three regions with different properties: non- or low-conducting gas at $x < 0$ (region 1) and at $x > l$ (region 3), high-conducting gas at $0 < x < l$ (region 2). Let undisturbed distributions be the following:

$$p_0(x,z)=\begin{cases} p_1 & at \quad x\le 0 \\ p_1+E_* B\sigma_0 x & at \quad 0<x<l \\ p_1+E_* B\sigma_0 l & at \quad x\ge l\end{cases} \qquad \rho_0(x,z)=\begin{cases} \gamma p_0/a_1^2 & at \quad x\le 0 \\ \gamma p_0/a_2^2 & at \quad 0<x<l \\ \gamma p_0/a_3^2 & at \quad x\ge l\end{cases}$$

$$T_0(x,z)=\begin{cases} T_1 & at \quad x\le 0 \\ T_2 & at \quad 0<x<l \\ T_3 & at \quad x\ge l\end{cases} \qquad \sigma_0(x,z)=\begin{cases} \sigma_1 & at \quad x\le 0 \\ \sigma_2 & at \quad 0<x<l \\ \sigma_3 & at \quad x\ge l\end{cases}$$

where a is sound velocity, T is temperature and γ is the specific heat ratio assigned in all calculations to be equal to 1.25, $T_2 > T_1$, $T_2 > T_3$., $\sigma_2 > \sigma_1$, $\sigma_2 > \sigma_3$. It is assumed that the Joule dissipation is negligibly small in the background flow and may be disregarded in studies of the linear stage of the instability development. The last assumption is valid when the energy dissipated by Joule heating during instability development is much less than the internal energy of high-conducting plasma. To determine initial parameters in region 3 it is supposed isentropic relation between them and parameters in region 1. It is also assumed that conductivity is pressure and temperature independent, i.e. remains constant in every fluid particle during the instability development.

In these assumptions the dispersion equation has been solved analytically. Its solution obtained in terms of hypergeometric functions is omitted here due to very cumbersome expressions. The dependence of the instability increment n on the wave number k has been calculated numerically. In connection with computational difficulties occurring at k>20 m^{-1} the simplified dispersion relation has been used in calculations for large values of k. The simplification has been made taking into account the inequality $\rho n^2 << k^2\gamma p$, performed at k>2 m^{-1}.

2.3. Results of linear analysis

The dependences of $n(k)$ have been calculated at the following set of the background flow parameters: in the case of compressible fluid p_1=1 MPa, a_1=a_3=800 m/s, a_2=1200 m/s, σ_1=σ_3=0, σ_2=500 Sm/m, B= 4 T, E^*=-1200 V/m, in the incompressible case the values of p_1, ρ_1, σ_1, σ_2, B and E_* are the same, the value of ρ_2 is equal to ρ_0 at $x=0$ from the compressible background flow. The analysis of results has showed that at layer thickness $l > 10^{-3}$ m its influence on the instability increment is negligible in wide range of background flow parameters both in the incompressible and compressible cases. In all calculations layer thickness has been assigned to be equal to 0.1 m.

The dispersion curves built for the cases of compressible and incompressible fluids are given in figure 1. One can see that at $k < 0.63$ or at disturbance wave length $\lambda > 10$ m ($\lambda=2\pi/k$) the dispersion equation for compressible fluid has three solutions. Two of them tend to zero at $k \rightarrow 0$ but the third monotonically increases with reduction of k and in the limit approaches to finite value. The physical meaning of the upper branch of the solution (curve 2 in figure 1) is not clear completely. It is seems possible that this solution peculiarity is connected with the aforementioned constancy of conductivity in high-conducting plasma. In consequence of this some accidental expansion of the high-conducting layer leads to an increase of integral ponderomotive force acting on the layer. A force balance disturbance results in further layer expansion. If so, the upper branch of dispersion curve may disappear at inserting in model of real dependence of electric conductivity on thermodynamic parameters. This question will be considered later, for the moment the upper branch of solution is considered as unphysical.

In the incompressible case the increments are less and, correspondingly, the instability develops faster. This explains by the infinitely large sound velocity in the incompressible fluid. It is necessary to note the difference between increments quickly decreases with the growth of k and at $k > 50$ becomes less than 1%.

The dispersion curve related to the compressible case in wide range of k is presented in figure 2. One can see that the dispersion curve is rather smooth in the conjugation point at k=20 m^{-1}. It is shown that the aforementioned simplification of the dispersion relation is valid.

To evaluate the maximum instability mode taking place due to viscous effects the approximation of incompressible fluid has been used. The estimation has been obtained on the basis of the classic study performed in [8]. For an MHD flow with a plane discontinuity one can apply a ponderomotive force instead of a gravity force. This leads to the following expression of the wave length of the mode of maximum instability:

$$\lambda_{\max} \approx 4\pi \cdot \nu^{2/3}\left((\rho_1 + \rho_2)/|f_{pm1} - f_{pm2}|\right)^{1/3}, \quad (4)$$

where ν is kinematic viscosity. Substituting typical magnitudes of parameters in (4) one can obtain that λ_{max} is less than (10^{-2} - 10^{-3}) m. Taking into account this estimation the wave length λ~10^{-1} m has been chosen as basic for inviscid computational modelling of the nonlinear stage of instabilities development. One can suppose that in the vicinity of basic wave length the inviscid model is valid for the disturbances evolution description.

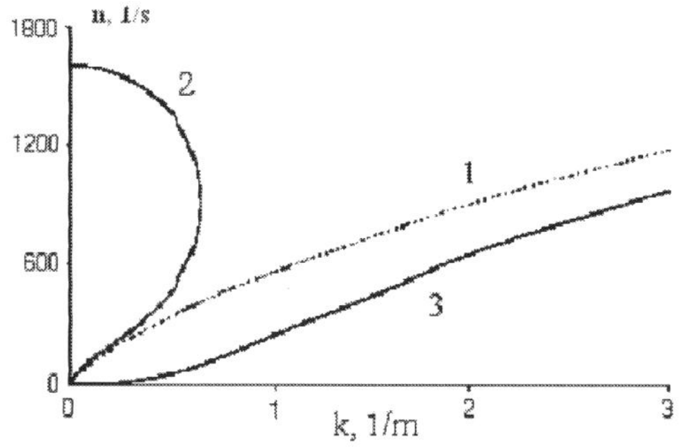

Figure 1. The dispersion curves for incompressible (1) and compressible (2, 3) fluids

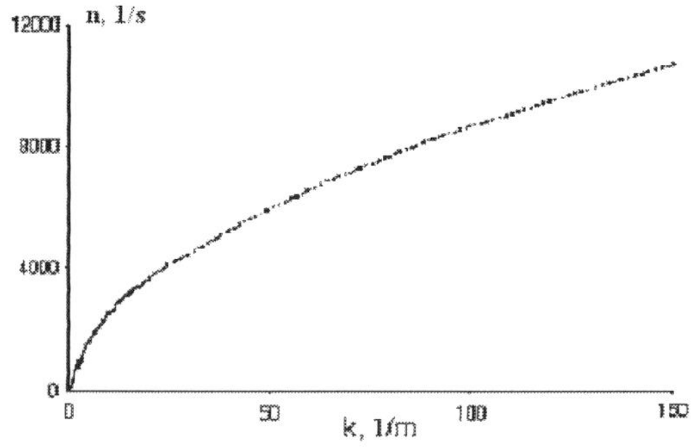

Figure 2. The dispersion curve 3 for compressible fluid in wide range of k

3. Nonlinear analysis: problem formulation

A model MHD flow in the *XZ*-plane is formed by two semi-infinite regions of non-conducting gas ($x \le 0$ and $x \ge l$) separated by high-conducting layer of thickness l. The viscosity, the heat conductivity and the induced magnetic fields are neglected, the perfect non-radiative gas approximation is used. It is assumed that electric field **E** and current density **j** are directed normally to the aforementioned plane. The velocity of the undisturbed flow is uniform and has *x*-component u_0 only, the flow inhomogeneities in y-direction are absent. The distributions of E and B are uniform and constant throughout the whole process. The flow is considered in the system of co-ordinates moving with velocity u_0.

With these assumptions the flow can be described by a two-dimensional system of non-steady-state gasdynamic equations taking into account momentum and energy body sources of MHD origin:

$$\frac{\partial \mathbf{F}}{\partial t} + \frac{\partial \mathbf{G}}{\partial t} + \frac{\partial \mathbf{H}}{\partial t} = \mathbf{R}, \qquad (5)$$

$$\mathbf{F} = \begin{vmatrix} \rho \\ \rho u \\ \rho w \\ \rho e \end{vmatrix}, \quad \mathbf{G} = \begin{vmatrix} \rho u \\ \rho u^2 + p \\ \rho u w \\ u(\rho e + p) \end{vmatrix}, \quad \mathbf{H} = \begin{vmatrix} \rho w \\ \rho u w \\ \rho w^2 + p \\ w(\rho e + p) \end{vmatrix}, \quad \mathbf{R} = \begin{vmatrix} 0 \\ S_\lambda jB \\ 0 \\ S_\lambda jE \end{vmatrix},$$

where the total energy density $e = c_v T + (u^2 + w^2)/2$. The system of equations is written here in the dimensionless form. For normalization of the variables the following parameters are used: the sound velocity a_{1d} for velocity components and sound velocity (indices "1" and "*d*" specify the left non-conducting region and the dimension form of a parameter, correspondingly), ρ_{1d} for density, $\rho_{1d} a_{1d}^2$ for pressure, a_{1d}^2 for energy density, the assigned wave length of disturbance λ_d for spatial co-ordinates, λ_d / a_{1d} for time, B_d for induction of magnetic field, σ_d for electrical conductivity (σ_d is the conductivity in high-conducting gas near upstream surface of layer, the dependence of conductivity on thermodynamic parameters used in this work is the same as in [7]), $\sigma_d a_{1d} B_d$ for current density, $a_{1d} B_d$ for electrical field. The parameter S_λ having meaning of Stuart number is determined as $\lambda_d \sigma_d B_d^2 / \rho_{1d} a_{1d}$. The system of equations is closed by dimensionless equation of state in form $p = \rho T$.

The initial distributions of variables are assigned by the following way. The undisturbed state of the flow has been determined from condition that ponderomotive force within the layer is balanced by pressure gradient, in the non-conducting regions pressure is uniform. At $x > 0$ transversal velocity component w is zero, at $x < 0$ w has a constant value varied in different versions of calculations from zero to 0.2. As a rule, it was supposed that $u_0 = a_{10}$. It is supposed that parameters in the left and right non-conducting regions are related isentropically, temperature within high-conducting layer T_{20} is uniform.

To study instability development these distributions are disturbed. Two types of disturbances have been used (figure 3). The first type (figure 3a) consists of the variation of the temperature distribution in the high-conducting layer in accordance with the expression $T = T_2(1 + T_a e^{-x/d} \sin(2\pi z))$, where values of T_a, d are assigned (hereafter *x*-axis is horizontal and *z*-axis is vertical).

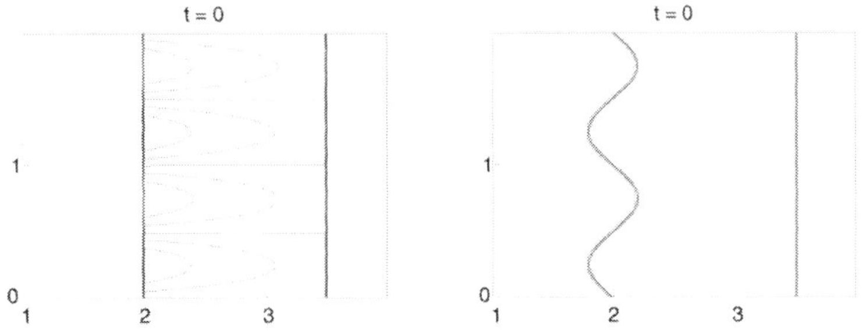

a) b)

Figure 3. The disturbance types of initial flow structure (a – the temperature field, b – the interface shape)

Journal of Physics: Conference Series **216** (2010) 012011 doi:10.1088/1742-6596/216/1/012011

The second type of disturbances (figure 3b) results in distortion of the interface between upstream non-conducting region and high-conducting layer. The distorted configuration of the interface is given by a sinusoid with assigned amplitude.

On the left and right boundaries of calculation region «soft» conditions ($\partial/\partial x = 0$) have been used, on the lower and upper boundaries the periodical conditions have been realized. The modified Godunov-Kolgan method described in [4] has been used in calculations.

4. Results and discussion

Within the framework of this formulation five versions of calculations have been carried out. They have been obtained at the following set of initial dimensionless data fixed for all versions: $\gamma = 1.2$, $\rho_1 = 1$, $p_1 = 0.833$, $T_1 = 0.833$, $T_2 = 2.315$, $\rho_3 = 0.836$, $p_3 = 0.672$, $T_3 = 0.804$, $B = 1$, $E_* = -0.4$, $\lambda = 1$, vertical size of computational region L_y is equal to 2. The dimensional thickness of the high-conducting layer l_{hd} has been fixed and equal to 0.15 m, accordingly, its dimensionless value has been changed with alternation of the dimensional wave length.

The initial three versions of calculations have been conducted to study features of the RTI (i.e. at $w_1 = 0$, and KHI is absent). The first version is characterized by $S_\lambda = 0.268$ corresponding to the following dimension parameters: $\lambda_d = 0.1$ m, $\sigma_d = 500$ Sm/m, $B_d = 4$ T, $\rho_{1d} = 0.926$ kg/m^3. The horizontal region size L_x is equal to 4, high-conducting layer is bounded by co-ordinates $x_{h1} = 2$ and $x_{h2} = 3.5$. The first type of the flow disturbance has been used with parameters $T_a = 0.2$ and $d = 1$. The results obtained are shown in figure 4 for two moments of time.

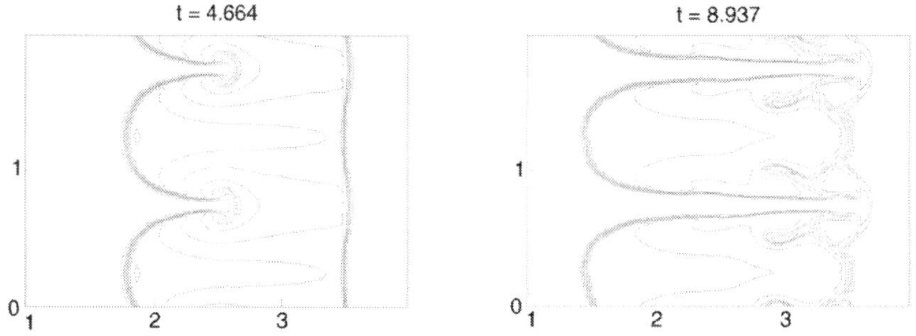

Figure 4. The level lines of temperature for two moments of time (the 1st version of calculations)

One can see a rather classical structures characterized by occurrence of «jets» with extended «heads» of more dense non-conducting medium and «bubbles» of less dense high-conducting gas. The dimensional break time of the layer τ_{db} is near to $1.2 \cdot 10^{-3}$ s.

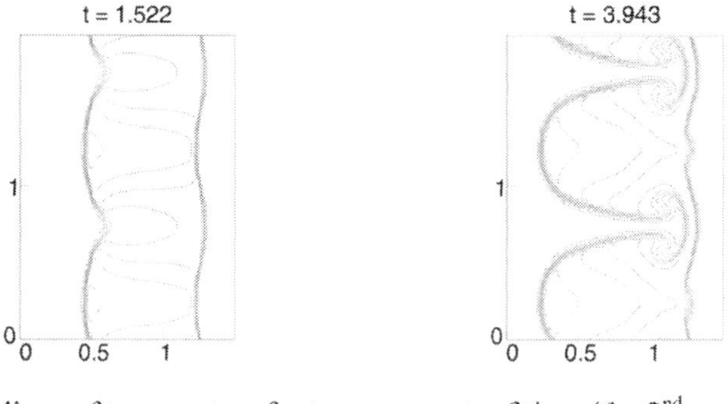

Figure 5. The level lines of temperature for two moments of time (the 2nd version of calculations)

The analysis of these two versions reveals weak variation of the width of jets, the speed of head extension and so on. It is interesting that the values of τ_{db} observed in these calculations are very close.

To obtain the third version of solution the second type of the flow disturbance has been used with amplitude of interface distortion being equal to 0.2. The other determining parameters are the same as in the first version. The results are illustrated by figure 6.

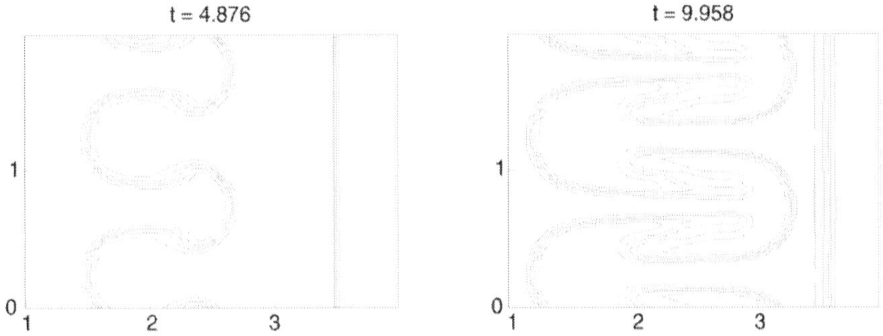

Figure 6. The level lines of temperature for two moments of time (the 3rd version of calculations)

It is likely that the observable difference of this version from the previous ones is due to more strong initial disturbance, practical absence of the linear stage and fast transition to slower nonlinear stage of the instability development. Apparently this fact has resulted in a rather higher τ_{db} (about $2 \cdot 10^{-3}$ s in the dimensional form).

The presence of transversal velocity in the left non-conducting region leads to fundamental changes of results (in the 4th and 5th versions of calculations the transversal velocity w_l has been assigned to be equal to 0.2, the first and second types of initial disturbance have been used correspondingly). As viewed in figures 7 and 8, in this case the development of disturbances is specific and essentially differ from the solutions corresponding to the individual development of the RTI discussed above and the KHI with well known effect of "cat eyes" creation.

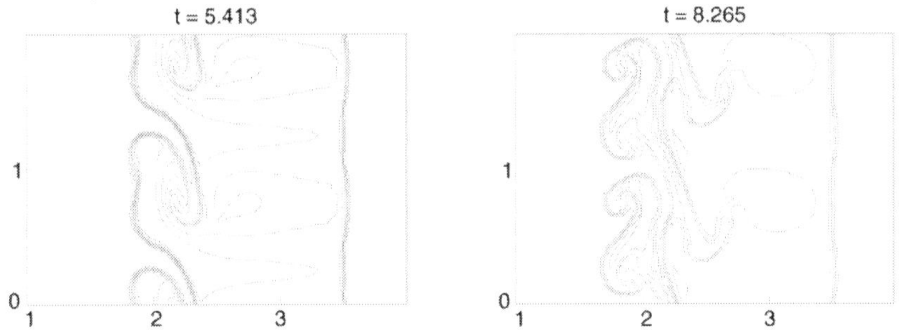

Figure 7. The level lines of temperature for two moments of time (the 4th version of calculations)

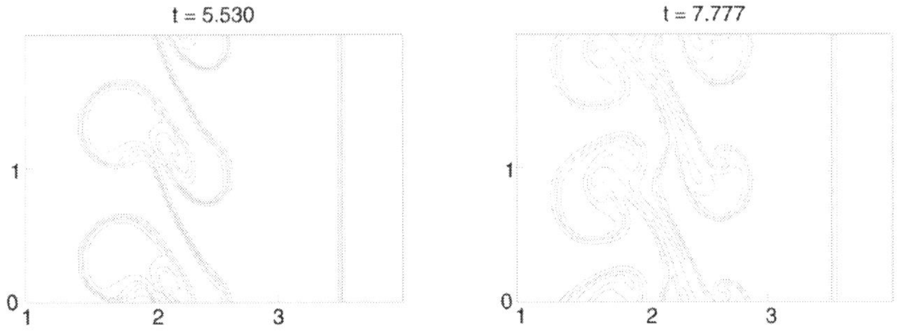

Figure 8. The level lines of temperature for two moments of time (the 5th version of calculations)

The destruction of the disturbed surface of the high-conducting layer is accompanied by formation and subsequent separation of its fragments. Although explicitly expressed break of the layer by the non-conducting gas is not observed it is clear that layer-by-layer destroy of the plasmoid surface introduces new aspects to the problem of the plasmoid stability. On the other hand, this process is relatively slow. In calculations the reduction of the layer in half required about $1.5 \cdot 10^{-3}$ s.

5. Conclusions

The study conducted has shown the following. First, the increments of the MHD instability of RT type arising from nonlinear analysis are one order of magnitude less than those predicted by linear theory. Second, at the same type of the initial disturbances the time of the break of high-conducting layer rather weakly depends on wave length. Third, the presence in upstream non-conducting region of velocity component parallel to the interface provokes the interaction of the MHD instabilities of the Raylegh-Taylor and Kelvin-Helmholtz types. Manifestation of this interaction introduces essential alterations in the development of the RTI.

It is necessary to note that the effect of destruction of the surface of the high-conducting layer with separation of its fragments may induce difficulties in the solution of the stability problem considered. On the other hand, this process is relatively long in time. The estimation of its importance requires to compare the characteristic times (for instance, of this process and of a residence of plasmoid in a channel of some device). Apparently this question calls for special consideration.

6. Acknowledgments

The initial stage of this work was sponsored by the International Science Foundation (ISF) of Grants MFT 000, MFT-300.

References

[1] Burns J O, Norman M L and Clarke D A 1991 *Science* **253** 522
[2] Li-Xin Li, Ramesh Narayan 2004 *Astrophys. J.* **601** 414
[3] Yang Zhang, Ning Ding 2006 *Phys. Plasmas* **13** 022701
[4] Bityurin V A, Likhachev A P 1992 *Progress in Astronautics and Aeronautics* **148** 373 (Washington: AIAA)
[5] Inogamov N A, Demyanov A Yu and Son E Ye 1999 *Hydrodynamics of Stirring* (Moscow: MIPT Press)
[6] Yabe T, Hoshino H and Tsuchiya T 1991 *Phys. Rev. A* **44** 2756
[7] Likhachev A P, Medin S A 1995 *Magnetohydrodynamics* **31** 374
[8] Chandrasekhar S 1961 *Hydrodynamic and Hydromagnetic Stability* (Oxford: Clarendon Press)

Journal of Physics: Conference Series **216** (2010) 012012 doi:10.1088/1742-6596/216/1/012012

Numerical simulations of interactions between shock wave and gas-liquid-air interfaces

X X Lu, L Li[1], X B Ren , X F Yan , Y C Dong

School of Aerospace, Tsinghua University, Beijing, 100084, P. R. China

Email: rockysys07@bbn.cn

Abstract. A gas-liquid-air model and problems of interactions between the shock wave and the interfaces during the expansion process are rather familiar in the study of liquid explosive dispersal and underwater explosion near air-water surface. It involves large density discontinuity and large pressure gradient at the same time which brings difficulties for the numerical simulations. In this paper, we use the level-set methods to capture the interface and the modified ghost fluid method (MGFM) to treat the interface conditions. Strong shock-interface interactions and influences of the disturbance on the two interfaces are investigated. The initial fragmentation of the liquid is studied based on the analyses of the fluid pressure distribution and the variation of the interface, combining the conditions of the cavitations occurrence in the liquids.

1. Introduction

The initial period of liquid explosive dispersal and under water explosion near air-water surface have similar characteristics. Both of these two kinds of problems involve complex physical phenomena which can be characterized by propagation of shock wave in multi-medium flow consisting of gas, liquid and air [1-3]. When the shock wave produced by explosion passes through the gas-liquid-air interfaces, strong refractive shock wave and rarefaction wave will be formed due to the large density discontinuity between gas and liquid. The large density discontinuity between gas and liquid and large pressure gradient corresponding to the shock wave coexist in these problems and make their numerical simulations very difficult.

In fact, for the aforementioned strong shock-interface interaction problems, two kinds of problems should be solved. One of them is to capture the moving material interface. The other is to treat the interface conditions of flow. Due to that interface problems involving external physics arise in various areas of science, a lot of methods to track the interface have been developed, such as front tracking, phase-field methods, the volume of fluid (VOF) approach and level set method [3-7]. Among of them, the level set method devised by Osher and Sethian has had major successes in this area as a simple and versatile method for computing and analyzing the motion of an interface in two or three dimensions. Mulder et al [8] firstly applied the level set method to two phase inviscid compressible flow but the method suffered from spurious pressure oscillations at[2] the interface. This problem was overcome by the development of the ghost fluid method (GFM) which can couple the level set representation of

[1] To whom any correspondence should be addressed.

discontinuities to finite difference calculations of compressible flows [9]. The main appealing features of the GFM are maintenance of a sharp interface without smearing and to make the interface "invisible" during computations and the computations are carried out as for a single-medium manner such that its extension to multi-dimensions becomes fairly straightforward. But the single-medium manner across the interface in the GFM may cause numerical inaccuracy when there is a strong shock wave interacting with the interface. To enhance its accuracy, Liu et al [10] developed a modified GFM (MGFM) with a predicted more reasonable ghost fluid status. The MGFM was further modified to apply to nearly cavitating flow and found to be quite robust and can provide relatively reasonable results [11]. However, as other popular high resolution schemes, the MGFM may also give inaccurate numerical results even over a very fine mesh when applied to gas flows with an initial high density ratio as well as a high pressure ratio [12].

In this paper, strong shock impacting on gas-liquid-air interfaces is investigated. Here the level-set methods are used to capture the moving material interfaces and the modified ghost fluid method (MGFM) to treat the interface conditions. In section 2, the formulation of gas-liquid-air problem is given. Its one dimension case is discussed in section 3. In section 4, 2D models corresponding to the initial period of liquid explosive dispersal and under water explosion near air-water surface are investigated respectively by numerical simulation and compared with experimental results. Some conclusions are given in section 5.

2. Mathematical formulations and numerical methods

2.1. Governing equations for the fluid

The governing equations for compressible fluid are the conservative form of the Euler equations. The Euler equations for 2-D compressible flows are written as

$$\frac{\partial \boldsymbol{U}}{\partial t}+\frac{\partial \boldsymbol{F(U)}}{\partial x}+\frac{\partial \boldsymbol{G(U)}}{\partial y}=0 \tag{2.1}$$

where $\boldsymbol{F(U)}=\left[\rho u, \rho u^2+p, \rho uv, (E+p)u\right]^{\mathrm{T}}$, $\boldsymbol{G(U)}=\left[\rho v, \rho uv, \rho v^2+p, (E+p)v\right]^{\mathrm{T}}$, $\boldsymbol{U}=\left[\rho, \rho u, \rho v, E\right]^{\mathrm{T}}$. Here, ρ is the density; u and v are the velocity components in the respective x and y directions; p is the pressure and E is the total energy per unit volume which is the sum of internal energy and kinetic energy. In this work, a second-order Total Variation Diminishing (TVD) scheme is applied to solve the Euler equations.

2.2. Equations of state (EOS)

In the present work, we are interested in the interaction with the strong shock wave and the interfaces with large density ratio. Two kinds of EOS are considered. The first one is the $\gamma-$law for perfect gases, which is written as

$$p=(\gamma-1)\rho e \tag{2.2}$$

We let $\gamma=1.4$ for all gases for simplicity. The second one is the Tait's EOS for water, which is written as

$$p=B(\frac{\rho}{\rho_0})^N-B+A \tag{2.3}$$

where $N=7.15$, $B=3.31\times10^8\,\mathrm{Pa}$, $A=10^5\,\mathrm{Pa}$.

2.3. Level set equation

In the level set approach, the interface is defined implicitly as an iso-surface of a smooth function ϕ, which is defined as signed distance function here. The transport of the interface can simply be described by

$$\frac{\partial\phi}{\partial t}+u\frac{\partial\phi}{\partial x}+v\frac{\partial\phi}{\partial y}=0 \tag{2.4}$$

In this work, the fifth-order weighted essentially non-oscillatory (ENO) scheme is used to discretize the spatial derivatives with the third-order Runge-Kutta time discretization. In order to ensure that ϕ remains a distance function, re-initialization is performed by solving a Hamilton-Jacobi equation

$$\frac{\partial\phi}{\partial\tau}+S(|\nabla\phi|-1)=0 \tag{2.5}$$

where S is a modified sign function (see [13] for details).

2.4. The MGFM algorithm

In a GFM-based algorithm, a given band of points as ghost cells is defined in the vicinity of the interface and the ghost fluid status is defined. Each medium with its ghost nodes is solved independently. By combining the solution for each medium according to the new interface location, the overall solution for the whole domain at the next time step is then obtained.

We applied MGFM method to treat the interface conditions. In this algorithm, equations below are solved for the status on the interface:

$$\frac{p_I-p_L}{W_L}+(u_I-u_L)=0 \tag{2.6a}$$

$$\frac{p_I-p_R}{W_L}-(u_I-u_R)=0 \tag{2.6b}$$

$$W_L^2=\rho_{IL}\rho_L\frac{p_I-p_L}{\rho_{IL}-\rho_L} \tag{2.6c}$$

$$W_R^2=\rho_{IR}\rho_R\frac{p_I-p_R}{\rho_{IR}-\rho_R} \tag{2.6d}$$

where subcripts '*I*', '*IL*', '*IR*' refer to the interface, the left side of the interface and the right side of the interface, respectively (see [10] for details). For closure of (2.6a)-(2.6d), the relationship of density and pressure based on the energy jump condition, $e_1-e_0=\frac{1}{2}(p_1+p_0)(\frac{1}{\rho_0}-\frac{1}{\rho_1})$, are applied.

The status (p_I, u_I, ρ_{IL}, ρ_{IR}) on the interface is obtained after solving the Riemann problem above via iteration and then assigned to the ghost nodes near the interface. Then the isobaric technique is applied to the proximate real fluid nodes to the interface to fix the real fluid density to suppress the possible "overheating" and also assign density for the ghost fluid.

3. One dimensional problem

Some shock tube-like problems are considered here. Let us consider a shock wave impacting on a gas-water interface from the left side and transporting in the water region and then transmitting to another gas medium, as shown in Figure 1. We are interested in the distribution of pressure and density behind the shock wave. Assuming that the pressure and velocity of each media to be uniform initially, the governing equation is the 1-D Euler equation and the initial condition is given as

$$\boldsymbol{U}\big|_{t=0} = \begin{cases} \boldsymbol{U}_1 & x < x_s \\ \boldsymbol{U}_{10} & x_s < x < x_{gw} \\ \boldsymbol{U}_{20} & x_{gw} < x < x_{wg} \\ \boldsymbol{U}_{30} & x > x_{wg} \end{cases} \tag{3.1}$$

where $\boldsymbol{U}_{10}, \boldsymbol{U}_{20}, \boldsymbol{U}_{30}$ are the initial states in medium 1, 2 and 3 before the shock wave, and $\boldsymbol{U}_1$ is the status behind the incident shock and satisfies the Rankine-Hugoniot conditions.

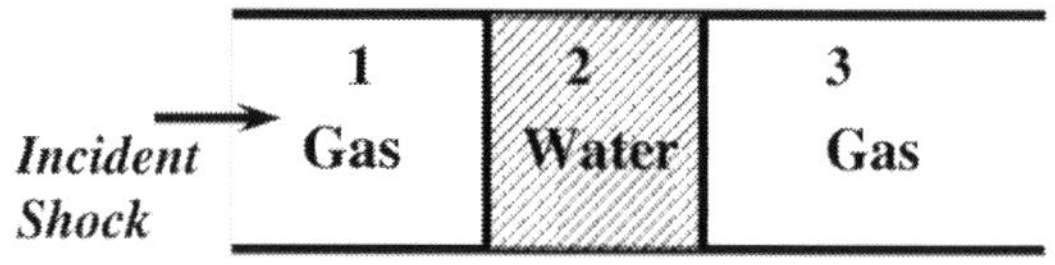

Figure 1. Schematic diagram for 1-D problem

Distribution of pressure and density under different strength of the incident shock and different initial density ratio will be discussed. The interface between left gas region and the water region (named F1) is initially located at $x_{gw} = 0$, where the interface between the water region and the right gas region (named F2) is initially located at $x_{wg} = 0.5$. The initial position of the incident shock front is the same as F1, that is, $x_s = x_{gw} = 0$. The calculations are done with CFL=0.5 and 501 uniform mesh points in a domain with dimensionless length 5.

Though the MGFM may give inaccurate numerical results even over a very fine mesh when applied to gas flows with an initial high density ratio as well as a high pressure ratio, it can provide helpful results here because that refer to [12], the calculation error mainly concentrates in the disability to provide the transmitted shock location accurately after it transfers a long distance while the value of the discontinuity is mostly correct. Therefore, analysis of the characteristic of the interaction between shock wave and the gas-liquid-air interfaces is still valuable.

The pressure and density profile for three different strength of incident shock under several time steps are shown in Figure 2 and Figure 3, where strength of the incident shock $p_1/p_{10} = 1000{,}5000{,}100000$ are considered; other initial flow parameters are $p_{10} = p_{20} = p_{30} = 10^5\,Pa$, $\rho_{10} = 0.001 g/cm^3$, $\rho_{20} = 1.0 g/cm^3$, $\rho_{30} = 0.001 g/cm^3$. The short horizontal lines in Figure 2 indicate the location of the water region for different time steps. From Figure 2, the pressure behind the reflected shock wave is about 7.5 times higher than the incident shock pressure. The water regions are compressed and pushed forward. When the transmitted shock in the water region meets the water-air interface, a strong rarefaction wave is reflected and the pressure and density of the water drop greatly and approach to the initial states. Comparison among these figures shows that characteristics under different strength of incident shock are similar and the differences only lie in the values of the pressure and density.

Figure 4 shows the pressure and density distributions for incident shock strength $p_1/p_{10} = 5000$ and $\rho_{10}/\rho_{20} = 0.1$. Other initial parameters are $\rho_{20} = 1.0 g/cm^3$, $\rho_{30} = 0.001 g/cm^3$, $p_{10} = p_{20} = p_{30} = 10^5\,Pa$. The high density of the upwind gas increases the velocity behind the shock and prevents the reflecting shock wave to propagate toward left side. Moreover, the refractive shock strength is only 4.4 times higher than the incident shock wave. Figure 5 shows the pressure and density distributions for incident shock strength $p_1/p_{10} = 5000$ and $\rho_{30}/\rho_{20} = 0.1$. Other initial parameters are $\rho_{10} = 0.001g/cm^3$, $\rho_{20} = 1.0g/cm^3$, $p_{10} = p_{20} = p_{30} = 10^5 Pa$. The distributions of pressure and density are similar with the cases shown in Figure 2 and 3 except that at the right water-gas interface, the transmitted shock wave is strengthened and the rarefaction wave is weakened greatly. Comparison between Figure 4 and 5 shows that the density of the left side gas influences the characteristics of the flow field more evidently than that of the right side gas.

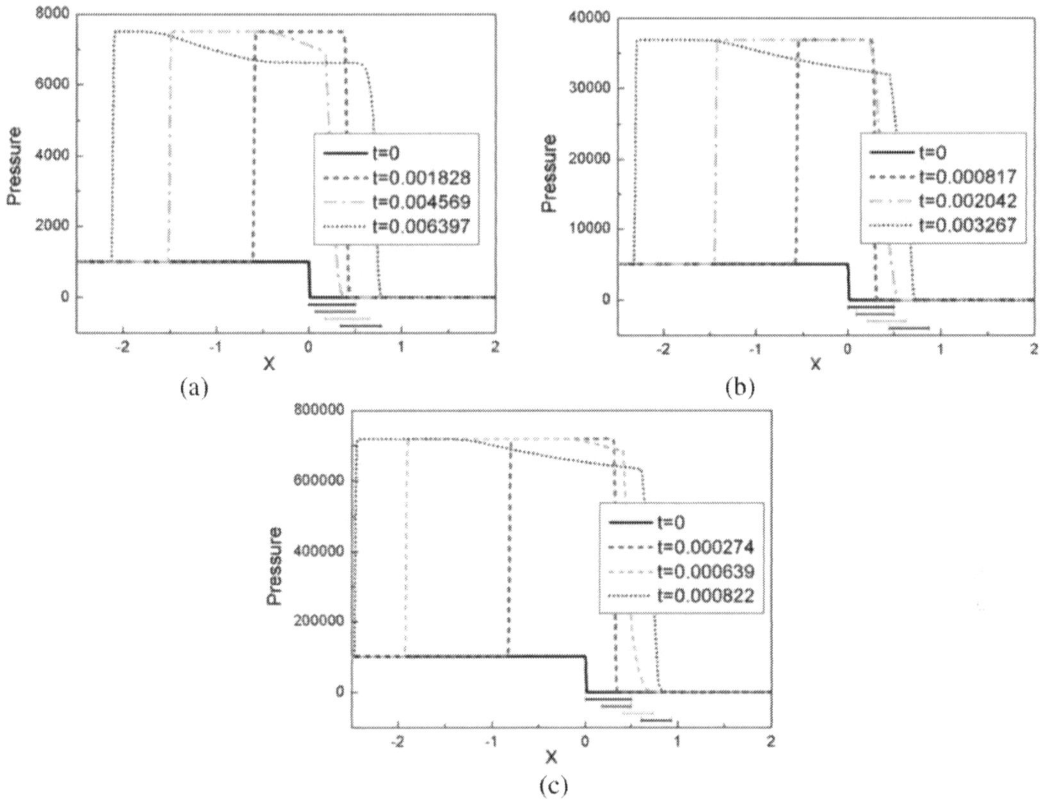

Figure 2. Pressure profiles for different strength of incident shock. (a) Strength of incident shock: 1000. (b) Strength of incident shock: 5000. (c) Strength of incident shock: 100000

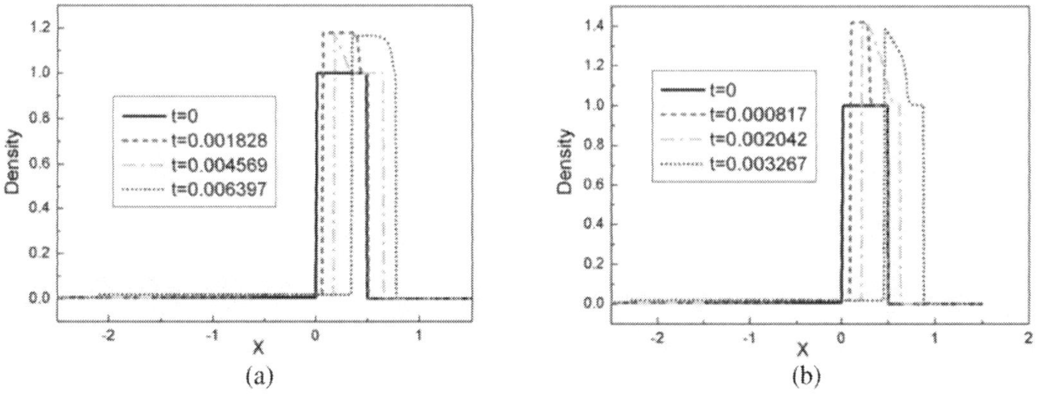

Journal of Physics: Conference Series **216** (2010) 012012 doi:10.1088/1742-6596/216/1/012012

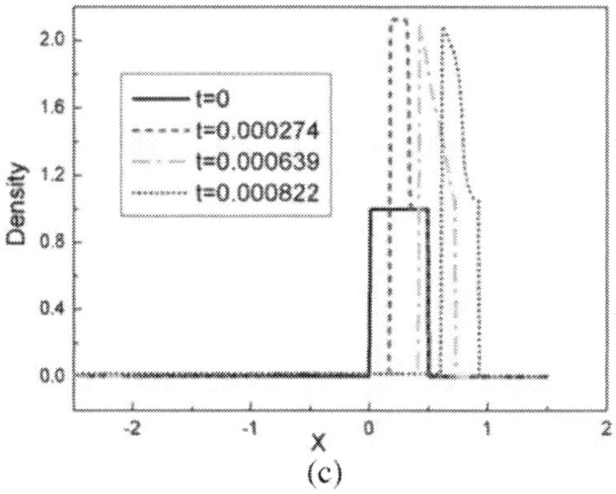

Figure 3. Density profiles for different strength of incident shock. (a) Strength of incident shock: 1000. (b) Strength of incident shock: 5000. (c) Strength of incident shock: 100000

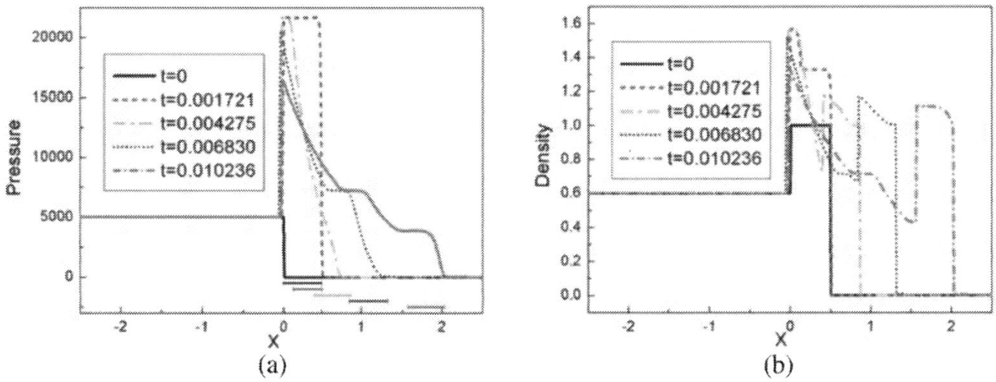

Figure 4. Results of several time steps for $\rho_{10} / \rho_{20} = 0.1$. (a) Pressure profile. (b) Density profile.

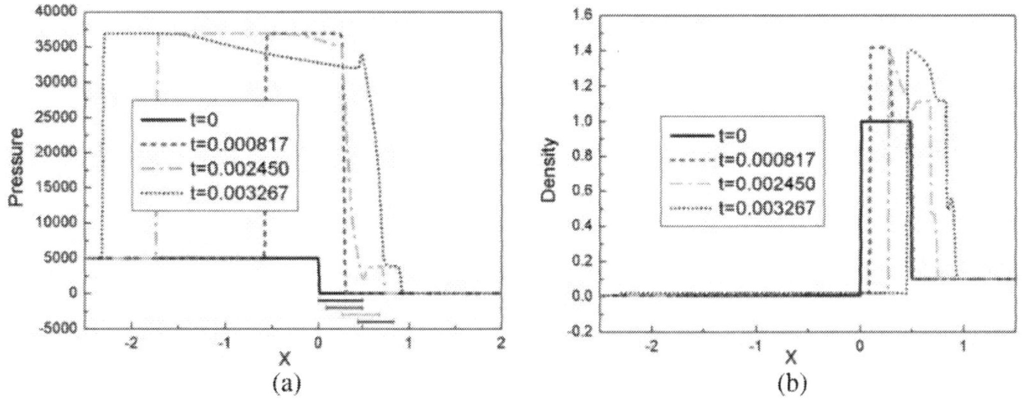

Figure 5. Results of several time steps for $\rho_{30} / \rho_{20} = 0.1$. (a) Pressure profile. (b) Density profile.

4. Two dimensional problem

4.1. Liquid explosive dispersion

The interaction of shock wave and liquid during the liquid explosive dispersion is analyzed. The initiation of the detonator is represented by the high pressure gas in the centre of the water ring, which is the gas region in Figure 6(a). Water and the air are of the standard atmosphere pressure. The initial pressure of the explosion gas is 1000atm.The inner and outer radii of the water ring are 3cm and 8cm respectively (see Figure 6(a)).

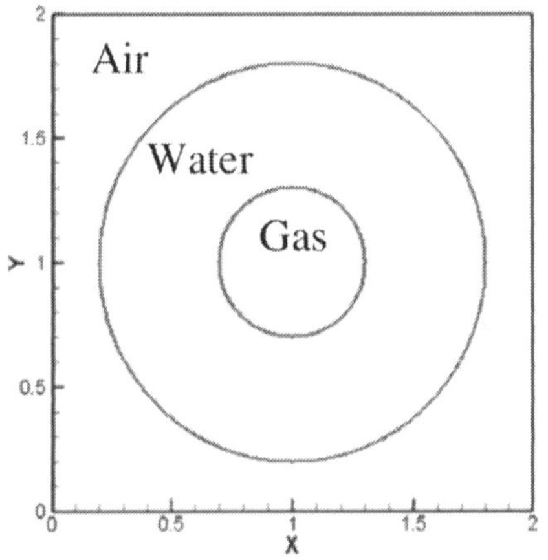

(a)

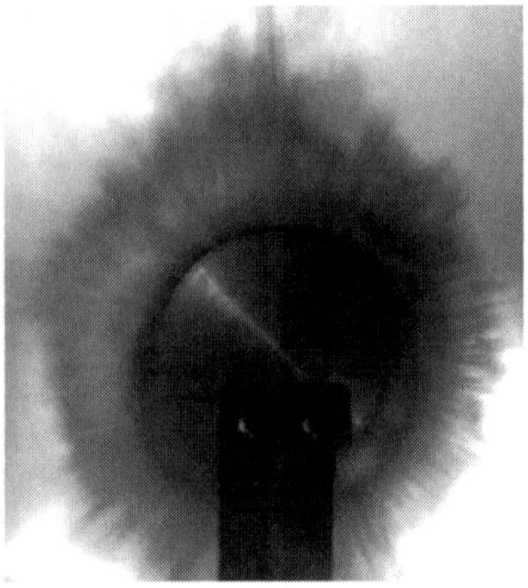

(b)

Figure 6. (a) A 2-D model for the initial period of a liquid explosive dispersion. (b) Experiment result obtained from a quasi-2-D liquid explosion dispersion

Pressure profile and the contour of density for time t=0.000044s are shown in Figure 7. The colour of deep blue in the left figure indicates a region with very low pressure (about 22pa, which is the low point for the Tait's EOS in effect) and the corresponding region in the figure of density contour is the outer region of the water. In this region, the pressure is lower than the saturation vapour pressure of water and the water in this region is hard to keep the continuous liquid state. This region is where cavitations are likely to happen. On the other hand, geometry of the water ring hasn't been changed markedly yet at this time, so if there's a solid shell outside the water ring, the shell should not have been broken yet. Thus, Figure 7 can equivalently show the status in a liquid explosion process just before the liquid jets out the crust, which means that as soon as the detonating occurs, the liquid's characteristics have changed before the crust of the device of liquid explosive dispersion is broken. Cavitations and gasification may occur in the outer part of the liquid. Figure 6(b) is obtained from an experiment of liquid explosive dispersion 2ms after initiation of detonator and shows how the liquid jet out the thin crust [14]. It can be seen that the outer part of the water region is bright, where the property of water is obviously changed. We thought that cavitations or gasification has occurred in this region. This shows that the experimental results are consistent with the numerical results.

It should be pointed out that simulation broke down shortly after this time because the EOS for water doesn't suit for the state of the region with very low pressure and a new EOS needs to be applied to this problem. However present simulation can point out the region where the pressure is low enough for the cavitations to occur.

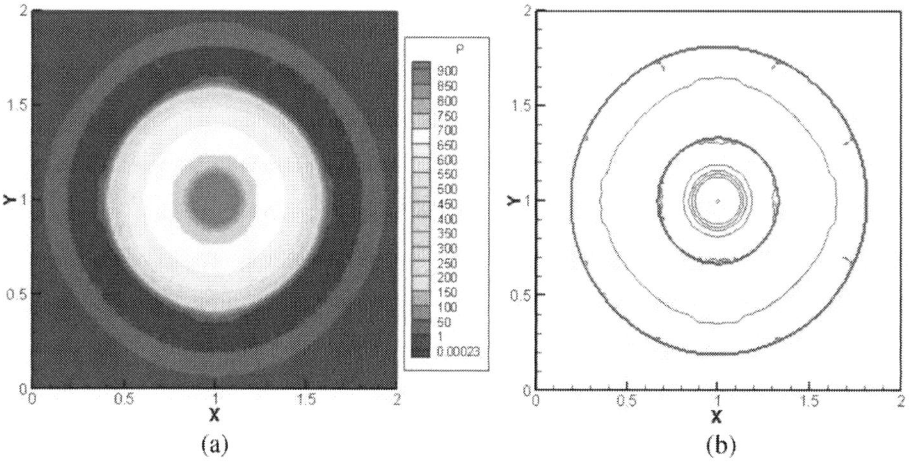

Figure 7. Numerical results for the 2-D model in Figure 6(a). (a) Pressure distribution. (b) Density contour. (The scale for x and y is 10cm)

4.2. Underwater explosion near the water-air surface

Another case with interaction of shock wave and gas-liquid-air interfaces is under water explosion problem near air-water surface. Ballhaus and Holt investigated an underwater spherical blast problem and proposed an analytical solution. Here the same underwater explosion problem near the water-air surface is investigated using MGFM method and compared with the analytical results in [2]. Moreover, influence of disturbance of water-air surface is discussed.

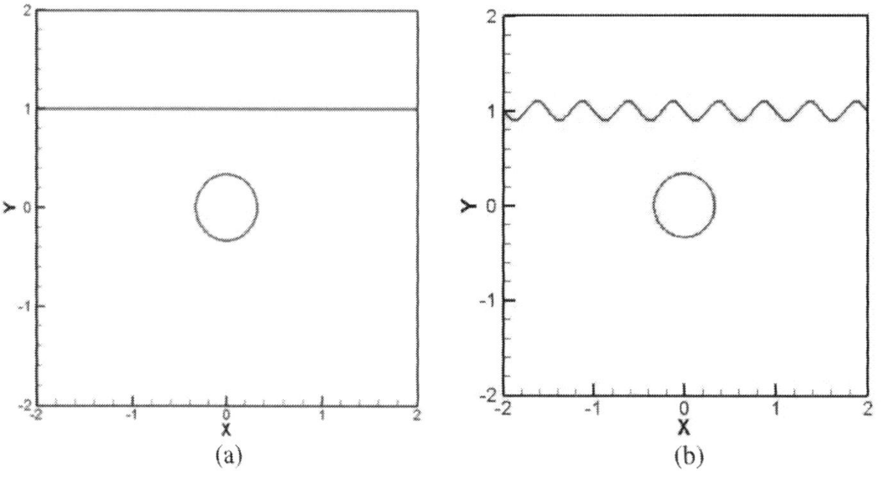

Figure 8. Underwater explosion initial conditions. (a) Undisturbed water-air surface. (b) Disturbed water-air surface.

Figure 8 demonstrates the underwater explosion initial conditions (the initial position of the water surface and the gas sphere) with respect to undisturbed and disturbed water-air surface. The same parameters as that in [2] are used in the following analysis. The explosion centre is 1 meter below the water surface and at time t=0, the explosion produces a spherical shock wave that begins to propagate outward. The radius of the initial explosion gas sphere is 1/3 meter and initial pressure is 9000atm. Figure 9 shows the distributions of pressure and the density contours for time t=0.0010312s and

0.0011456s with respect to the undisturbed water-air surface. The shape and position of the water-air surface and the gas sphere is displayed in the density contour. In the figure of pressure profile, the color of deep blue indicates a region with very low pressure where cavitations are likely to happen. It can be seen that the low pressure region increases with the increasing of time. Figure 10 shows refractive wave series and the position of the water-air surface based on the analytical results in [2].The region where pressure is lower than the initial value is labeled with dots. It shows that the low pressure region occurs firstly at the intersection of the refraction wave and the water-air surface. The region becomes larger with the increasing of time and new low pressure region generates in the middle of the water-air surface. This is consistent with the numerical results basically.

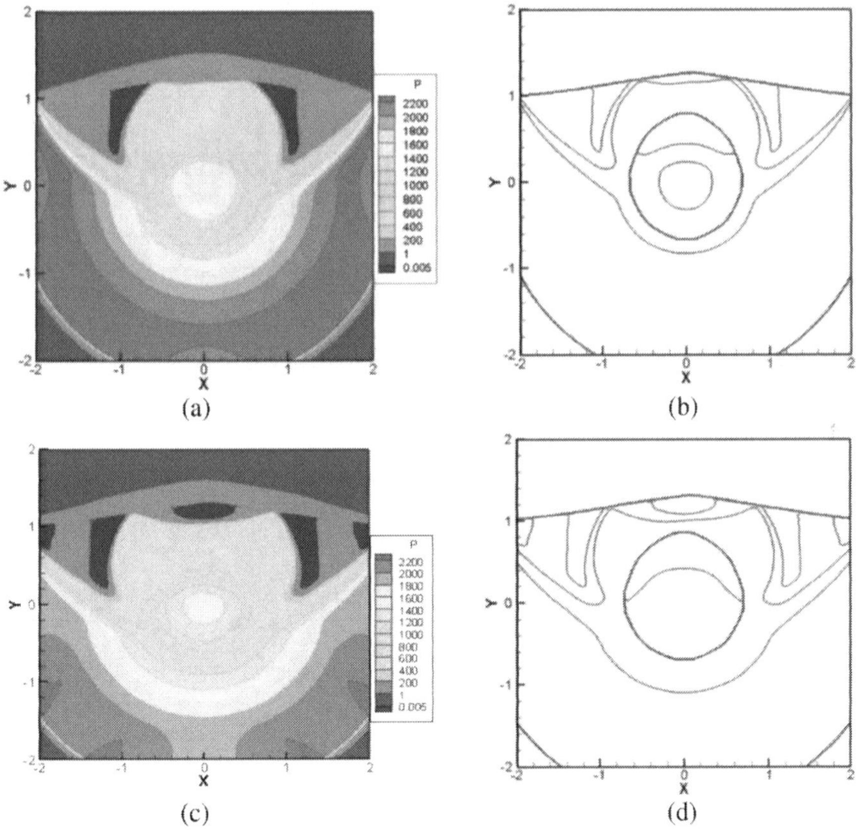

Figure 9. Distributions of pressure and the density contours for the undisturbed surface. (a) Pressure field at time t=0.0010312s. (b) Density contour at time t=0.0010312s. (c) Pressure field at time t=0.0011456s. (d) Density contour at time t=0.0011456s.

Journal of Physics: Conference Series **216** (2010) 012012 doi:10.1088/1742-6596/216/1/012012

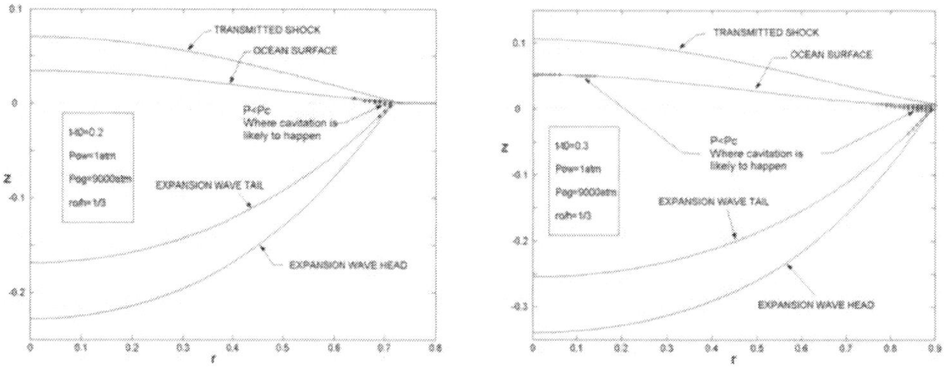

Figure 10 Refractive wave series and the position of the water-air surface based on [2]

The distributions of pressure and the density contours for time t=0.00057338s and 0.0011456s with respect to the disturbed water-air surface are shown in Figure 11. It can be found by comparison with Figure10 that when the water-air surface are disturbed, low pressure regions occur at more early time and distribute in wider range. In addition, the flow field becomes more irregular. The results show that the disturbance of water-air surface can make the cavitations generate more easily and the flow become more complex.

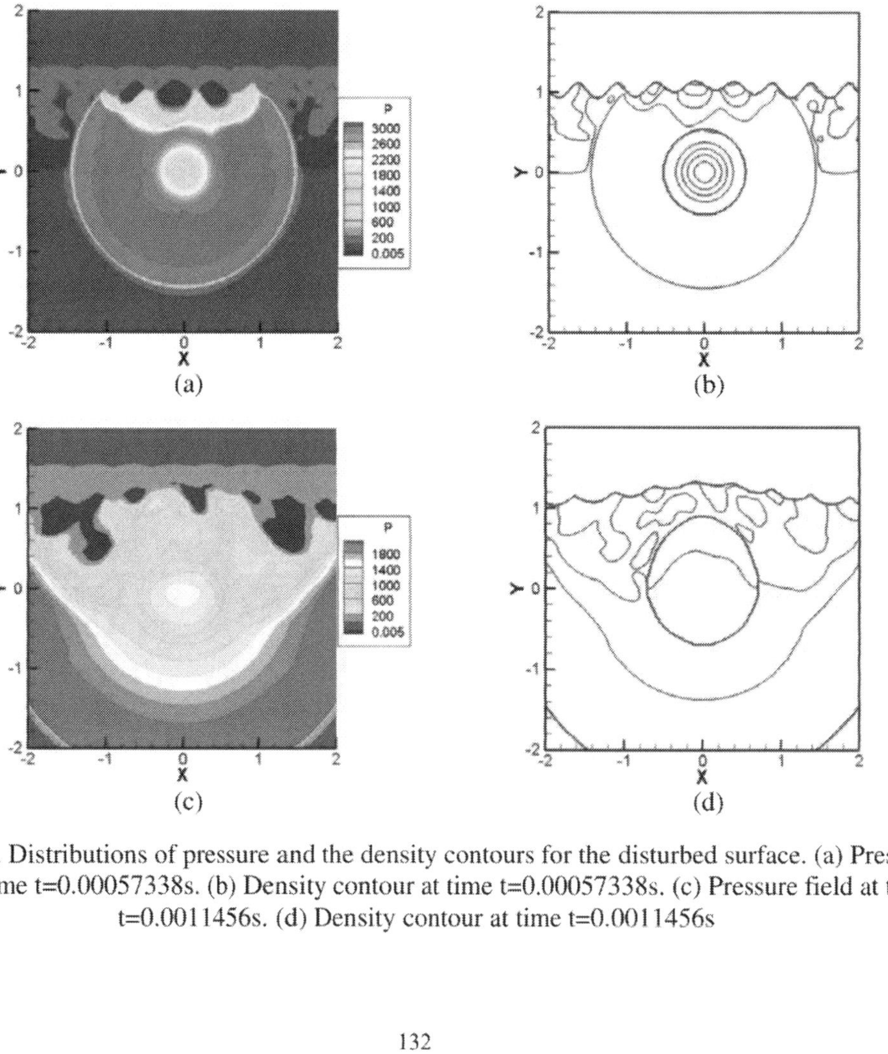

Figure 11. Distributions of pressure and the density contours for the disturbed surface. (a) Pressure field at time t=0.00057338s. (b) Density contour at time t=0.00057338s. (c) Pressure field at time t=0.0011456s. (d) Density contour at time t=0.0011456s

5. Conclusions

Interactions of strong shock wave with gas-liquid-air interfaces are investigated in this paper. Here the level-set and MGFM methods are used to capture the moving material interfaces and treat the interface conditions, respectively. Firstly distributions of pressure and density under different strength of the incident shock and different initial density ratio are discussed for one dimensional problem. The results show that characteristics under different strength of incident shock are similar and the differences only lie in the values of the pressure and density. Moreover the density of the left side gas influences the characteristics of the flow field more evidently than that of the right side gas. Secondly two typical 2D models corresponding to liquid explosive dispersal and under water explosion near air-water surface are investigated respectively. Numerical results are in good agreement with corresponding experimental or analytical results. The two typical problems have similar characteristics. Both of them have low pressure region which may generate cavitations or gasification. In addition, the disturbance of water-air surface can make the flow become more complex. The future work is to introduce more reasonable model to treat the two phase flow problem caused by cavitations.

Acknowledgments

The work in this paper was supported by National Science Foundation Committee of China, No.10572149 and Allied Key-projects by China Academy of Engineering Physics and National Science Foundation Committee of China (NSAF), No.10676120.

References

[1] Gardner D R 1990 Near-field dispersal modeling for liquid fuel-air-explosive DE91000079 (SAND-90-0686)
[2] Ballhaus W F Jr and Holt M 1974 Interaction between the ocean surface and underwater spherical blast waves *Phys. Fluids* **17** 1068-79
[3] Holt M 1977 Underwater explosions *Ann. Rev. Fluid Mech.* **9** 187-214
[4] Brackbill J U, Kothe D B, and Zemach C 1992 A continuum method for modeling surface tension *J. Comput. Phys.* **100** 335-54
[5] Unverdi S O and Tryggvason G 1992 A front-tracking method for viscous, incompressible, multi-fluid flows *J. Comput. Phys.* **100** 25 -37.
[6] Kobayashi R 1993 Modeling and numerical simulations of dendritic crystal growth, *Physica D* **63** 410-23
[7] Osher S and Fedkiwy R P 2001 Level set methods: an overview and some recent results *J. Comput. Phys.* **169** 463–502
[8] Mulder W, Osher S, and Sethian J A 1992 Computing interface motion in compressible gas dynamics *J. Comput. Phys.* **100** 209-28
[9] Fedkiw R P, Aslam T, Merriman B and Osher S 1999 A non-oscillatory Eulerian approach to interfaces in multi-material flows(the Ghost Fluid Method) *J. Comput. Phys.* **152** 457–92
[10] Liu T G, Khoo BC and Yeo K S 2003 Ghost fluid method for strong shock impacting on material interface *J. Comput. Phys.* **190** 651–81
[11] Liu T G, Khoo B C and Wang C W 2005 The ghost fluid method for compressible gas– water simulation *J. Comput. Phys.* **204** 193-221
[12] Tang H Z and Liu T G 2006 A note on the conservative schemes for the Euler equations, *J. Comput. Phys.* **218** 451-59
[13] Osher S and Fedkiwy R P 2003 *Level set methods and dynamic implicit interfaces* (New York: Springer)
[14] Li L, Cui J, Ren X B, Dong Y C, Lu X X, Wang J S, Wang H Y, Lv X 2009 Experimental Investigations on the liquid dispersion flow during explosive disseminations process, *Chinese Science Bulletin* (Submitted)

Journal of Physics: Conference Series **216** (2010) 012013 doi:10.1088/1742-6596/216/1/012013

Observation of the instability in the shear layer behind a paraboloidal-nose cylinder

Fumihiko Mikami, Ryohei Toyota, Nobuhide Nishikawa

Division of Mechanical Engineering, Graduate School of Engineering, Chiba University, Chiba 263-8522, Japan

E-mail: `fmikami@faculty.chiba-u.jp`

Abstract. Flow visualization was used to investigate experimentally the evolution process of the instability in the separated shear layer behind an axisymmetric paraboloidal-nose cylinder at high angles of attack. The appearance of instability is observed as uniformly spaced vortex loops showing wavy folds of a shear layer. The distance between the counter-rotating primary and secondary vortices varies with time and the vortex loops always appear first in the region between the primary and secondary vortex at the time and place where these vortices approach closest together. The second vortex loops appear in the downstream of the first ones delayed by the appearance of the first vortex loops. It is suggested that the short wavelength perturbations of the preceding vortex loops advect on the outer surface of the primary vortex along the fluid path accompanied by the dominating vortex loops in downstream region.

1. Introduction

A flow over a paraboloidal-nose cylinder at incidence provides the fundamental example of the simple body flow field seen in most high angle of attack flows and provides the model flow of three-dimensional separated flows. Three-dimensional flow separation about a slender body at incidence has been the subject of many studies, including topological structures for a body of revolution such as a slender cone [1], a round-nosed body [2], and a prolate spheroid [3, 4]. The structures of the symmetric separated flow about a paraboloidal-nose cylinder at incidence with main, primary, and secondary vortices on the lee side has been well established in experiments and is illustrated in figure 1. The feeding shear layers roll up to form a pair of counter-rotating main vortices.

It was shown by Riley & Lowson [5] that steady sub-vortical structures appear in the free shear layer over delta wings. They demonstrated that the smoke filaments formed near the leading edge of the wing spiral into the core of the main vortex as they pass downstream. They suggested that the steady sub-vortices are due to a local three-dimensional Kelvin-Helmholtz instability of the free shear layer.

In the preset study, we investigate experimentally the instability of the flow behind a paraboloidal-nose cylinder with appearance of vortex loops in the feeding shear layer encircling the main vortices. This instability is unsteady in contrast to the steady sub-vortical structures over delta wings, and little data on this kind of instability were available in previous works. The present measurements were accomplished using flow visualizations together with PIV measurements in a towing system in a settled air chamber.

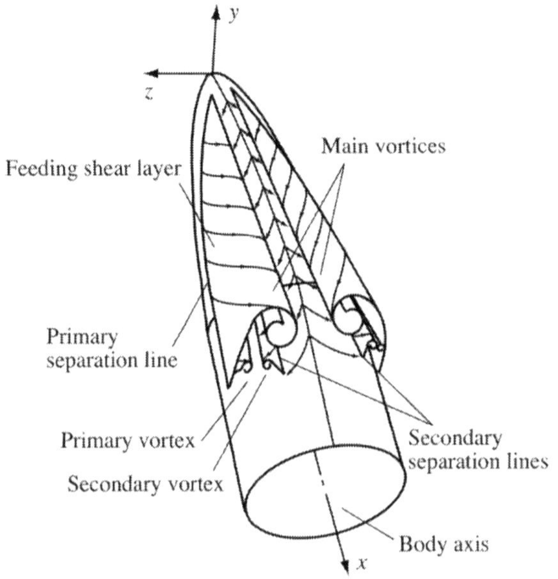

Figure 1. The vortical flow field on the lee side of a slender body at high angle of attack.

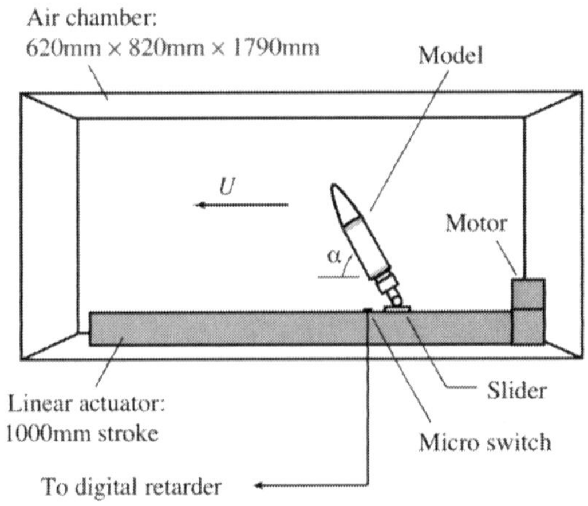

Figure 2. Experimental setup.

2. Experimental setup

The model of a 180mm length cylinder with diameter $D = 75$mm and a paraboloidal nose length 140mm ($1.87D$) was tested in this experiments.

The flow facility used in all the experiments described herein is a towing system in a settled air chamber with a rectangular cross section shown in figure 2. The dimensions of the chamber are 620mm in width × 820mm in height and a length of 1790mm. A linear actuator enables positioning of a model with the stroke of 1000mm at a maximum speed of 1.5 m/s. The slider of the linear guide, on which a model is fixed at an angle of attack α, is driven by the electric motor programmed as a foregoing acceleration with $1.9D$ in stroke and a steady traveling motion with $9.4D$ followed by a deceleration with $1.9D$.

In the following, we will adopt a coordinate system (the same as for the figure shown in section 1) with the x-axis along the body axis, the y-axis along the symmetric plane, and the z-axis perpendicular to them, where the origin is taken at the nose apex.

The evolution of the wake was then measured by a particle image velocimetry (PIV) using cross-sectional and longitudinal cuts produced by the sheet illumination of the Nd:YAG laser. Glycol/water fog droplets (nominally 5 μm diameter, created with a commercial theatrical fog generator) were seeded into the air chamber and the visualized image was recorded with a CCD camera (1008×1018pixel, 8bit). The cross-sectional cuts of the flow along planes parallel to yz-plane performed at various downstream locations, x/L, will be referred to as 'front views'. While the longitudinal cuts along planes parallel to xz-plane performed at various locations, y/R ($2R = D$, where R is the radius of the cylinder), will be referred to as 'top views'.

To visualize the three-dimensional evolution of the wake, we employed a titanium tetrachloride ($TiCl_4$) to produce a dense white smoke (titanium dioxide), as a result of a chemical reaction, due to the presence of moisture in the air. The titanium tetrachloride was applied on a portion of the nose surface, which allowed us to visualize integrated streaklines of a white smoke developing in the wake.

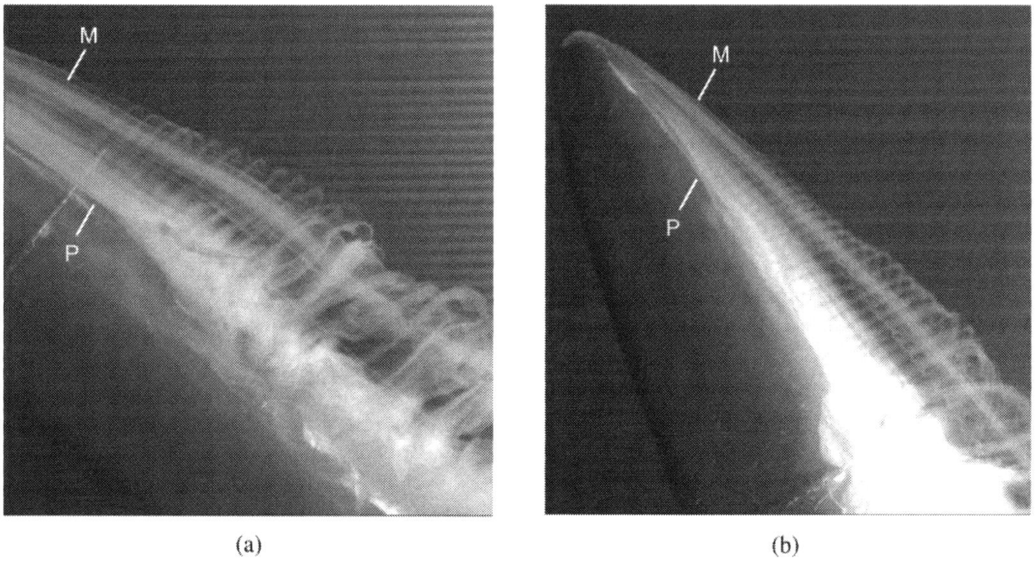

Figure 3. Visualization of vortex loops behind an axisymmetric paraboloidal-nose cylinder. $Re = 7200$. Titanium tetrachloride was applied near the nose apex and on the windward symmetry line. Main vortex (M) and primary vortex (P) are seen in the picture. (a) $\alpha = 40°$, (b) $\alpha = 70°$.

3. Results

The $TiCl_4$ visualization was used to study and characterize the instability in the feeding shear layer. A CCD camera (the same as for the PIV measurements described in section 2) and a stroboscope (pulse duraton 22 μs) were used to capture snapshots of the flow. Titanium tetrachloride was applied on the portion of the body surface near the nose apex and on the windward symmetry line of the nose. The test speed was $U = 1.5$ m/s and Reynolds number was $Re = 7200$ ($Re = UD/\nu$, where ν is the kinematic viscosity of the air).

For $\alpha = 40°$ and $70°$ (figures 3a, 3b), the main vortex and the primary vortex (denoted in the picture by the letters 'M' and 'P', respectively) are seen in the picture with vortex loops, which are encircling the main vortex and extend to the cylinder part of the body from the end of the nose at $\alpha = 40°$ and from the middle of the nose at $\alpha = 70°$. The regularity of the vortex loops is quite apparent in the visualization shown in figure 3, where the spacing of the loops is almost uniform. Further downstream, there appear slight irregularity in the instantaneous vortex-loop shapes with turbulence.

To characterize the structure of the vortex loops, we took several pictures at $\alpha = 40°$ by changing the amounts and the places to apply titanium tetrachloride. Figure 4 shows the close up view of the vortex loops. In this picture, the most outer sheet of white smoke estimates the feeding shear layer and we observed the wavy folds of the shear layer around the main vortex. The waves extend downstream while increasing in magnitude. Figure 5 was taken at a different location, where the main, primary, and secondary vortices (denoted in the picture by the letters 'M' , 'P', and 'S', respectively) are seen in this picture. Observe that at the border region between the primary and the secondary vortices, a serial pattern of vortices appear with a constant spacing. The above observations support our assumption that each vortex loop consists of a vortex tube.

To characterize the temporal behavior of the vortex loops, we took sequence of pictures using

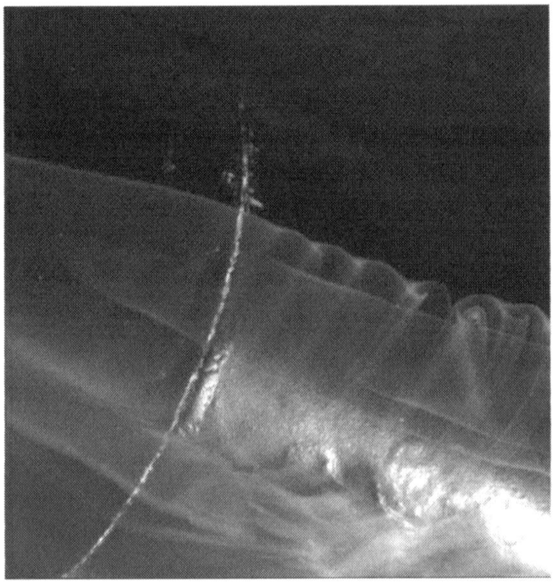

Figure 4. Close-up view of the vortex loops showing the wavy folds of a shear layer. $Re = 7200$, $\alpha = 40°$.

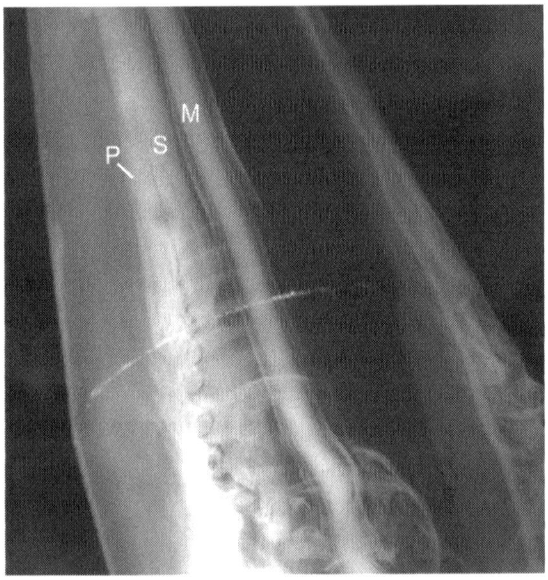

Figure 5. Footprints of the vortex loops between the primary and the secondary vortices. $Re = 7200$, $\alpha = 40°$.

a high-speed camera moving with the model. The pictures were taken at a frame rate of 240 frames per second with an exposure-time of 1/2000 seconds. A metal halide lamp was used for illumination. Figure 6 presents a representative sequence of $TiCl_4$ visualizations showing the vortex loops that appeared intermittently around $x/L = 1.0$ at $\alpha = 40°$. In these pictures, the main flow is from top to bottom, and the primary and the secondary vortices (denoted in the first frame of the pictures by the letters 'P' and 'S') are seen. Note that the distance between the primary and the secondary vortices varies with time and the vortex loops first appear between these vortices at the time and place where these two vortices approach closest together. The newly created vortex loops (denoted in the pictures by the letters A_1 and B_1) are so weak that the surface of the primary and the secondary vortices adjacent to the vortex loops are almost flat.

At the downstream region of these vortex loops, the ripples appear on the upper surface of the primary vortex, delayed by the appearance of the first vortex loops. The ripples propagate downstream while increasing in magnitude, forming new vortex loops (denoted in the picture by the letters A_2 and B_2) . The delayed vortex loops are originated not from the region between the primary and the secondary vortices but from the outward surface of the primary vortex, where the delayed loops are stronger in magnitude than the preceding ones. It is considered that the vortex loops shown in figure 3 correspond to the delayed loops, since the preceding loops are so weak that they are overshadowed by the delayed ones.

Figure 7 shows vorticity distributions at $\alpha = 40°$ obtained using a particle image velocimetry and measured at various cross sections, with ω^* non-dimensionalized by the velocity of the free stream U and the diameter of the cylinder D as $\omega^* = (D/U)\omega$. The locations of the primary and the secondary vortices (denoted in the figure by the letters 'P', and 'S' respectively) are obtained from the front views of averaged vorticity distributions from 50 sets of data in various cross sections shown in figure 7a, and are depicted in figures 7c and 7d (data at $x/L = 0.93$ and 1.25 are not presented in figure 7a). In figure 7c, preceding vortex loops and delayed vortex loops (denoted in the figure by the letters 'L_1' and 'L_2' respectively) are clearly seen in the

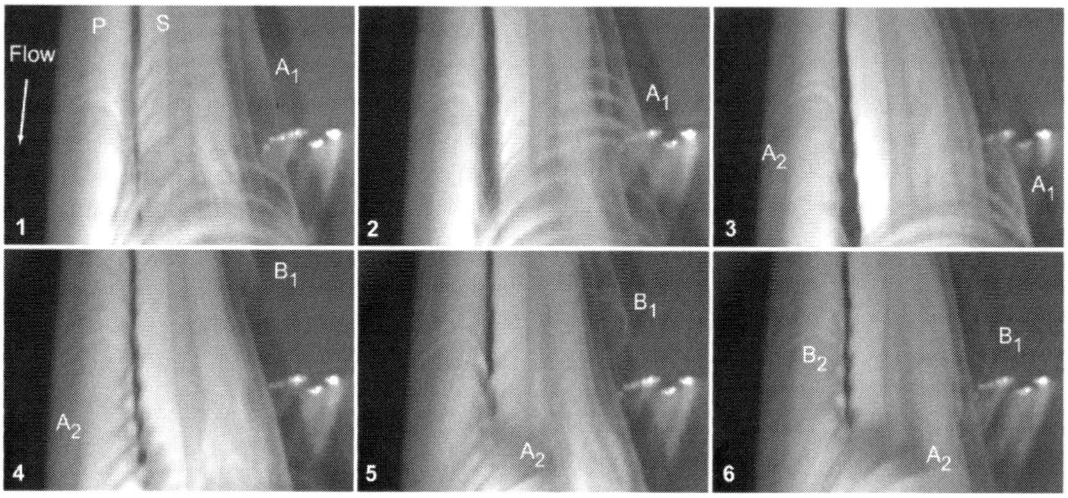

Figure 6. Vortex loop formation around $x/L = 1.0$. $Re = 7200$, $\alpha = 40°$. Time interval between frames is 8.33 ms. 'P' and 'S' represent the primary and the secondary vortex, respectively. The vortex loops A_1 and B_1 first appear on the edge between the primary and the secondary vortices, accompanied by ripples in the downstream region near the upper surface of the primary vortex. The ripples grow along the shear layer, forming stronger new vortex loops A_2 and B_2.

longitudinal cross section at $y/R = 1.07$ that is above the primary and the secondary vortices corresponding to the line A-A in figure 7a. Note that preceding loops L_1 are located between the primary and the secondary vortices, while delayed loops L_2 are located in the outward side of the primary vortex. This observation is consistent with the behavior of the vortex loops described above in figure 6. Note also that the preceding loops extends to downstream region in parallel with the delayed ones. In other words, the preceding loops coexist with the delayed ones forming a double loop structure. In such a case, however, the preceding loops are difficult to observe by visualizations, since they are overshadowed by the delayed loops. The vortex tubes constituting the delayed loops are observed also in the other orthogonal cross section at $x/L = 1.43$ in figure 7b, where its direction of rotation is counter clock-wise in the right half of the front view and clock-wise in the right half of the top view. This result confirms that each delayed vortex loop consists of a vortex tube.

To confirm further that we have observed the change in the distance between the primary and the secondary vortices accompanied by the formation of the preceding vortex loops, we visualized the vortex cores on the lee side of the body at $\alpha = 40°$ using $TiCl_4$. In figure 8, a pair of parallel primary and secondary vortices are seen behind the body, where symmetric deformation of the vortex pair appeared in the nose region with an axial wavelength of about $4b \sim 5b$ (b is the distance between the vortex pair). This instability of the counter-rotating vortex pair is due to the Crow instability [6], which cause the temporal change of the distance between the primary and the secondary vortices. Further downstream, the breakdown of the primary and the secondary vortices are occurred as can be seen in figure 8. This vortex breakdown is due to the deceleration of the axial flow behind the cylinder, which causes disorder of the vortex loops as shown in figure 3.

Under the effect of the primary and the secondary vortices approaching each other, we found that the preceding vortex loops appear in the region between the two vortices. This evidence favors the hypothesis that observed vortex loops are a result of the short wavelength instability of

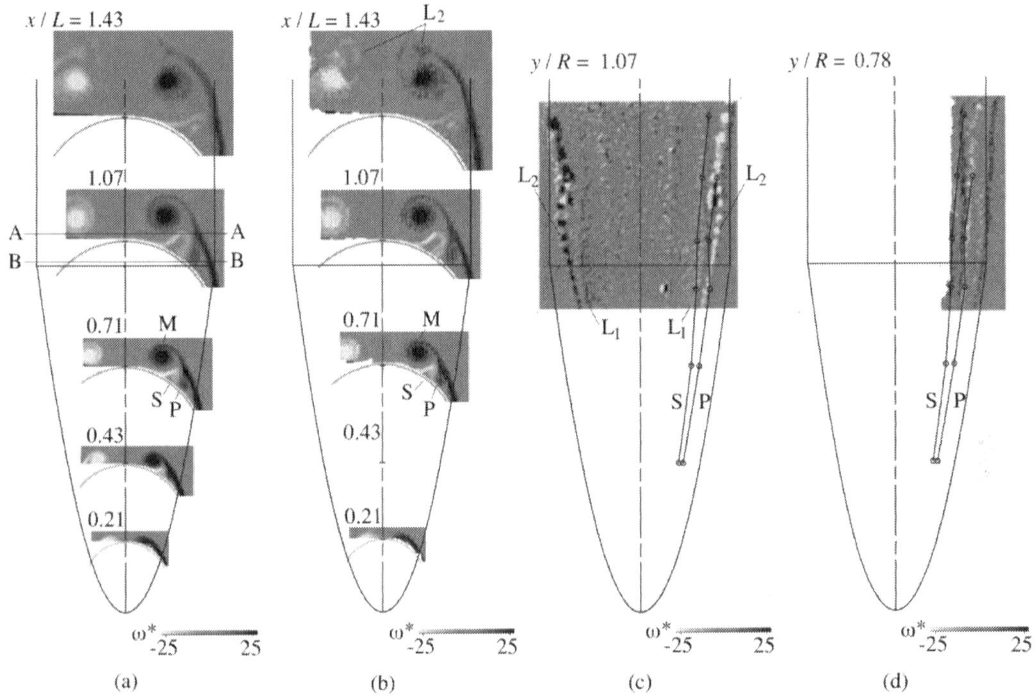

Figure 7. Vorticity distributions. $Re = 7200$, $\alpha = 40°$. (a) front views for averaged vorticity from 50 sets of data, (b) front views for instantaneous vorticity, (c) and (d) top views for instantaneous vorticity in the cross sections corresponding respectively to the lines A-A and B-B in (a).

a counter-rotating vortex pair as shown in figure 9. Leweke & Williamson [7], studying the short wavelength instability, show that the amplitude of the short wavelength instability is larger in the regions where the large-scale Crow instability brings the two vortices closer together. However, it must be considered that the down-wash flow between the primary and the secondary vortices cannot lift up the short wavelength perturbations as vortex tubes. It would be interesting to address the link between the preceding vortex loops and the three-dimensional separation type of foci or spiral nodes [2]. It would be natural that the short wavelength perturbations affect the skin-friction lines and may explain the observed preceding vortex loops.

The formation of the preceding vortex loops has an effect on the development of the ripples in the downstream region. We found, in figure 6, that the second vortex loops appeared in the downstream of the preceding loops delayed by the appearance of the preceding loops. These vortex loops originate from the ripples on the upper surface of the primary vortex, where the strong cross flow lifts up these ripples towards the main vortex along the feeding shear layer. It is important to mention that the organized structure and the wavelength of the preceding loops are retained at the ripples. This result can be explained by assuming that the upstream short wavelength perturbations are advected along the path of the flow on the surface of the primary vortex.

4. Conclusions

We have studied the development of the instability in the shear layer as an appearance of vortex loops. Flow visualization of the temporal properties of the vortex loops revealed that the

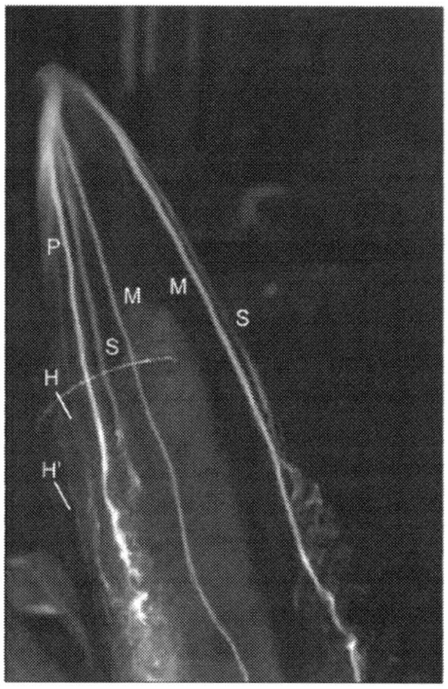

Figure 8. Visualization of vortex cores. $Re = 7200$, $\alpha = 40°$. Higher order vortices (represented by H and H') are seen in the picture in addition to main (M), primary (P) and secondary (S) vortices.

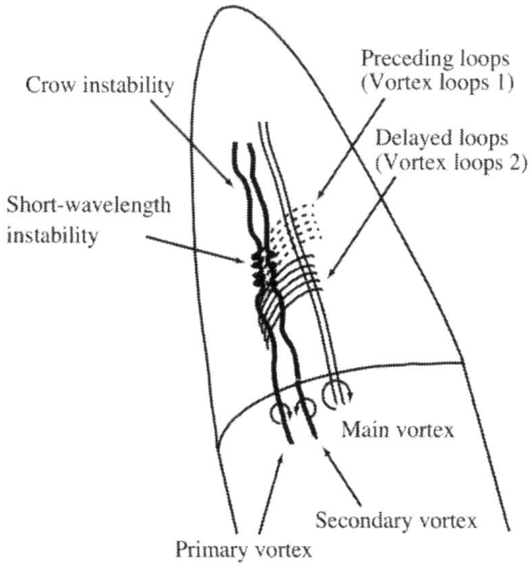

Figure 9. Vortex skeleton model of a flow behind an axisymmetric paraboloidal-nose cylinder at high angles of attack. Vortices in the other side of the symmetric plane are not shown.

preceding vortex loops always grow first, accompanied by the delayed vortex loops dominating in the wake. We have also shown that the distance between the primary and the secondary vortices varies with time possibly by the Crow instability of a counter-roating vortex pair and that the preceding loops appear intermittently at the time and place where the primary and the secondary vortices approach closest together. Furthermore, we found that the preceding loops appear in the region between the primary and the secondary vortex, while the delayed loops are originated from the ripples on the upper surface of the primary vortex. This result was further confirmed by the PIV measurements of vorticity distributions. The PIV measurements also revealed that the preceding loops and the delayed loops can coexist in the downstream region. It appears to indicate that the delayed vortex loops are due to the short wavelength perturbations of the preceding vortex loops advected on the surface of the primary vortex.

References

[1] Peake D J, Fisher D F and McRae D S 1982 *AIAA J.* **20** 1338
[2] Tobak M and Peake D J 1982 *Ann. Rev. Fluid Mech.* **14** 61
[3] Wang K C, Zhou H C, Hu C H, and Harrington S 1990 *Proc. R. Soc. Lond.* A **429** 73
[4] Chesnakas C J and Simpson R L 1996 *Trans. ASME J. Fluids Eng.* **118** 268
[5] Riley A J and Lowson M V 1998 *J. Fluid Mech.* **369** 49
[6] Crow S C 1970 *AIAA J.* **8** 2172
[7] Leweke T and Williamson C H K 1998 *J. Fluid Mech.* **360** 85

Journal of Physics: Conference Series **216** (2010) 012014 doi:10.1088/1742-6596/216/1/012014

Forced and thermocapillary convection in silicon Czochralski crystal growth in semispherical crucible

F Mokhtari [1,2], A Bouabdallah[2], M Zizi[2], S Hanchi[3] and A Alemany[4]

[1] Physics Department, Faculty of Science, University of Mouloud Mammeri, Tizi Ouzou, Algeria.
[2] LTSE Laboratory, University of Science and Technology USTHB. BP 32 Elalia, Babezzouar, Algiers, Algeria.
[3]UER Mécanique/ E.M.P B.P 17, Bordj El Bahri, Algiers, Algeria
[4] Laboratoire EPM, CNRS, Grenoble, France

E-mail: abouab2002@yahoo.fr

Abstract. In order to understand the influence of a semispherical crucible geometry combined with different convection modes as a thermocapillary convection, natural convection and forced convection, induced by crystal rotation, on melt flow pattern in silicon Czochralski crystal growth process, a set of numerical simulations are conducted using Fluent Software. We solve the system of equations of heat and momentum transfer in classical geometry as cylindrical and modified crystal growth process geometry as cylindro-spherical. In addition, we adopt hypothesis adapted to boundary conditions near the interface and calculations are executed to determine temperature, pressure and velocity fields versus Grashof and Reynolds numbers. The analysis of the obtained results led to conclude that there is advantage to modify geometry in comparison with the traditional one. The absence of the stagnation regions of fluid in the hemispherical crucible corner and the possibility to control the melt flow using the crystal rotation enhances the quality of the process comparatively to the cylindrical one. The pressure field is strongly related to the swirl velocity.

1. Introduction

The silicon crystals used in the technology of semiconductors are mainly produced by the Czochralski technique. During the Czochralski process three modes of heat transfer are present, namely, conduction in melt and crystal, natural convection due to the temperature gradient between the crucible wall and the melt/crystal interface. In addition, forced convection caused by crystal and/or crucible rotation combined to thermocapillary convection which is known as Marangoni convection induced by surface tension gradients on the free surface of the melt. In some cases the heat transfer by radiation between various surfaces of the furnace is also present.

The effect of natural convection on the growth in the traditional system (cylindrical crucible) was studied by several researchers; Kobayashi N [1] and Langlois W E [2] were the first to do it. Turbulent natural convection in large melt volumes have been studied in recent works [3-6]. According to Kumar V et al. [7] the effect of surface tension gradients was neglected in several studies on silicon growth because of the weakness of Prandtl number (Pr=0.0113). They showed that the Marangoni convection influences the thermal and dynamic field in the melt. K.Kakimoto and

H.Ozoe [8] carried by using three-dimensional numerical simulations and found that the surface tensions influence the flow and the oxygen transport in melt. By simulating the three-dimensional flow in an oxide Czochralski melt, Basu B et al. [9] have found that the crystal rotation affects the quality and the stability of the growing crystal, the centrifugal force due to the rotating crystal leads to a stabilising effect on the three dimensional flow caused by natural convection. They have found a critical value of Richardson number ($Ri=Gr/Re^2=235$) beyond it the flow become periodic. According to D.T.J.Hurle [10] the rotating crystal acts as a centrifugal fan sucking up fluid axially spinning it up in an Ekman layer and ejecting it tangentially. The radially outward flow due to the rotating crystal meets the radially inflowing fluid driven by the hot crucible wall and a down flow occurs at some radial distance which depends upon the relative strength of the crystal flow and the buoyancy-driven convection. Z.Galazka and H.Wilke [11] have showed that during the growth of YAG, the interface convexity decreases when the crystal rotation rate increases, it becomes almost flat for high Reynolds number values. Crochet M J et al [12] studied numerically the effect of forced convection induced by rotation on the Czochralski crystal growth of GaAs in a cylindrical crucible for different Reynolds number values. Liu L and Kakimoto K [13] analysed the effect of crystal rotation on the growth interface shape in the presence of a transverse magnetic field, they found that the melt-crystal interface changes from an obvious 3D shape when the crystal is not rotating to an almost 2D shape with increase in rotation rate.

Our work consists in studying numerically the combined effects of crucible geometry, thermocapillary convection and crystal rotation on silicon Czocharlski crystal growth. We show the advantage of the modified geometry, and we establish the relation between the swirl velocity and the corresponding pressure field for different Reynolds number values.

2. Modelling

Anselmo A P et al. [14] and Tavakoli M H et al. [15] showed that a cylindrical crucible with a curved bottom has advantages for crystallization. Therefore, it is proposed recently by Mokhtari F et al. [16,17,18] to modify the device as a cylindro-spherical system (figure 1).

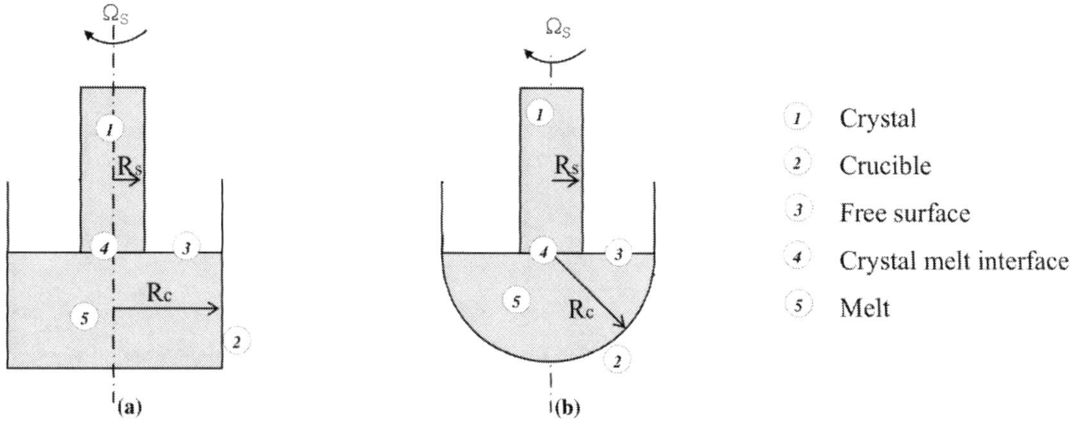

Figure 1. Traditional (a) and modified (b) Czochralski crystal growth systems

To facilitate the procedure of modelling related to the considered problem, we admit some assumptions; the molten silicon is assumed to be a viscous, Newtonian and incompressible fluid satisfying the Boussinesq assumption. We assume the solid liquid interface and the free surface to be flat initially. The thermo physical properties of the fluid are constant except for the density variation in the buoyancy force term. The flow is symmetric in the axial direction. Fixed temperatures are imposed on the walls of the melt crucible and the crystal melt interface. Considering the preceding assumptions, the resolution of the problem will be done in two-dimensional geometries, limited to axisymmetric sections of the crucible represented by the schematic diagrams (figure 2).

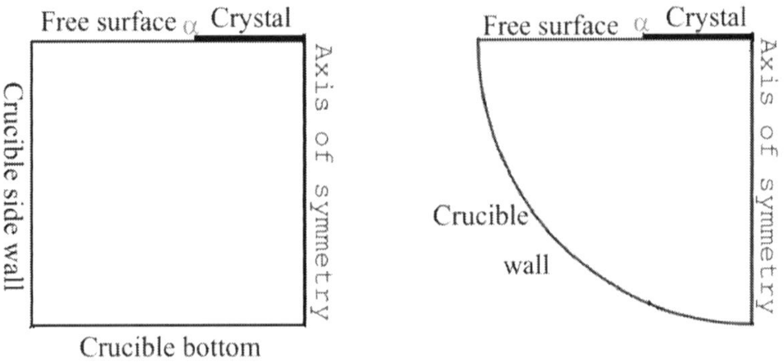

(a) **(b)**

Figure 2. Computational domains for mathematical model: (a) traditional (b) modified

The dimensionless governing equations for the fluid motion and temperature field derived under the above assumptions are

Continuity

$$\frac{\partial V_r}{\partial r} + \frac{V_r}{r} + \frac{\partial V_z}{\partial z} = 0$$

Momentum

$$\frac{\partial V_r}{\partial t} + V_r \frac{\partial V_r}{\partial r} + V_z \frac{\partial V_r}{\partial z} = -\frac{\partial P}{\partial r} + \frac{1}{r}\frac{\partial}{\partial r}\left(r\frac{\partial V_r}{\partial r}\right) - \frac{V_r}{r^2} + \frac{\partial^2 V_r}{\partial z^2}$$

$$\frac{\partial V_\varphi}{\partial t} + V_r \frac{\partial V_\varphi}{\partial r} + V_z \frac{\partial V_\varphi}{\partial z} + \frac{V_r V_\varphi}{r} = \frac{1}{r}\frac{\partial}{\partial r}\left(r\frac{\partial V_\varphi}{\partial r}\right) + \frac{1}{r^2}\frac{\partial^2 V_\varphi}{\partial \varphi^2} + \frac{2}{r^2}\frac{\partial V_r}{\partial \varphi} + \frac{\partial^2 V_\varphi}{\partial z^2}$$

$$\frac{\partial V_z}{\partial t} + V_r \frac{\partial V_z}{\partial r} + V_z \frac{\partial V_z}{\partial z} = -\frac{\partial P}{\partial z} + \frac{1}{r}\frac{\partial}{\partial r}\left(r\frac{\partial V_z}{\partial r}\right) + \frac{\partial^2 V_z}{\partial z^2} \text{-Gr.T}$$

Energy

$$\frac{\partial T}{\partial t} + V_r \frac{\partial T}{\partial r} + V_z \frac{\partial T}{\partial z} = \frac{1}{\Pr}\left\{\frac{1}{r}\frac{\partial}{\partial r}\left(r\frac{\partial T}{\partial r}\right) + \frac{\partial^2 T}{\partial z^2}\right\}$$

Where Grashof number $Gr = \frac{g\beta R_c^3 (T_c - T_f)}{\upsilon^2}$ characterises natural convection effect. Prandtl number $\Pr = \frac{\upsilon}{\alpha}$ measures the relative importance of heat transfer by conduction and convection.

The non dimensional boundary conditions are as follows

Crucible wall and bottom $V_r = V_\varphi = V_z = 0$ $T = 1$

Crystal melt interface $V_r = V_z = 0$ $V_\varphi = r.\text{Re}$ $T = 0$

Where $\text{Re} = \frac{\Omega_S R_C R_S}{\upsilon}$ is Reynolds number associated to the rotating crystal.

Axis of symmetry $\partial V_z / \partial r = 0$ $V_r = 0$ $\partial T / \partial r = 0$

Free surface: $V_z = 0$ $\frac{\partial V_r}{\partial z} = \frac{Ma}{\Pr}\frac{\partial T}{\partial r}$

The last one is the Marangoni condition, it supposes that the flow perpendicular to the free surface is null, and that the shear stresses at the free surface are balanced by the surface tension gradients [19],

$$\mu\frac{\partial V_r}{\partial z}=-\frac{\partial \sigma}{\partial r}$$

Where $Ma=-\frac{\partial \sigma}{\partial T}\frac{R_C\left(T_C-T_f\right)}{\mu\alpha}$ is Marangoni number measuring the effect of thermocapillary convection.

The properties used in simulation are dimensionless parameters. The Prandtl number characterizing silicon is 0.0113. We propose to maintain the Grashof number to 10^5, the Marangoni number value corresponding is Ma=2000 [18] and we vary Reynolds number in the range of 1000 to 15000.
The problem is solved using the finite volume package Fluent. The steady segregated axisymmetric swirl solver was used with the first order upwind discretization for convection, SIMPLE algorithm for pressure-velocity coupling and the PRESTO (PREssure STaggering Option) scheme for pressure interpolation. The convergence is handled by monitoring residuals of continuity, momentum and energy equations.

3. Results and discussions

The aim of this work is based on the numerical study of the geometry effect combined with convection effect with its different forms: natural, thermocapillary (Marangoni convection) and forced (induced by crystal rotation) on flow pattern, temperature and pressure field in silicon Czochralski crystal growth process. By carrying out the convergence test of grid and residuals, we find a grid of 200*200 and residuals of 10^{-6} We maintain the Grashof number value to 10^5, thus the corresponding Marangoni number value is approximately Ma=2000, we modify the crystal rotation rate by increasing Reynolds number value. For the different analyzed cases we represent the corresponding isotherms, isobars, streamlines and azimuthal velocity contours.

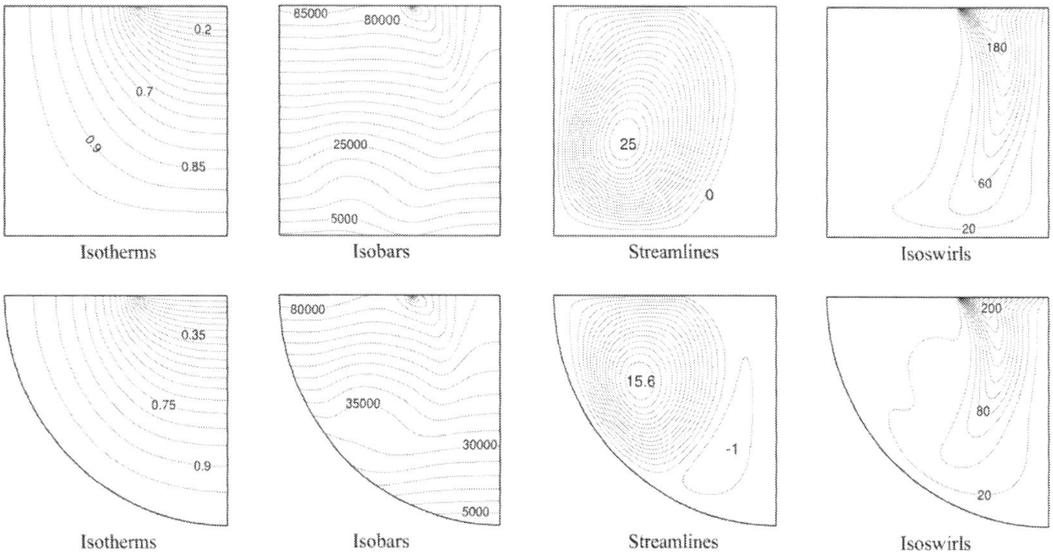

Figure 3. Isotherms, isobars, streamlines, isoswirls for Gr=10^5, Ma=2000, Re=1000

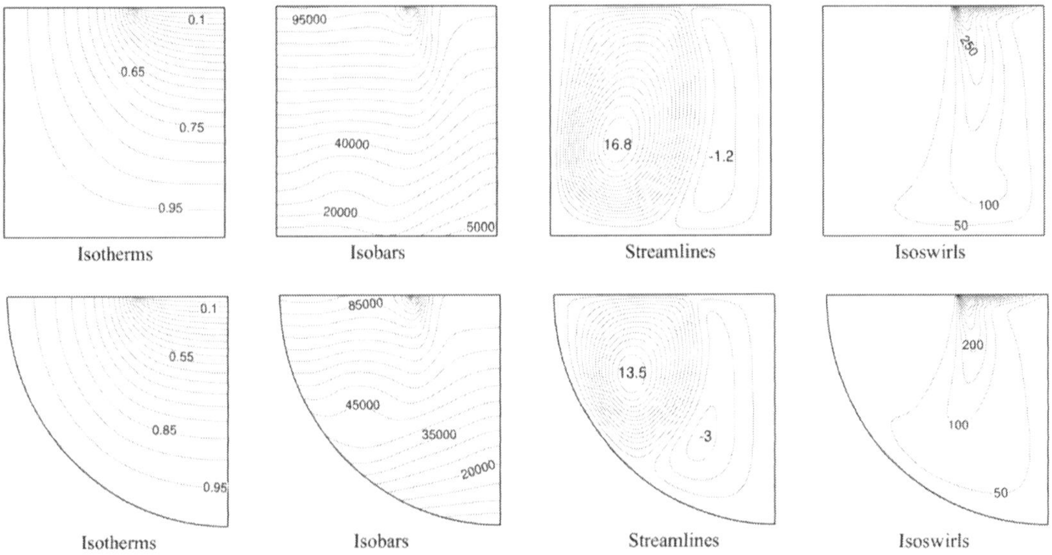

Figure 4. Isotherms, isobars, streamlines, isoswirls for Gr=10^5, Ma=2000, Re=2000

For low Reynolds number values (Re<1000), the melt flow in the crucible is induced by buoyancy and thermocapillary effects so that the convection modes present are natural and Marangoni convection, one anticlockwise vortex is present in the melt. Beyond Re=1000 a second vortex, induced by crystal rotation, begins to form in the melt, it is clockwise. Its intensity and volume become higher by increasing the crystal rotation rate.

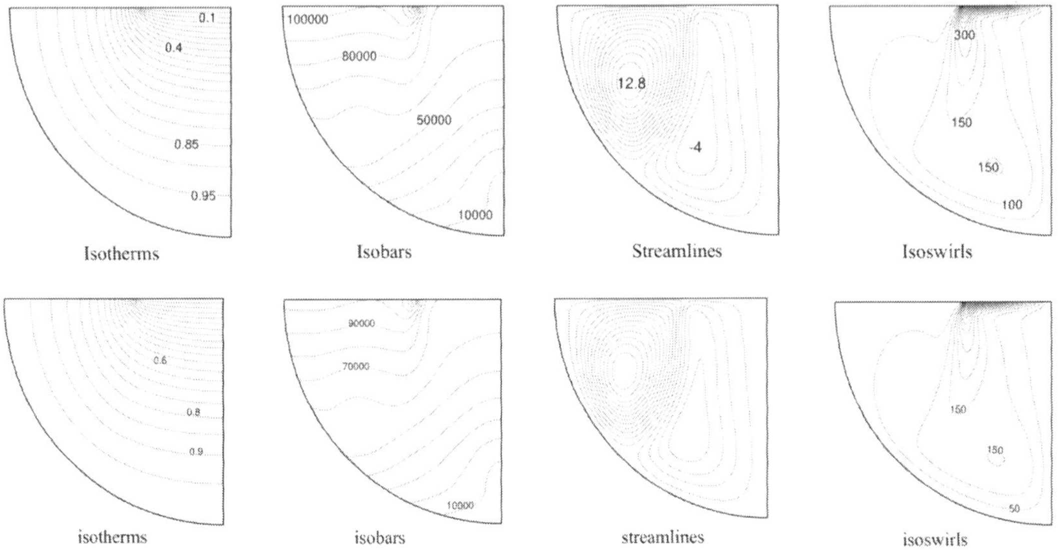

Figure 5. Isotherms, isobars, streamlines, isoswirls for Gr=10^5, Ma=2000, Re=3000

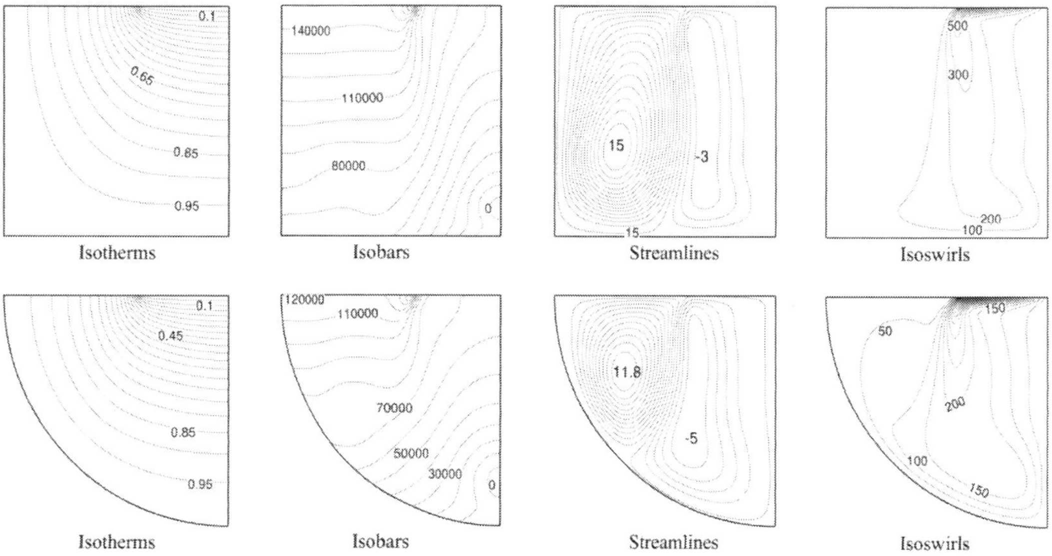

Figure 6. Isotherms, isobars, streamlines, isoswirls for Gr=10^5, Ma=2000, Re=4000

For Reynolds number values Re∈[1000, 5000] corresponding to figures. 3-7 the natural and Marangoni convection dominate, the anticlockwise vortex becomes bigger and more intense. For Re≈5500 the two vortices have almost the same intensity and occupy the same volume in the melt, this Reynolds number value is corresponding to the transition from the natural and thermocapillary mode toward the forced mode. Beyond this critical value, that's the forced vortex (clockwise vortex) which is more intense, it takes the upper part of the crucible whereas the anticlockwise vortex goes down to the crucible bottom.

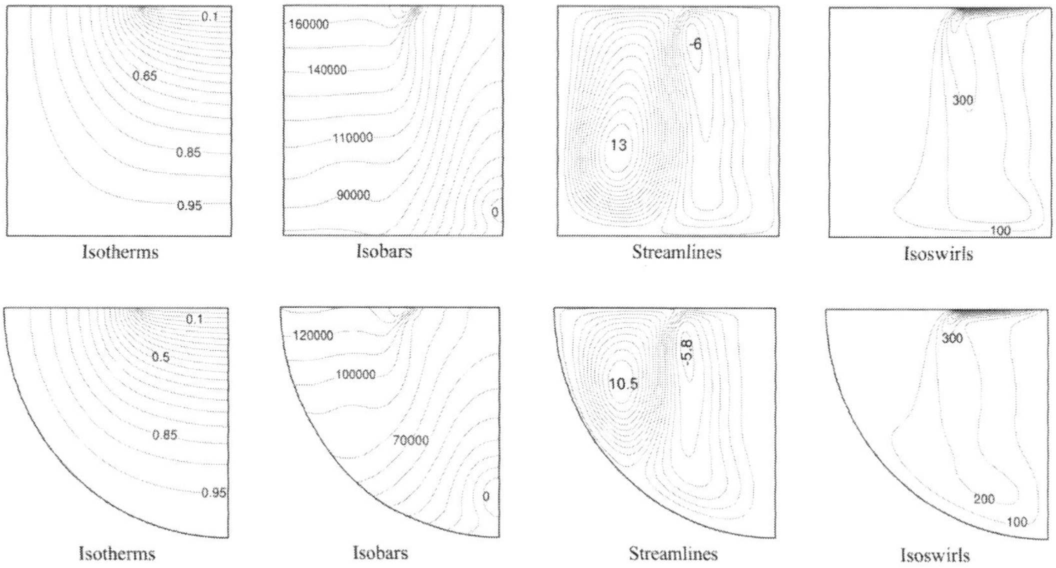

Figure 7. Isotherms, isobars, streamlines, isoswirls for Gr=10^5, Ma=2000, Re=5000

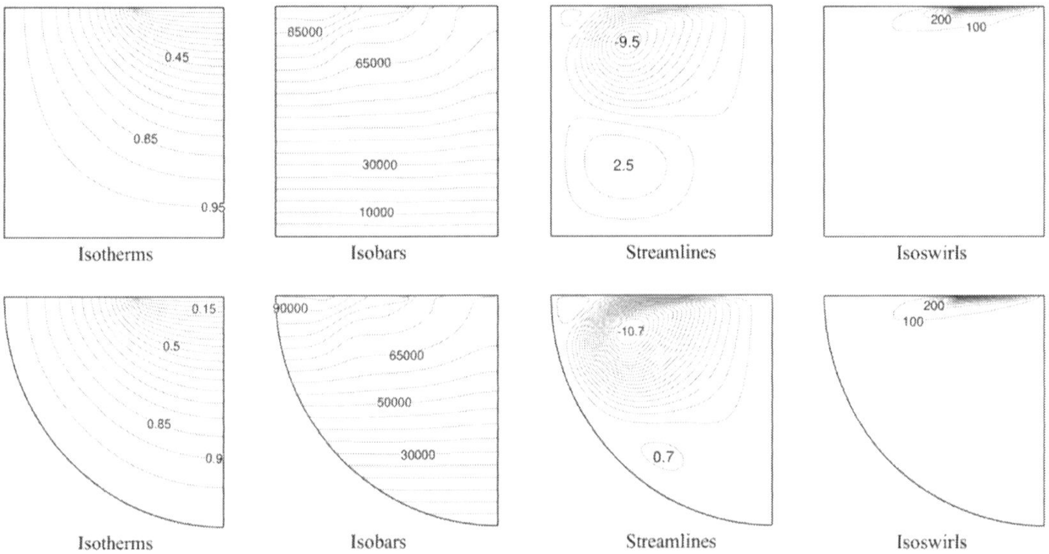

Figure 8. Isotherms, isobars, streamlines, isoswirls for Gr=10^5, Ma=2000, Re=6000

By following the center location of both vortices, we note that for low rotation velocities (Re<3000) the center of natural and Marangoni vortex is above. For Reynolds number Re=4000 (figure 6), they are almost in the same level, but beyond Re=5000 (figure 7), the center of forced vortex moves upward whereas the other remains in the lower part of the crucible.

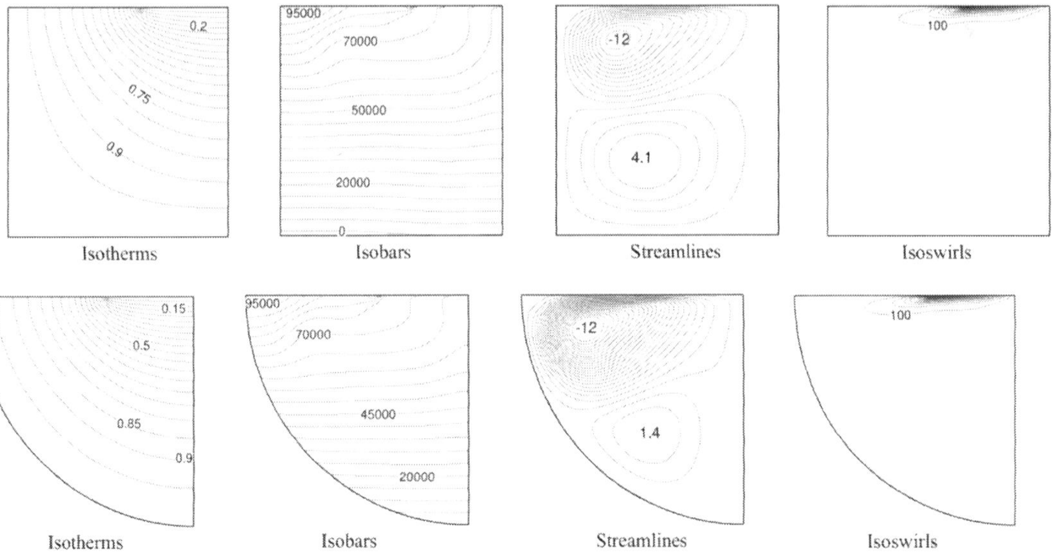

Figure 9. Isotherms, isobars, streamlines, isoswirls for Gr=10^5, Ma=2000, Re=7000

Unlike isotherms, the isobars undergo significant distortions under the rotation effect and their intensity increases. For Re=3000, 4000 and 5000 corresponding respectively to figures. 5, 6 and 7, the incline of isobars in the melt zone under the crystal is in the same phase with the presence of a second vortex under the crystal (this vortex is induced by forced convection), when this vortex moves to the free surface (Re=6000, figure 8) the isobars become parallel to the free surface and perpendicular to the symmmetry axis.

Journal of Physics: Conference Series **216** (2010) 012014 doi:10.1088/1742-6596/216/1/012014

In all studied cases we note the existence of a node in isobars at the point α (figure 2), where three phases are present simultaneously: gas (argon), liquid (silicon melt) and solid (crystal). This node, which is induced by Marangoni convection, has been noted in a precedent work [18]. By comparing streamlines and isotherms obtained for both considered geometries, we note the absence of stagnation region or region of died fluid in the hemispherical crucible and the isotherms follow the symmetrical geometry (spherical). The intensity of the natural and thermocapillary vortex is weaker comparing with the cylinder.

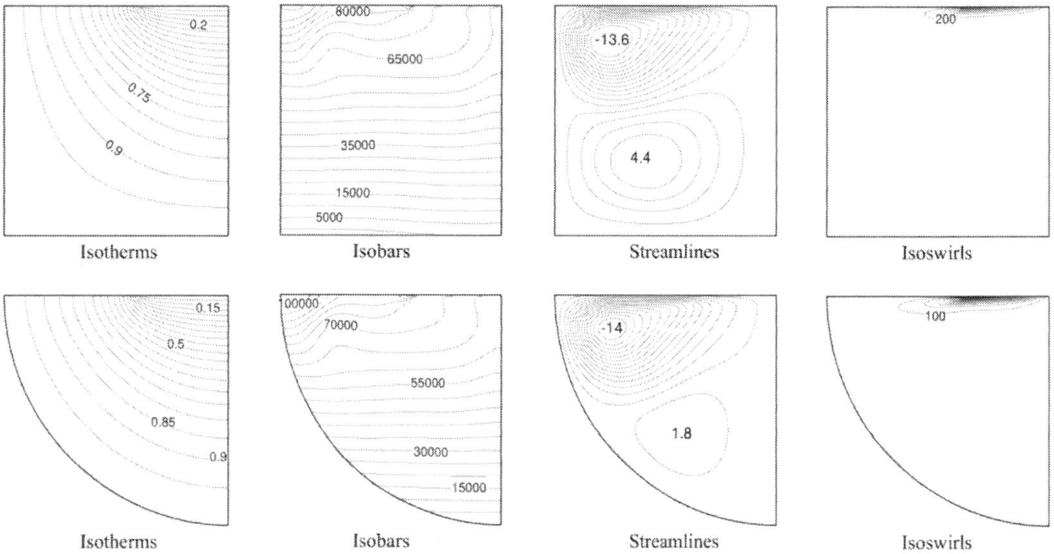

Figure 10. Isotherms, isobars, streamlines, isoswirls for $Gr=10^5$, Ma=2000, Re=8000

For the Reynolds number value Re=1000 corresponding to figure 3 the forced vortex appears in the hemispherical crucible but it is absent in the cylinder. We have showed in a precedent work [18] that the Richardson number $Ri= Gr/Re^2$ corresponding to the transition from natural mode to forced mode is more important for the cylinder when the Marangoni effect is not considered. The presence of the forced vortex is also accompanied with the deformation of isobars especially in the part of melt under the crystal with increasing the pressure everywhere in the melt.

The contour lines of azimuthal velocity occupy more space in the hemispherical crucible than in the cylindrical one. In figure 5 corresponding to Reynolds number value Re=3000, a small node is formed in the contours of azimuthal velocity in the lower part of the melt, beyond this Reynolds number value a dynamic boundary layer forms at the crystal bottom in both geometries, it is the Ekman layer [10]. But the interaction of the Ekman layer on the melt flow is still unknown.

For Re=6000 (figure 8) , the forced vortex of the hemispherical case is bigger than the one in the cylindrical crucible, it remains small traces of the natural and thermocapillary vortex, thus we can conclude that the hemispherical geometry helps us to better control the melt flow in the traditional one even with thermocapillary convection.

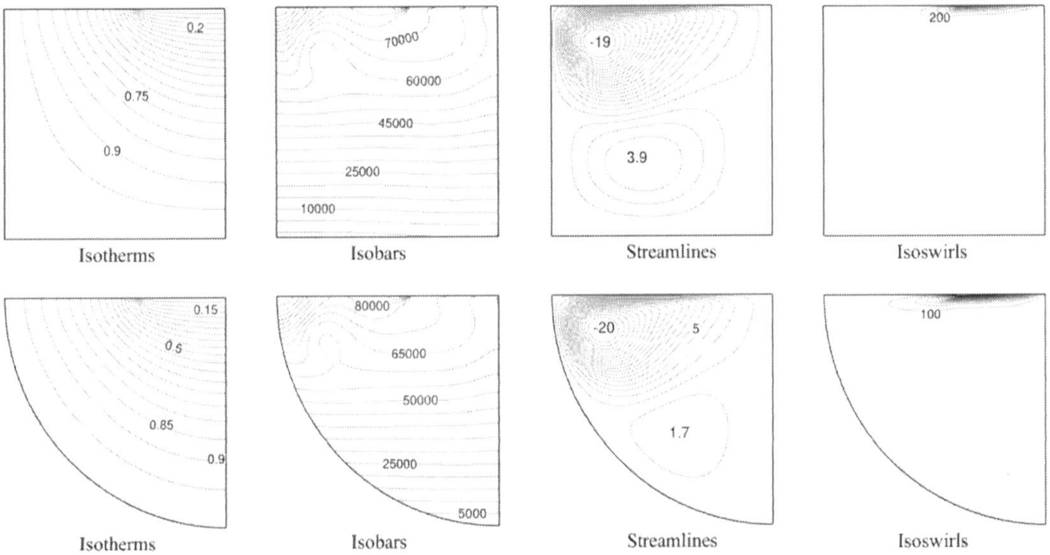

Figure 11. Isotherms, isobars, streamlines, isoswirls for Gr=10^5, Ma=2000, Re=10000

Beyond Re=6000, le center of the forced vortex begins to move toward the crucible wall, its intensity increases more and more, the natural and thermocapillary vortex begins to rebuild but remaining always in the lower part of the melt. The movement of the center of forced vortex is accompanied with the deformation of isobars in the upper part of the melt under the free surface.

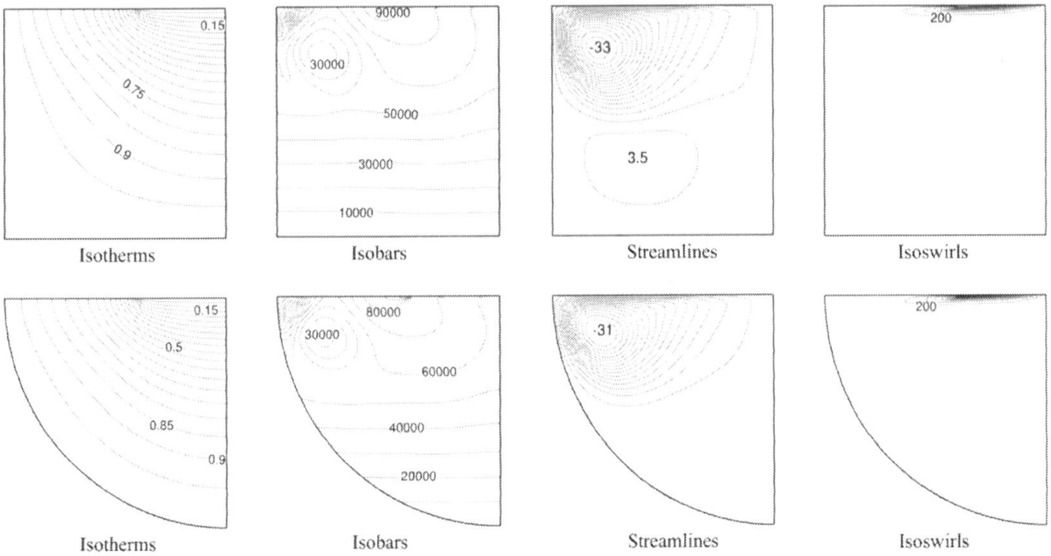

Figure 12. Isotherms, isobars, streamlines, isoswirls for Gr=10^5, Ma=2000, Re=15000

According to Kobayashi S [20] oxygen is the major impurity in Czochralski silicon crystals, the oxygen comes from a continuous corrosion of the walls of the silica crucible by molten silicon [10]. This oxygen is transported by the natural vortex, one part is evaporated at the free surface if there is a contact between this vortex and the free surface, the remaining part is transferred to the forced vortex. The later assigns it to the crystal by intermediate of the growth interface. Thus the contact area between both vortices has an important role and can provide some information about the quality of the grown crystal.

For high Reynolds number values, Re=7000, 8000 and 10000 (figures. 9, 10, 11), the natural vortex in the cylindrical crucible is more intense and more voluminous than the one in the hemispherical crucible. For Re=15000 (figure 12), the natural vortex disappears in the hemispherical crucible but it remains in the cylindrical one. Thus we establish that the contact area between the two vortices is greater in the cylindrical system compared to the hemispherical one, this difference between the two geometries gives more advantage for the modified geometry because the more the contact area is smaller less impurities can be transferred to the forced vortex and then to the crystal.

4. Conclusion

Under the effect of natural, thermocapillary and forced convection, we tried to show the effect of crucible geometry on temperature, velocity and pressure fields in silicon melt of a Czochralski crystal growth system. The analysis of the obtained results enables us to conclude the advantage of the modified geometry in comparison with the traditional one. The absence of the stagnation region or region of died fluid in the hemispherical crucible corner and the facility to control the melt flow, by using the crystal rotation, and then the pressure field favors the hemispherical crucible shape to the cylindrical one.

The pressure field is strongly related to the swirl velocity, a clockwise vortex induced by forced convection is present in the melt. The distortion of isobars follows the location, the shape and the center position of this vortex. For low Reynolds values (Re<6000), the forced vortex is situated under the crystal, the isobars deformation or their inclination is also situated under the crystal. For high Reynolds number values (Re>6000) the forced vortex is under the free surface and the isobars deform in the same region i.e. under the free surface and their deformation follows the movement of the forced vortex center. The obtained results for high rotation rate values enable us to note that the contact area between the natural and forced vortexes is small in the hemispherical crucible in comparison with the cylindrical one. Consequently this helps us to minimize impurities in the grown crystal.

5. References

[1] Kobayashi N 1978 *J. Cryst. Growth* **43** 357
[2] Langlois W E 1979 *J. Cryst.Growth* **46** 743
[3] Lukanin D P, Kalaev V V, Makarov Yu N, Wetzel T, Virbulis J and Amon W V 2004 *J. Cryst. Growth* **206** 20
[4] Jana S, Dost S, Kumar V and Durst F 2006 *Int. J. Eng. Sci.* **44** 554
[5] Kalaev V V 2007 *J. Cryst.Growth* **303** 203
[6] Raufeisen A, Jana S, Breuer M Botsch T and Durst F 2007 *J. Cryst. Growth* **303** 146
[7] Kumar V, Basu B, Enger S, Brenner G and Durst F 2003 *J. Cryst. Growth* **255** 27
[8] Kakimoto K and Ozoe H 2000 *J. Cryst. Growth* **212** 429
[9] Basu B, Enger S, Brenner G and Durst F 2001 *J. Cryst. Growth* **230** 148
[10] Hurle D T J 1983 *J. Cryst. Growth* **65** 124
[11] Galazka Z and Wilke H 2000 *J. Cryst. Growth* **216** 389
[12] Crochet M J, Wouters P J, Geyling FT and Jordan A S 1983 *J. Cryst. Growth* **65** 153
[13] Liu L and Kakimoto K 2005 *J. Cryst. Growth* **275** 1521
[14] Anselmo A P and Prasad V 1993 *J. Cryst. Growth* **131** 247
[15] Tavakoli M H, Wilke H and Crnogorac N 2007 *Cryst. Res. Tech.* **42** 1252
[16] Mokhtari F, Bouabdallah A, Zizi M, and Alemany A 2006 *Mag. Hyd. Dyn.* **42** 451
[17] Mokhtari F, Bouabdallah A, Zizi M, Hanchi S and Alemany A 2009 *Mag. Hyd. Dyn.* **45** 339
[18] Mokhtari F, Bouabdallah A, Zizi M, Hanchi S and Alemany A 2009 *Cryst. Res. Tech.* **44** 787
[19] Denn M D *Linear Stability Surface Tension-Driven Transport* (New Jersey: Prentice-Hall, Inc Englewood Cliffs) 173
[20] Kobayashi S 1997 *J. Cryst. Growth* **117** 163

The Third International Symposium on Bifurcations and Instabilities in Fluid Dynamics IOP Publishing
Journal of Physics: Conference Series **216** (2010) 012015 doi:10.1088/1742-6596/216/1/012015

Instabilities of non-isothermic ultra-thin two-layer films

Akexander A. Nepomnyashchy[1,2], **I. B. Simanovskii**[1]

[1] Department of Mathematics, Technion - Israel Institute of Technology, 32000, Haifa, Israel
[2] Minerva Center for Nonlinear Physics of Complex Systems, Technion - Israel Institute of Technology, 32000, Haifa, Israel

E-mail: yuri11@inter.net.il

Abstract. Evolution of ultra-thin bilayer films under the joint action of the van der Waals forces and Marangoni stresses in the presence of an inclined temperature gradient, is investigated. The problem is solved by means of a linear stability theory and nonlinear simulations. The linear analysis leads to the discovery of the phenomenon of "isotropization". In the nonlinear region, the formation of ordered structures that consist of equally sized droplets moving with equal velocities and located at equal distances from each other, is revealed. The influence of the gravity on the instabilities of non-isothermic two-layer films is studied.

1. Introduction

In the past few decades, the development of microfluidics and nanotechnology led to a significant progress in the exploration of thin film flows. The dynamics of ultra-thin (but still macroscopic) films, with the thickness less than 100 nm, is of a special interest. In the case of ultra-thin films, it is necessary to take into account the long-range intermolecular forces (first of all, van der Waals forces) acting between molecules of the liquid and substrate [1]. Depending on the sign of the Hamaker constant, these forces can either stabilize the film or create a longwave instability leading to the film rupture through the formation of holes [2].

The wetting properties of the interfaces can be significantly changed by the creation of two-layer films, which reveal much richer dynamics than one-layer films [3]. The nonlinear evolution of two-layer liquid films in the presence of van-der-Waals interactions and other effects, such as evaporation, adsorption of surfactants, thermocapillary and solutocapillary effects, action of an electrostatic field, and Rayleigh-Taylor instability, has been studied [4], [5], [6], [7], [8]. Besides a rupture, the development of instabilities may lead to the formation of different morphologies including holes, droplets and even labyrinthine structures. In the present work, we have found the phenomenon of a spontaneous ordering of a droplet system under the action of an imposed inclined temperature gradient.

2. Formulation of the Problem

Consider a system of two superposed layers of immiscible liquids with different physical properties. The bottom layer rests on a solid substrate, the top layer is in contact with the adjacent gas phase. The temperature of the solid substrate is $T_s + Ax$, the temperature of the gas near the interface is assumed to be $T_g + Ax$. All the variables referring to the bottom layer are marked by subscript 1, and all the variables referring to the top layer are marked by subscript 2. The equilibrium coordinates of interfaces are H_i^0, $i = 1, 2$. The deformable interfaces are described by equations $z = H_1(x, y, t)$ (liquid-liquid interface) and $z = H_2(x, y, t)$ (liquid-gas interface). The ith fluid has density ρ_i, kinematic viscosity ν_i, dynamic viscosity $\eta_i = \rho_i \nu_i$, thermal diffusivity χ_i and heat conductivity κ_i. The surface tension coefficients on the lower and upper interfaces, σ_1 and σ_2, are linear functions of temperature T: $\sigma_1 = \sigma_1^0 - \alpha_1 T$, $\sigma_2 = \sigma_2^0 - \alpha_2 T$. The effect of gravity is neglected. For a heat flux on the liquid-gas interface we use an empirical condition,

$$\kappa_2 \frac{\partial T_2}{\partial x_i} n_{2i} = -q(T_2 - T_g - Ax), \tag{1}$$

where q is the heat exchange coefficient which is assumed to be constant.

The system of equations and boundary conditions governing the problem is rather complicated. However, in the case of thin film flows, when the fluid system is thin in one direction and extended in other directions, the nonlinear model governing three-dimensional flows with a deformable interface can be drastically simplified by means of long-wavelength expansions. The longwave approach is based on the assumption that the characteristic spatial scales in the directions x and y are much larger than that in the direction z. The details of the longwave approach applied to thermocapillary flows can be found in review papers [12], [13]. In the framework of the leading order approximation, which is known as lubrication approximation, pressures P_i, $i = 1, 2$, do not depend on the transverse coordinate z.

The van der Waals forces manifest themselves as external normal stresses ("disjoining pressures") imposed on each interface [1]. These pressures can be calculated using the potential energy of the van der Waals interactions

$$\Psi(H_1, H_2) = -\frac{A_{sg} - A_{s2} - A_{g1}}{12\pi H_2^2} - \frac{A_{s2}}{12\pi H_1^2} - \frac{A_{g1}}{12\pi (H_2 - H_1)^2},$$

where A_{sg}, A_{s2} and A_{g1} are Hamaker constants for the interactions between the solid substrate and the gas phase across the two liquid layers, between the solid substrate and *an infinite layer* of the liquid 2 across liquid 1, and between the gas and *an infinite layer* of the liquid 1 across liquid 2, correspondingly (see [7]). These Hamaker constants can be calculated on the basis of Lifshitz theory [1].

In the present paper, we consider solely the case where $A_{sg} - A_{s2} - A_{g1} > 0$, $A_{s2} < 0$, and $A_{g1} < 0$. That means that the effective interaction between the liquid 2/gas interface and the liquid 1/substrate interface is attractive, and it creates a van der Waals instability. However, the effective interactions between two other pairs of interfaces are repulsive, and they prevent the rupture of the layers.

Let us formulate the mathematical model in the non-dimensional form. The natural vertical length scale is the initial thickness of the lower layer, H_1^0. We fix the horizontal length scale, L^*, as

$$L^* = (H_1^0)^2 \sqrt{6\pi\sigma_1^0 / |A_{sg}|}, \tag{2}$$

which is convenient for the analysis of the instability induced by intermolecular forces [7]. We choose

$$\tau^* = \frac{\eta_1 (L^*)^4}{\sigma_1^0 (H_1^0)^3} \tag{3}$$

as a time scale and

$$p^* = \frac{\sigma_1^0 H_1^0}{(L^*)^2} \tag{4}$$

as a pressure scale.

Governing equations look as follows:

$$h_{1\tau} + \nabla \cdot \mathbf{q}_1 = 0,\ h_{2\tau} + \nabla \cdot \mathbf{q}_2 = 0, \tag{5}$$

$$\mathbf{q}_1 = f_{11}\nabla p_1 + f_{12}\nabla p_2 + \mathbf{q}_1^T,\ \mathbf{q}_2 = f_{21}\nabla p_1 + f_{22}\nabla p_2 + \mathbf{q}_2^T, \tag{6}$$

where $h_j = H_j/H_1^0$, $p_j = P_j/p^*$, $j = 1, 2$,

$$f_{11} = -\frac{1}{3}h_1^3,\ f_{12} = -\frac{1}{2}h_1^2(h_2 - h_1), f_{21} = \frac{1}{6}h_1^3 - \frac{1}{2}h_1^2 h_2,$$

$$f_{22} = (h_2 - h_1)\left[h_1^2\left(\frac{1}{2} - \frac{\eta}{3}\right) + h_1 h_2\left(-1 + \frac{2\eta}{3}\right) - \frac{\eta}{3}h_2^2\right].$$

The contributions of disjoining pressures are included:

$$p_1 = -\nabla^2 h_1 - \sigma\nabla^2 h_2 + w_1(h_1, h_2), \tag{7}$$

$$p_2 = -\sigma\nabla^2 h_2 + w_2(h_1, h_2), \tag{8}$$

$$w_1 = \frac{a_0 - a_1 - a_2}{h_2^3} + \frac{a_1}{h_1^3}, \tag{9}$$

$$w_2 = \frac{a_0 - a_1 - a_2}{h_2^3} + \frac{a_2}{(h_2 - h_1)^3}. \tag{10}$$

The non-dimensional expressions for the fluxes generated by the thermocapillary effect are:

$$\mathbf{q}_1^T = \frac{M_\perp}{2}h_1^2\nabla[d(Bih_1 - \alpha\kappa)] - \frac{M_\parallel}{2}(1+\alpha)h_1^2\mathbf{e}_x. \tag{11}$$

$$\mathbf{q}_2^T = \frac{M_\perp}{2}\{-h_2^2\nabla(d\eta\alpha\kappa) + (2h_2 - h_1)h_1\nabla\{d[Bih_1 - \tag{12}$$

$$\alpha\kappa(1-\eta)]\}\} - \frac{M_\parallel}{2}[\eta\alpha h_2^2 + (1 + \alpha - \eta\alpha)h_1(2h_2 - h_1)]\mathbf{e}_x.$$

Here

$$M_\perp = \frac{\alpha_1(T_s - T_g)}{\sigma_1^0}\left(\frac{L^*}{H_1^0}\right)^2, \tag{13}$$

$$M_\parallel = \frac{\alpha_1 A L^*}{\sigma_1^0}\left(\frac{L^*}{H_1^0}\right)^2 \tag{14}$$

are the modified transverse and longitudinal Marangoni numbers, correspondingly,

$$Bi = \frac{qH_1^0}{\kappa_2} \tag{15}$$

is the Biot number, and

$$d = [\kappa + Bi(1-\kappa)h_1 + Bi\kappa h_2]^{-1}, \tag{16}$$

$\eta = \eta_1/\eta_2$, $\kappa = \kappa_1/\kappa_2$, $\sigma = \sigma_2^0/\sigma_1^0$, $\alpha = \alpha_2/\alpha_1$,

$$a_0 = \mathrm{sign}(A_{sg}),\ a_1 = \frac{A_{s2}}{|A_{sg}|},\ a_2 = \frac{A_{g1}}{|A_{sg}|}. \tag{17}$$

The system of equations (5) contains ten nondimensional parameters: $M_\perp$, $M_\parallel$, Bi, σ, α, η, κ, a_0, a_1, and a_2.

First, let us consider the linear stability of a parallel thermocapillary flow with flat interfaces $h_1 = 1$, $h_2 = h$. The growth rate $\lambda(\mathbf{k})$ of a disturbance with a wavevector $\mathbf{k} = (k_x, k_y)$ is determined by the dispersion relation $det(N - \lambda I) = 0$, where I is a 2×2 unit matrix, and

$$N = k^2 A + k^4 B + M_\perp k^2 C + ik_x M_\| D. \tag{18}$$

The 2×2 matrices in the right-hand side of (18), which do not depend on the wavevector $\mathbf{k}$, describe the influence of van der Waals forces (A), surface tension (B), transverse heating (C) and longitudinal heating (D). In the absence of the thermocapillary flow ($M_\| = 0$), the growth rate $\lambda = \lambda(k^2)$ does not depend on the orientation of the wavevector. In the case $\beta = k_x M_\| / M_\perp k^2 = O(1)$ the growth rate depends significantly on the wavevector orientation. Let us emphasize that this is a peculiarity of a two-layer system, because in the case of a one-layer system D would be a number rather than a matrix, and the term containing k_x does not influence the growth rate. However, an asymptotic analysis of the dispersion relation in the limit $\beta \gg 1$ shows that in the *leading orders* $\lambda \sim ik_x M_\| \lambda_1 + k^2 \lambda_2 + \ldots$, where λ_1 and λ_2 are real coefficients that do not depend on $\mathbf{k}$. Thus, we come to the following paradoxical situation. The longitudinal component of the temperature gradient is the only factor that violates the rotational symmetry of the problem. However, when this component is sufficiently large, the isotropy of the problem is partially restored (the real part of the growth rate is isotropic in the leading order, while the imaginary part is not isotropic), except the region of nearly longitudinal structures that have small β even for large $M_\|$.

The investigation of the film evolution under the joint action of the thermocapillary stresses and intermolecular forces governed by full nonlinear equations (5)-(12) has been fulfilled by means of nonlinear simulations. Evolution equations (5) have been discretized by central differences for spatial derivatives and solved using an explicit scheme. Periodic boundary conditions have been applied on the boundaries of the computational region. Initial conditions for h_j, $j = 1, 2$ have been chosen in such a way that the mean value of $h_1(X, Y, 0)$ was equal to 1 and the mean value of $h_2(X, Y, 0)$ was equal to h, where $h > 1$. Small random deviations of $h_j(X, Y, 0)$ from their mean values were imposed using a code creating pseudo-random numbers. The computations have been performed in the region 240×240 using the grid 400×400.

3. Numerical Results

We have observed numerous types of film dynamics.

A typical scenario of the film decomposition due to the instability induced by the van der Waals forces includes the creation of a disordered array of droplets connected by a thin two-layer film. These droplets are subject to coarsening. Under the action of a transverse heating, $M_\perp \neq 0$, the dependence of the liquid-liquid interface hight on the distance from the droplet center becomes non-monotonic, so that this interface resembles a "volcano". In the absence of a longitudinal heating, $M_\| = 0$, the droplets do not move; the coarsening takes place due to the coalescence of large droplets caused by their growth, and the disappearance of small droplets. In the presence of a longitudinal component of the temperature gradient, the shape of the liquid-liquid interface beneath the droplets is strongly asymmetric: the crater-like depression is shifted to the front part of the droplet, while in its back part a high elevation is observed. The coarsening of droplets is accelerated by their drift due to the thermocapillary flow.

Surprisingly, under certain conditions we observe the formation of ordered structures that consist of equally sized droplets moving with equal velocities and located at equal distances from each other. Let us emphasize that the ordered droplets system is not a result of a casual monodispersity of droplets due to a specific choice of initial conditions, but a result of a natural evolution of the systems to its stable configuration. The same ordered system of droplets has been obtained from different initial conditions. Note that the deformation of the gas/liquid interface is much stronger than that of the liquid/liquid interface.

Outside the region of parameters corresponding to perfectly ordered structures, we observe a dynamical regime of "splashes" which is characterized by the coexistence of several "immortal" big droplets and a number of small droplets that are ejected by big droplets, grow with time and finally coalesce with big droplets.

Some more regimes have been found in simulations, e.g., oblique traveling waves and modulated longitudinal structures.

Under the action of the gravity forces, the ordered system of droplets is destroyed and transforms through intermediate unstable stages to a new stable configurations of the vertical stripes or the inclined stripes.

In conclusion, we have considered the evolution of a two-layer film under a joint action of thermocapillary effect and intermolecular forces in the case of an inclined temperature gradient. The governing equations have been derived in the lubrication approximation. It is found that for sufficiently large values of the horizontal temperature gradient, the real part of the linear growth rate does not depend on the direction of the wavenumber, except the case of nearly longitudinal disturbances (phenomenon of "isotropization"). Numerous types of nonlinear evolution regimes have been observed, among them ordered systems of droplets, "splashes", oblique waves, and modulated longitudinal structures. The influence of the gravity on the instabilities of non-isothermic two-layer films is investigated.

The research was partially supported by the European Union via the FP7 Marie Curie scheme [PITN-GA-2008-214919 (MULTIFLOW)].

References

[1] Israelachvili J N 1992 *Intermolecular and Surface Forces* (Academic Press, New York)
[2] Thiele U 2003 *Eur. Phys. J. E* **12** 409
[3] Pototsky A, Bestehorn M, Merkt D and Thiele U 2004 *Phys. Rev. E* **70** 025201
[4] Danov K D *et al.* 1998 *Chem. Eng. Sci.* **53** 2809
[5] Danov K D *et al.* 1998 *Chem. Eng. Sci.* **53** 2823
[6] Paunov V N *et al.* 1998 *Chem. Eng. Sci.* **53** 2839
[7] Fisher L S and Golovin A A 2005 *J Coll. Interf. Sci* **291** 515
[8] Merkt D, Pototsky A, Bestehorn M and Thiele U 2005 *Phys. Fluids* **17** 064104
[9] Bandyopadhyay D, Gulabani R and Sharma A 2005 *Ind. Eng. Chem. Res* **44** 1259
[10] Pototsky A, Bestehorn M, Merkt D and Thiele U 2005 *J. Chem. Phys* **122** 224711
[11] Limat L *et al.* 1992 *Physica D* **61** 166
[12] Davis S H 1987 *Ann. Rev. Fluid Mech.* **19** 403
[13] Oron A, Davis S H and Bankoff S G 1997 *Rev. Mod. Phys* **69** 931

Stability of MHD shear flows: Application to space physics

Michael S. Ruderman[1] **and Nikolai A. Belov**[2]

[1]Department of Applied Mathematics, University of Sheffield, Hounsfield Road, Hicks Building, Sheffield S3 7RH, United Kingdom

[2]Institute for Problems in Mechanics, Russian Academy of Sciences, Vernadsky Ave. 101, Moscow 117527, Russia

E-mail: M.S.Ruderman@sheffield.ac.uk

Abstract. Shear slows of magnetised plasmas are routinely observed in the solar atmosphere, in planetary magnetospheres, and in interplanetary space. They are also ubiquitous elements of models of remote astrophysical objects like the interacting stellar winds in binary stellar systems. Studying stability of such flows is paramount for understanding physical processes in space. The simplest shear flow is a tangential magnetohydrodynamic (MHD) discontinuity. We start our review from considering the instability of tangential MHD discontinuity (called the Kelvin-Helmholtz (KH) instability) first in incompressible plasmas, and then taking the compressibility into account. We introduce the notion of the absolute and convective instabilities. The physical behaviour of absolutely unstable flows is qualitatively different from that of convectively unstable flows. Studying the absolute and convective instabilities is based on the analysis of asymptotic behaviour (for large time) of the solution of the initial value problem. The initial value problem for a tangential discontinuity is ill-posed: the instability increment is unbounded. This implies that the absolute and convective instabilities of tangential discontinuities cannot be studied. To obtain a well-posed problem we have either to take dissipation into account, or to consider a continuous velocity profile. In both cases we obtain a surprising result: the account of either of these two effects decreases the threshold value of the velocity jump needed for instability. This phenomenon is related to negative energy waves. We show that, in both cases, the instability is the so-called negative energy instability rather than the KH instability. Finally we consider two examples of the theory application: the heliopause stability and the stability of the Earth's magnetopause.

1. Introduction

Shear slows of magnetised plasmas are routinely observed in the solar atmosphere, in planetary magnetospheres, and in interplanetary space. They are also ubiquitous elements of models of remote astrophysical objects like accretion disks near neutron stars and black holes, and the interacting stellar winds in binary stellar systems. Hence, studying stability of such flows is paramount for understanding physical processes in space.

The simplest shear flow is a planar tangential discontinuity. This is a flow where all flow parameters are constant at the two sides of a particular plane, and have jumps across the plane. In addition, the magnetic field and velocity are tangential to the plane.

The standard approach to studying stability of shear flows is the normal mode analysis. This analysis is based on the linearised magnetohydrodynamic (MHD) equations. Perturbations of all variables are taken to be proportional to $\exp(-i\omega t)$, where t is the time. As a result we obtain an eigenvalue problem for ω. The flow is unstable if there is at least one eigenvalue with the positive imaginary part, and stable otherwise. The instability that a tangential discontinuity can suffer from is called the Kelvin-Helmholtz (KH) instability.

It turns out that the normal mode analysis is not sufficient to describe the flow behaviour for large time. If the normal mode analysis predicts stability, then no other investigation is necessary. However if it predicts instability, then the behaviour of perturbations at large time can be quite different dependent on the nature of the instability. To study the behaviour of perturbations at large time we need to introduce the notion of absolute and convective instabilities.

The instability is called absolute if an initial perturbation of the flow grows exponentially with time at any fixed spatial position. It is called convective if an initial perturbation is exponentially decays with time at any spatial fixed position. The schematic picture of time evolution of an initial perturbation in absolutely and convectively unstable flows is shown in Fig. 1. We see that in the lower panel the initial perturbation is convected so quickly that it decays at any spacial position in spite that the perturbation amplitude increases. The instability character, absolute or convective, depends on the reference frame. In the lower panel the instability is absolute in the primed reference frame moving together with the growing perturbation.

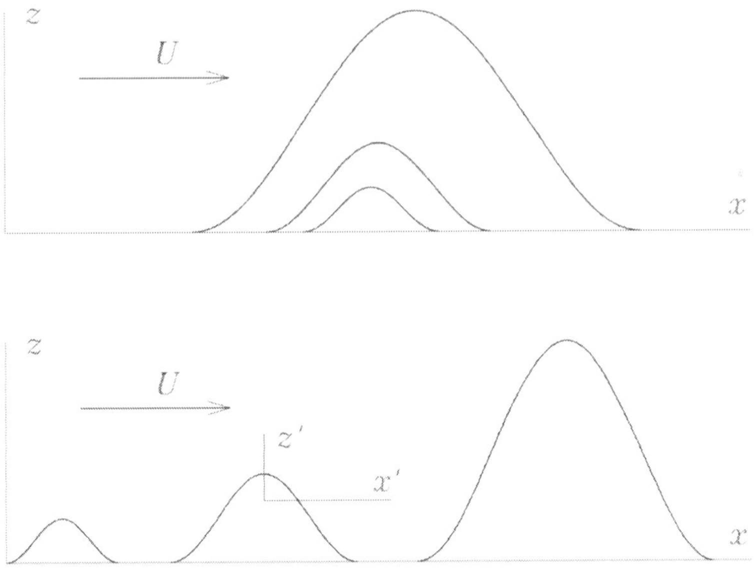

Figure 1. The schematic picture of time evolution of an initial perturbation in absolutely and convectively unstable flows. We can see that the initial perturbation grows at any spatial position in the upper panel. In the low panel it grows but simultaneously it is convected so quickly that it decays at any spatial position. However it grows at any spatial position in the primed reference frame moving together with the perturbation.

To find out if the instability is absolute or convective we have to solve the initial value problem and then to evaluate the asymptotic behavior at a fixed spatial position as $t \to \infty$. The initial value problem for a tangential discontinuity is ill-posed: the perturbation increment is unbounded because, for a perturbation harmonic in space, it is proportional to the wavenumber. This implies that we cannot study the absolute and convective instabilities of tangential discontinuities.

There are two ways to alleviate this difficulty and obtain a well-posed problem: either to take dissipation into account, or to consider a continuous velocity profile connecting two regions with constant but different flow velocities. In both cases the normal mode analysis gives a surprising from the first sight result: the account of either of these two effects not only makes the initial value problem well-posed, but also decreases the threshold value of the velocity jump needed to make the flow unstable. This phenomenon is related to the existence of negative energy waves. It can be shown that in both cases the instability is the so-called negative energy instability rather than the Kelvin-Helmholtz instability.

In this paper we give a brief review of linear stability of MHD shear flows and then describe its applications to the stability of the Earth's magnetopause and heliopause. In the next section we present the linearised MHD equations used in this paper and write down the linearised boundary conditions at the tangential discontinuity. In Sect. 3 we consider the Kelvin-Helmholtz instability of tangential discontinuities. In Sect. 4 we describe negative energy instability of tangential discontinuities in dissipative fluids. This kind of instability is also called the dissipative instability. In Sect. 5 we study the negative energy instability of shear flows with continuous velocity profile. Since this instability is related to the wave energy resonant absorption it is also called the resonant instability. In Sect. 6 we apply the general results concerning the stability of MHD shear flows to the heliopause stability. In Sect. 7 a similar application is made to the stability of the Earth's magnetopause. Sect. 8 contains the summary and conclusions.

2. Governing equations and boundary conditions

In what follows we use the MHD equations, which are the mass conservation, momentum, induction and entropy equations,

$$\frac{\partial \rho}{\partial t} + \nabla \cdot (\rho \boldsymbol{v}) = 0, \tag{1}$$

$$\rho\left(\frac{\partial \boldsymbol{v}}{\partial t} + (\boldsymbol{v}\cdot\nabla)\boldsymbol{v}\right) = -\nabla p + \frac{1}{\mu_0}(\nabla\times\boldsymbol{B})\times\boldsymbol{B} + \boldsymbol{F}_{\rm vis} + \rho\boldsymbol{g}, \tag{2}$$

$$\frac{\partial \boldsymbol{B}}{\partial t} = \nabla\times(\boldsymbol{v}\times\boldsymbol{B}) + \lambda\nabla^2\boldsymbol{B}, \tag{3}$$

$$\frac{\partial}{\partial t}\left(\frac{p}{\rho^\gamma}\right) + \boldsymbol{v}\cdot\nabla\left(\frac{p}{\rho^\gamma}\right) = (\gamma-1)\rho^{-\gamma}\left(\mathcal{L}_{\rm vis} + \frac{j^2}{\sigma} - \nabla\cdot\boldsymbol{q}\right). \tag{4}$$

Here ρ is the density, p the pressure, $\boldsymbol{v}$ the velocity, $\boldsymbol{g}$ the gravity acceleration, and $\boldsymbol{B}$ the magnetic field; μ_0 is permeability of free space, λ the coefficient of magnetic diffusion, and γ the ratio of specific heats. $\boldsymbol{F}_{\rm vis}$ is the viscous force. The most general expression for it is given by Braginskii [1],

$$\boldsymbol{F}_{\rm vis} = \nabla\cdot\mathbf{P}, \tag{5}$$

where $\mathbf{P}$ is the viscosity tensor determined by

$$\mathbf{P} = \eta_0\mathbf{P}_0 + \eta_1\mathbf{P}_1 + \eta_2\mathbf{P}_2 - \eta_3\mathbf{P}_3 - \eta_4\mathbf{P}_4, \tag{6}$$

$$\mathbf{P}_0 = \left(\boldsymbol{bb} - \tfrac{1}{3}\mathbf{I}\right)Q, \tag{7}$$

$$\mathbf{P}_1 = \nabla\boldsymbol{v} + (\nabla\boldsymbol{v})^T - \boldsymbol{bW} - \boldsymbol{Wb} + (\boldsymbol{bb}-\mathbf{I})\nabla\cdot\boldsymbol{v} + (\boldsymbol{bb}+\mathbf{I})\boldsymbol{b}\cdot\nabla(\boldsymbol{b}\cdot\boldsymbol{v}), \tag{8}$$

$$\mathbf{P}_2 = \boldsymbol{bW} + \boldsymbol{Wb} - 4(\boldsymbol{bb})\boldsymbol{b}\cdot\nabla(\boldsymbol{b}\cdot\boldsymbol{v}), \tag{9}$$

$$\mathbf{P}_3 = \tfrac{1}{2}\{(\boldsymbol{b}\times\nabla)\boldsymbol{v} + [(\boldsymbol{b}\times\nabla)\boldsymbol{v}]^T + \nabla(\boldsymbol{b}\times\boldsymbol{v}) + [\nabla(\boldsymbol{b}\times\boldsymbol{v})]^T - \boldsymbol{b}(\boldsymbol{b}\times\boldsymbol{W}) - (\boldsymbol{b}\times\boldsymbol{W})\boldsymbol{b}\}, \tag{10}$$

$$\mathbf{P}_4 = \boldsymbol{b}(\boldsymbol{b}\times\boldsymbol{W}) + (\boldsymbol{b}\times\boldsymbol{W})\boldsymbol{b}, \tag{11}$$

$$Q = 3\boldsymbol{b}\cdot\nabla(\boldsymbol{b}\cdot\boldsymbol{v}) - \nabla\cdot\boldsymbol{v}, \qquad \boldsymbol{W} = \nabla(\boldsymbol{b}\cdot\boldsymbol{v}) + \boldsymbol{b}\cdot\nabla\boldsymbol{v}. \tag{12}$$

In Eqs. (6)–(12) $\eta_0, \dots, \eta_4$ are the coefficients of viscosity, I is the unit tensor, $\boldsymbol{b} = \boldsymbol{B}/B$ is the unit vector in the magnetic field direction, the symbol $\boldsymbol{ab}$ indicates the dyadic product of vectors $\boldsymbol{a}$ and $\boldsymbol{b}$ (in particular, one of the vectors can be ∇), and the superscript 'T' indicates a transposed tensor.

In the entropy equation $\mathcal{L}_{\rm vis}$ is the rate of mechanical energy dissipation due to viscosity given by

$$\mathcal{L}_{\rm vis} = \mathsf{P} : \nabla \boldsymbol{v} = P_{ij}\frac{\partial v_i}{\partial x_j}. \tag{13}$$

It can be shown that $\mathcal{L}_{\rm vis} > 0$. The quantity $\boldsymbol{j}$ is the electric current density related to the magnetic field by

$$\boldsymbol{j} = \frac{1}{\mu_0}\nabla \times \boldsymbol{B}, \tag{14}$$

and σ is the conductivity related to λ by $\lambda = 1/(\mu_0\sigma)$. Finally, $\boldsymbol{q}$ is the thermal energy flux given by [1]

$$\boldsymbol{q} = -\kappa_\parallel \nabla_\parallel T - \kappa_\perp \nabla_\perp T - \kappa_\wedge \boldsymbol{b} \times \nabla T, \tag{15}$$

where $\kappa_\parallel$, $\kappa_\perp$ and $\kappa_\wedge$ are the parallel, perpendicular and skew coefficients of thermal conduction, $\nabla_\parallel = \boldsymbol{b}(\boldsymbol{b}\cdot\nabla)$, $\nabla_\perp = \nabla - \boldsymbol{b}(\boldsymbol{b}\cdot\nabla)$, and T is the temperature related to ρ and p by the ideal gas law,

$$p = \frac{k_B}{m}\rho T \tag{16}$$

with k_B and m being the Boltzmann constant and the mean mass per particle, respectively.

Let us now write down the boundary conditions at a tangential discontinuity. We assume that the equation of a perturbed discontinuity is $f(x, y, z, t) = 0$ in Cartesian coordinates x, y, z. If an elementary fluid volume is at this surface at the initial moment of time, then it remains on this surface forever. Let $x = x(t)$, $y = y(t)$, $z = z(t)$ be the equations of this volume trajectory. Then we have an identity $f(x(t), y(t), z(t), t) \equiv 0$. Differentiating this identity with respect to time we obtain

$$\frac{\partial f}{\partial x}\frac{dx}{dt} + \frac{\partial f}{\partial y}\frac{dy}{dt} + \frac{\partial f}{\partial z}\frac{dz}{dt} + \frac{\partial f}{\partial t} = 0.$$

Taking into account that the derivatives of $x(t)$, $y(t)$ and $z(t)$ are equal to the components of the elementary volume velocity, we rewrite this equation as

$$\boldsymbol{v}\cdot\nabla f + \frac{\partial f}{\partial t} = 0. \tag{17}$$

Since $\nabla f = \boldsymbol{n}|\nabla f|$ where $\boldsymbol{n}$ is the normal vector to the surface of tangential discontinuity, it follows from this equation that the normal component of the velocity is continuous at the discontinuity. Introducing the jump of a function $g(x, y, z, t)$ through the discontinuity,

$$[\![g(\boldsymbol{r})]\!] = \lim_{\varepsilon \to +0}[g(\boldsymbol{r} + \varepsilon\boldsymbol{n}) - g(\boldsymbol{r} - \varepsilon\boldsymbol{n})]$$

where $\boldsymbol{r} = (x, y, z)$ is the position vector, and the point $\boldsymbol{r}$ in this expression is on the surface of tangential discontinuity, we eventually write the kinematic boundary condition as

$$[\![v_n]\!] = 0, \tag{18}$$

where $v_n = \boldsymbol{n}\cdot\boldsymbol{v}$.

In the presence of magnetic field and viscosity the stress tensor is given by

$$\mathsf{T} = -\left(p + \frac{B^2}{2\mu_0}\right)\mathsf{I} + \frac{\boldsymbol{BB}}{\mu_0} + \mathsf{P}.$$

Since at the tangential discontinuity $\boldsymbol{n}\cdot\boldsymbol{B} = 0$, the stress at the surface of tangential discontinuity is

$$\boldsymbol{n}\cdot\mathbf{T} = -\left(p + \frac{B^2}{2\mu_0}\right)\boldsymbol{n} + \boldsymbol{n}\cdot\mathbf{P}.$$

The stress has to be continuous at the surface of tangential discontinuity, which gives us the dynamic boundary condition,

$$\left[\!\left[\left(p + \frac{B^2}{2\mu_0}\right)\boldsymbol{n} - \boldsymbol{n}\cdot\mathbf{P}\right]\!\right] = 0. \tag{19}$$

In an inviscid fluid this condition reduces to the continuity of the total pressure, kinetic plus magnetic,

$$\left[\!\left[p + \frac{B^2}{2\mu_0}\right]\!\right] = 0. \tag{20}$$

In what follows we study only the linear stability of MHD shear flows, so that we use the linearised equations. We assume that the equilibrium flow is stationary. In a dissipative fluid we, in general, need external forces and energy sources to maintain a stationary flow. We assume that such forces and energy sources are implicitly present. Linearising Eqs. (1)–(4) we obtain

$$\frac{\partial\rho'}{\partial t} + \nabla\cdot(\rho_0\boldsymbol{v}' + \boldsymbol{v}_0\rho') = 0, \tag{21}$$

$$\rho_0\left(\frac{\partial\boldsymbol{v}'}{\partial t} + (\boldsymbol{v}_0\cdot\nabla)\boldsymbol{v}' + (\boldsymbol{v}'\cdot\nabla)\boldsymbol{v}_0\right) + \rho'(\boldsymbol{v}_0\cdot\nabla)\boldsymbol{v}_0$$
$$= -\nabla p' + \frac{1}{\mu_0}[(\nabla\times\boldsymbol{B}')\times\boldsymbol{B}_0 + (\nabla\times\boldsymbol{B}_0)\times\boldsymbol{B}'] + \boldsymbol{F}'_{\text{vis}} + \rho'\boldsymbol{g}, \tag{22}$$

$$\frac{\partial\boldsymbol{B}'}{\partial t} = \nabla\times(\boldsymbol{v}_0\times\boldsymbol{B}' + \boldsymbol{v}'\times\boldsymbol{B}_0) + \lambda\nabla^2\boldsymbol{B}', \tag{23}$$

$$\frac{\partial}{\partial t}(p' - c_S^2\rho') + \rho_0^\gamma\boldsymbol{v}_0\cdot\nabla\left(\frac{p' - c_S^2\rho'}{\rho_0^\gamma}\right) + \rho_0^\gamma\boldsymbol{v}'\cdot\nabla\frac{p_0}{\rho_0^\gamma} + \gamma\rho_0^{\gamma-1}\rho'\boldsymbol{v}_0\cdot\nabla\frac{p_0}{\rho_0^\gamma}$$
$$= (\gamma - 1)\left(\mathcal{L}'_{\text{vis}} + \frac{2\boldsymbol{j}'\cdot\boldsymbol{j}_0}{\sigma} - \nabla\cdot\boldsymbol{q}'\right), \tag{24}$$

where c_S is the sound speed determined by $c_S^2 = \gamma p_0/\rho_0$, and the subscript '0' and the prime indicate an equilibrium quantity and the perturbation respectively. Under the assumption that T_0 is constant the expression for $\boldsymbol{q}'$ is given by Eq. (15) with T' substituted for T. We do not give the expressions for $\boldsymbol{F}'_{\text{vis}}$ and $\mathcal{L}'_{\text{vis}}$. They will be given later for particular cases. The linearized ideal gas law takes the form

$$p' = \frac{k_B}{m}\rho_0 T' + \frac{k_B}{m}\rho' T_0. \tag{25}$$

To derive the linearized boundary conditions we assume that the unperturbed tangential discontinuity coincides with the $z = 0$ plane, while the equation of the perturbed discontinuity can be written as $z = \zeta(x, y, t)$. Then, in particular, $w_0 = 0$, where w is the z-component of the velocity, and

$$\boldsymbol{n} = \frac{\boldsymbol{e}_z - \nabla\zeta}{\sqrt{1 + |\nabla\zeta|^2}},$$

where $\boldsymbol{e}_z$ in the unit vector in the z-direction. Then the linearized kinematic boundary condition obtained from Eq. (18) takes the form

$$[\![w' - \boldsymbol{v}_0 \cdot \nabla\zeta]\!] = 0. \tag{26}$$

To relate $\boldsymbol{v}'$ and ζ we use the linearized Eq. (17). As a result we obtain

$$w' = \frac{\partial\zeta}{\partial t} + \boldsymbol{v}_0 \cdot \nabla\zeta. \tag{27}$$

This equation is valid at both sides of the discontinuity, so that we can use it instead of Eq. (26). Note that both Eqs. (26) and Eq. (27) are valid not at $z = \zeta$, but at $z = 0$.

Using the fact that $\boldsymbol{n}' = -\nabla\zeta$ we write the linearised dynamic boundary condition obtained from Eq. (19) as

$$\begin{aligned}\Bigg[\!\!\Bigg[\left(p' + \frac{\boldsymbol{B}_0 \cdot \boldsymbol{B}'}{\mu_0}\right)\boldsymbol{e}_z - \boldsymbol{e}_z \cdot \mathbf{P}' - \left(p_0 + \frac{B_0^2}{2\mu_0}\right)\nabla\zeta + \mathbf{P}_0 \cdot \nabla\zeta \\ + \zeta\left[\boldsymbol{e}_z \frac{\partial}{\partial z}\left(p_0 + \frac{B_0^2}{2\mu_0}\right) - \boldsymbol{e}_z \cdot \frac{\partial \mathbf{P}_0}{\partial z}\right]\Bigg]\!\!\Bigg] = 0.\end{aligned} \tag{28}$$

This equation is also valid at $z = 0$. In an inviscid fluid this equation is simplified to

$$\left[\!\!\left[p' + \frac{\boldsymbol{B}_0 \cdot \boldsymbol{B}'}{\mu_0} + \zeta\frac{\partial}{\partial z}\left(p_0 + \frac{B_0^2}{2\mu_0}\right)\right]\!\!\right] = 0, \tag{29}$$

where we have taken into account that Eq. (20) is valid for the equilibrium quantities.

The linearized MHD equations and boundary conditions presented in this section will be used in what follows to study the stability of MHD shear flows.

3. Stability of MHD tangential discontinuities

In this section we study the stability of planar MHD tangential discontinuity in an ideal fluid. It is assumed that the unperturbed discontinuity coincides with the $z = 0$ plane, and the equilibrium quantities are constant above and below this plane, while they, in general, have jumps through the discontinuity. The equilibrium velocity and magnetic field are parallel to the xy-plane.

First we assume that the fluid is incompressible. In that case we use equation

$$\nabla \cdot \boldsymbol{v}' = 0 \tag{30}$$

instead of Eq. (21), and we do not need Eq. (24). Since the equilibrium quantities are independent of t, x and y, we can take perturbations of all quantities proportional to $\exp(i\boldsymbol{k}\cdot\boldsymbol{r} - i\omega t)$, where the wave vector $\boldsymbol{k}$ is parallel to the xy-plane. As a result we reduce the system of Eqs. (22), (23) and (30) to the system of ordinary differential equations. Eliminating all variables from this system in favor of w' we arrive at

$$\frac{d^2 w'}{dz^2} - k^2 w' = 0. \tag{31}$$

In addition we have

$$p' + \frac{\boldsymbol{B}_0 \cdot \boldsymbol{B}'}{\mu_0} = \frac{i\rho_0[(\omega - \boldsymbol{k}\cdot\boldsymbol{v}_0)^2 - (\boldsymbol{k}\cdot\boldsymbol{V}_A)^2]}{k^2(\omega - \boldsymbol{k}\cdot\boldsymbol{v}_0)}\frac{dw'}{dz}, \tag{32}$$

Journal of Physics: Conference Series **216** (2010) 012016 doi:10.1088/1742-6596/216/1/012016

where $\boldsymbol{V}_A = \boldsymbol{B}_0/\sqrt{\mu_0\rho_0}$ is the Alfvén velocity. Using this result we rewrite the dynamic boundary condition (29) as

$$\left[\!\left[\frac{\rho_0[(\omega - \boldsymbol{k}\cdot\boldsymbol{v}_0)^2 - (\boldsymbol{k}\cdot\boldsymbol{V}_A)^2]}{\omega - \boldsymbol{k}\cdot\boldsymbol{v}_0} \frac{dw'}{dz} \right]\!\right] = 0, \tag{33}$$

where we have taken into account that, in the absence of gravity, the derivative of the total pressure with respect to z is continuous as the tangential discontinuity. Using Eq. (27) we write the kinematic boundary condition as

$$\left[\!\left[\frac{w'}{\omega - \boldsymbol{k}\cdot\boldsymbol{v}_0} \right]\!\right] = 0. \tag{34}$$

It follows from Eq. (31) that

$$w' = \begin{cases} w_1 e^{kz}, & z < 0, \\ w_2 e^{-kz}, & z > 0. \end{cases} \tag{35}$$

Substituting this result in Eq. (33) and Eq. (34) we obtain the system of two linear homogeneous algebraic equations with respect to w_1 and w_2. This system has a non-trivial solution only if its determinant is zero. This gives the dispersion equation

$$\begin{aligned}(\rho_{01} + \rho_{02})\omega^2 \; - \; 2\boldsymbol{k}\cdot(\rho_{01}\boldsymbol{v}_{01} + \rho_{02}\boldsymbol{v}_{02})\omega + \rho_{01}[(\boldsymbol{k}\cdot\boldsymbol{v}_{01})^2 - (\boldsymbol{k}\cdot\boldsymbol{V}_{A1})^2] \\ + \; \rho_{02}[(\boldsymbol{k}\cdot\boldsymbol{v}_{02})^2 - (\boldsymbol{k}\cdot\boldsymbol{V}_{A2})^2] = 0,\end{aligned} \tag{36}$$

where the subscripts '1' and '2' refer to the regions $z < 0$ and $z > 0$ respectively. The discriminant of this equation is equal to

$$(\rho_{01} + \rho_{02})[\rho_{01}(\boldsymbol{k}\cdot\boldsymbol{V}_{A1})^2 + \rho_{02}(\boldsymbol{k}\cdot\boldsymbol{V}_{A2})^2] - \rho_{01}\rho_{02}[\boldsymbol{k}\cdot(\boldsymbol{v}_{01} - \boldsymbol{v}_{02})]^2. \tag{37}$$

When it is positive the roots of Eq. (36) are real and the perturbation is neutrally stable. However when the discriminant is negative, there are two complex conjugate roots and the perturbation grows exponentially. Hence, the instability criterion is

$$[\boldsymbol{k}\cdot(\boldsymbol{v}_{01} - \boldsymbol{v}_{02})]^2 > \frac{(\rho_{01} + \rho_{02})[\rho_{01}(\boldsymbol{k}\cdot\boldsymbol{V}_{A1})^2 + \rho_{02}(\boldsymbol{k}\cdot\boldsymbol{V}_{A2})^2]}{\rho_{01}\rho_{02}}. \tag{38}$$

We see that the instability criterion is independent of the magnitude of $\boldsymbol{k}$, however it does depend on its direction. The tangential discontinuity is unstable as soon as the instability criterion (38) is satisfied for at least one $\boldsymbol{k}$. In particular, when $\boldsymbol{B}_{01} \parallel \boldsymbol{B}_{02}$ and $\boldsymbol{v}_{01} - \boldsymbol{v}_{02}$ is not parallel to $\boldsymbol{B}_{01}$, all perturbations with $\boldsymbol{k} \perp \boldsymbol{B}_{01}$ are unstable for any value of $|\boldsymbol{v}_{01} - \boldsymbol{v}_{02}|$. On the other hand, when $\boldsymbol{B}_{01} \parallel \boldsymbol{B}_{02} \parallel (\boldsymbol{v}_{01} - \boldsymbol{v}_{02})$, Eq. (38) reduces to

$$|\boldsymbol{v}_{01} - \boldsymbol{v}_{02}|^2 > V_{\mathrm{KH}}^2 \equiv \frac{(\rho_{01} + \rho_{02})(\rho_{01}V_{A1}^2 + \rho_{02}V_{A2}^2)}{\rho_{01}\rho_{02}}. \tag{39}$$

The quantity V_{KH} is the so-called Kelvin-Helmholtz threshold velocity, and the instability that takes place when Eq. (39) is satisfied is the Kelvin-Helmholtz (KH) instability. The stability of an MHD tangential discontinuity in an ideal incompressible fluid was first studied by Syrovatskii [2] (see also Chandrasekhar [3]).

Later Fejer [4] generalized this analysis to take compressibility into account. He obtained the following dispersion equation written in the reference frame where $\boldsymbol{v}_{01} = 0$:

$$(\rho_1/m_1)[V_{A1}^2 k^2 \cos^2(\phi - \psi_1) - \omega^2] + (\rho_2/m_2)[V_{A2}^2 k^2 \cos^2(\phi - \psi_2) - (\omega - Uk\cos\phi)^2] = 0, \tag{40}$$

where $U = |\boldsymbol{v}_{02}|$, ψ_1 and ψ_2 are the angles between $\boldsymbol{k}$ and $\boldsymbol{B}_{01}$ and $\boldsymbol{B}_{02}$ respectively, and ϕ is the angle between $\boldsymbol{k}$ and $\boldsymbol{v}_{02}$. The quantities m_1 and m_2 are given by

$$m_1^2 = k^2 - \frac{\omega^4}{\omega^2(c_{S1}^2 + V_{A1}^2) - c_{S1}^2 V_{A1}^2 k^2 \cos^2(\phi - \psi_1)}, \tag{41}$$

$$m_2^2 = k^2 - \frac{(\omega - Uk\cos\phi)^4}{(\omega - Uk\cos\phi)^2(c_{S2}^2 + V_{A2}^2) - c_{S1}^2 V_{A1}^2 k^2 \cos^2(\phi - \psi_1)}. \tag{42}$$

The quantities m_1 and m_2 have to satisfy the conditions $\Re(m_1) > 0$ and $\Re(m_2) > 0$, where $\Re$ indicates the real part of a quantity.

Equation (40) can be reduced to the tenth-order algebraic equation. Written in the dimensionless form this equation contains 8 parameters, which makes its comprehensive analysis extremely difficult. Usually only particular cases with the reduced number of parameter are studied. One such particular case will be considered in Sect. 6.

If we rewrite the dispersion equation (36) or a more general equation (40) in terms of ω/k, then we obtain equations that are independent of k. Then it follows that the instability increment is proportional to k, and thus its is unbounded. This implies that the initial value problem for an MHD tangential discontinuity is ill-posed. In the next two sections we will discuss how to modify the equilibrium flow to make the initial value problem well-posed.

4. Negative energy waves and dissipative instabilities

4.1. Derivation of governing equation

In this section we consider the stability problem for an MHD tangential discontinuity in a viscous fluid. It is well known that tangential discontinuities cannot exist in viscous fluids: viscosity smears out the discontinuity. To overcome this difficulty Ruderman and Goossens [5] considered an MHD tangential discontinuity in a fluid viscous at one side of the discontinuity and inviscid at the other side. They also assumed that the fluid is incompressible. The third assumption that they made concerns the form of the viscosity tensor. The coefficients $\eta_0, \dots, \eta_4$ in Eq. (6) depend on the dimensionless parameter $\tau_i\omega_i$, where τ_i is the collisional time and ω_i the cyclotron frequency of the ions. When $\tau_i\omega_i \ll 1$ the viscosity tensor becomes isotropic, i.e. independent of the magnetic field direction, and reduces to a relatively simple form,

$$P_{ij} = \eta_0 \left(\frac{\partial v_i}{\partial x_j} + \frac{\partial v_j}{\partial x_i} \right). \tag{43}$$

Then

$$\boldsymbol{e}_z \cdot \mathsf{P}' = \eta_0 \left(\frac{\partial \boldsymbol{v}'}{\partial z} + \nabla w' \right).$$

Using this result, assuming that the fluid is viscous in the region $z < 0$ and inviscid in the region $z > 0$, and recalling that the equilibrium quantities are constant in $z < 0$ and $z > 0$, we reduce Eq. (28) to two boundary conditions at $z = 0$,

$$p_1' + \frac{\boldsymbol{B}_{01} \cdot \boldsymbol{B}_1'}{\mu_0} - 2\eta_0 \frac{\partial w'}{\partial z} = p_2' + \frac{\boldsymbol{B}_{02} \cdot \boldsymbol{B}_2'}{\mu_0}, \tag{44}$$

$$\frac{\partial \boldsymbol{v}_{\perp 1}'}{\partial z} + \nabla_\perp w_1' = 0, \tag{45}$$

where $\boldsymbol{v}_\perp' = \boldsymbol{v}' - w'\boldsymbol{e}_z$ and $\nabla_\perp = \nabla - \boldsymbol{e}_z \partial/\partial z$. We use the reference frame where the fluid in the region $z < 0$ is at rest, while the fluid in the region $z > 0$ is moving with the speed $U_0 > 0$ in the x-direction. It is assumed that $\boldsymbol{B}_{01}$ and $\boldsymbol{B}_{02}$ are also parallel to the x-direction. Then the boundary condition Eq. (27) gives

$$w_1' = \frac{\partial \zeta}{\partial t}, \qquad w_2' = \frac{\partial \zeta}{\partial t} + U_0 \frac{\partial \zeta}{\partial x}. \tag{46}$$

Now we take perturbations of all variables proportional to $\exp(i\boldsymbol{k}\cdot\boldsymbol{r} - i\omega t)$, where the wave vector $\boldsymbol{k}$ is parallel to the xy-plane. Then after long but straightforward calculation we obtain from the system of Eqs. (22), (23) and (30)

$$p_1' + \frac{B_{01}B_{x1}'}{\mu_0} = A_1(t,\boldsymbol{k})e^{kz}, \quad p_2' + \frac{B_{02}B_{x2}'}{\mu_0} = A_2(t,\boldsymbol{k})e^{-kz}. \tag{47}$$

In addition, we obtain the equation for w_2',

$$\frac{\partial^2 w_2'}{\partial t^2} + 2ik_xU_0\frac{\partial w_2'}{\partial t} - k_x^2(U_0^2 - V_{A2}^2)w_2' = \frac{k}{\rho_{02}}\left(\frac{\partial A_2}{\partial t} + iU_0k_xA_2\right). \tag{48}$$

Using the second equation in Eq. (46) we derive from Eq. (48) that

$$A_2 = \frac{\rho_{02}}{k}\left[\frac{\partial^2\zeta}{\partial t^2} + 2ik_xU_0\frac{\partial\zeta}{\partial t} - k_x^2(U_0^2 - V_{A2}^2)\zeta\right]. \tag{49}$$

Now we proceed to studying the motion in the region $z < 0$. Let us introduce the Reynolds number $R_e = \rho_{01}V_{A1}/k\eta_0$ and assume that $R_e \gg 1$. Then viscosity is only important in a thin boundary layer of thickness of the order of $R_e^{-1/2}/k$ near the discontinuity, while the motion outside the dissipative layer can be described by the ideal MHD equations. This analysis inspired us to use matched asymptotic expansions to study the motion in the region $z < 0$. To describe the motion in the boundary layer we introduce the stretching variable $Z = R_e^{1/2}z$. Then, using Eq. (47), we obtain the following equation valid in the boundary layer,

$$\frac{\partial^2 w_1'}{\partial t^2} + k_x^2V_{A1}^2w_1' = -\frac{k}{\rho_{01}}\frac{\partial A_1}{\partial t}e^{R_e^{-1/2}kZ} + \frac{\bar\eta_0}{\rho_{01}}\left(\frac{\partial^2 w_1'}{\partial Z^2} - R_e^{-1}k^2w_1'\right), \tag{50}$$

where $\bar\eta_0 = R_e\eta_0$. Using Eq. (30) to eliminate $\boldsymbol{v}_{\perp 1}'$ from Eq. (45) we obtain the boundary condition for w_1' valid at $Z = 0$,

$$\frac{\partial^2 w_1'}{\partial Z^2} - R_e^{-1}k^2w_1' = 0. \tag{51}$$

It immediately follows from Eqs. (46), (50) and (51) that

$$A_1 = -\frac{\rho_{01}}{k}\left(\frac{\partial^2\zeta}{\partial t^2} + k_x^2V_{A2}^2\zeta\right). \tag{52}$$

Outside the boundary layer we neglect viscosity and obtain

$$\frac{\partial^2 w_1'}{\partial t^2} + k_x^2V_{A2}^2w_1' = -\frac{k}{\rho_{01}}\frac{\partial A_1}{\partial t}e^{kz}. \tag{53}$$

It follows from Eqs. (52) and (53) that

$$w_1' = \frac{\partial\zeta}{\partial t}e^{kz}. \tag{54}$$

Let us look for the solution to Eqs. (50) in the form

$$w_1' = \left(\frac{\partial \zeta}{\partial t} + \bar{w}\right) e^{R_e^{-1/2} kZ}. \tag{55}$$

It follows from Eq. (46) that $\bar{w} = 0$ at $Z = 0$. In addition, it follows from the matching condition with the external solution given by Eq. (54) that $\bar{w} \to 0$ as $Z \to -\infty$. It is straightforward to show that $\bar{w} \to 0$ as $t \to \infty$ on the time scale kV_{A1}. In what follows we are only interested in slow processes with the time scale $R_e kV_{A1}$, so we can safely take $\bar{w} = 0$. Then we obtain from Eq. (55) that

$$\frac{\partial w_1'}{\partial z} = k\frac{\partial \zeta}{\partial t} \tag{56}$$

at $z = 0$. Substituting Eqs. (49), (52) and (56) in Eq. (44) we obtain the governing equation for ζ,

$$(\rho_{01} + \rho_{02})\frac{\partial^2 \zeta}{\partial t^2} + 2i\rho_{02}k_x U_0 \frac{\partial \zeta}{\partial t} + \rho_{02}k_x^2(U_c^2 - U_0^2)\zeta = -2\eta_0 k^2 \frac{\partial \zeta}{\partial t}, \tag{57}$$

where

$$U_c^2 = \frac{\rho_{01}V_{A1}^2 + \rho_{02}V_{A2}^2}{\rho_{02}} = \frac{\rho_{01}V_{\rm KH}^2}{\rho_{01} + \rho_{02}}. \tag{58}$$

Equation (57) is written for the Fourier transform of ζ with respect to x and y. Making the inverse Fourier transform we obtain the equation written in physical variables,

$$(\rho_{01} + \rho_{02})\frac{\partial^2 \zeta}{\partial t^2} + 2\rho_{02}U_0 \frac{\partial^2 \zeta}{\partial t \partial x} + \rho_{02}(U_0^2 - U_c^2)\frac{\partial^2 \zeta}{\partial x^2} = 2\eta_0 \nabla^2 \frac{\partial \zeta}{\partial t}. \tag{59}$$

4.2. Negative energy instability

Multiplying Eq. (59) by $\partial\zeta/\partial t$ and integrating the result with respect to x and y yields

$$\frac{dE}{dt} = -D, \tag{60}$$

where

$$E = \frac{1}{2}\iint \left[(\rho_{01} + \rho_{02})\left(\frac{\partial \zeta}{\partial t}\right)^2 + \rho_{02}(U_0^2 - U_c^2)\left(\frac{\partial \zeta}{\partial x}\right)^2\right] dx\, dy, \tag{61}$$

$$D = 2\eta_0 \iint \left[\left(\frac{\partial^2 \zeta}{\partial t \partial x}\right)^2 + \left(\frac{\partial^2 \zeta}{\partial t \partial y}\right)^2\right] dx\, dy. \tag{62}$$

Here the integration limits are $\pm\infty$ for perturbations decaying at infinity, and $\pm\frac{1}{2}L_x$, $\pm\frac{1}{2}L_y$ for periodic perturbations with the periods with respect to x and y equal to L_x and L_y respectively.

Consider the solution to Eq. (59) in the form $\zeta = \zeta_0 \Re[\exp(i\boldsymbol{k}\cdot\boldsymbol{r} - i\omega t)]$. Substituting this expression in Eq. (59) we obtain the dispersion equation

$$(\rho_{01} + \rho_{02})\omega^2 - 2\rho_{02}U_0\omega k_x + \rho_{02}(U_0^2 - U_c^2)k_x^2 = -2i\eta_0 k^2 \omega. \tag{63}$$

It is easy to see that the ratio of the right hand side of this equation to its left-hand side is of the order of $R_e^{-1} \ll 1$. This observation inspires us to use the regular perturbation method to find the approximate solution to Eq. (63). As a result we obtain

$$\omega = \omega_\pm \equiv C_\pm k \mp i\chi_\pm k^2, \tag{64}$$

Journal of Physics: Conference Series **216** (2010) 012016 doi:10.1088/1742-6596/216/1/012016

where

$$C_{\pm} = \frac{\rho_{02}U_0 \pm \sqrt{\rho_{01}\rho_{02}(V_{\rm KH}^2 - U_0^2)}}{\rho_{01} + \rho_{02}}, \quad \chi_{\pm} = \frac{2\eta_0 C_{\pm}}{\sqrt{\rho_{01}\rho_{02}(V_{\rm KH}^2 - U_0^2)}}. \tag{65}$$

Note that Eqs. (64) and (65) are only valid when $U_0 < V_{\rm KH}$, i.e. when there is no KH instability.

It follows from Eq. (65) that $C_+ > 0$, $\chi_+ > 0$, and the wave mode with $\omega = \omega_+$ is stable. The wave mode with $\omega = \omega_-$ is also stable when $C_- < 0$, however it is unstable when $C_- > 0$. This condition reduces to

$$U_0 > U_c. \tag{66}$$

This is the criterion for the onset of dissipative instability. Since, in accordance with Eq. (58), $U_c < V_{\rm KH}$, the dissipative instability occurs for the values of the shear velocity U_0 smaller than those needed for the onset of the KH instability. Hence, contrary to our intuition, dissipation destabilises tangential discontinuities.

The quantity E can be considered as the wave energy. Since $D > 0$, it follows from Eq. (61) that dissipation causes the wave energy to decrease. Let us calculate E for the wave modes with $\omega = \omega_{\pm}$. Since the imaginary part of ω is much smaller than its real part, we can take $\omega \approx C_{\pm}k$. Then we easily obtain

$$E_{\pm} = \pm 2\pi^2 k_x k_y^{-1} \zeta_0^2 C_{\pm} \sqrt{\rho_{01}\rho_{02}(V_{\rm KH}^2 - U_0^2)}. \tag{67}$$

Now we assume that ζ_0 is a slowly varying function of time. We see that $E_+ > 0$, so that the decrease of E_+ caused by dissipation results in the decrease of the wave amplitude ζ_0. When $U_0 < U_c$ we have $E_- > 0$, and the amplitude of the mode with $\omega = \omega_-$ also decreases due to dissipation. However $E_- < 0$ when $U_0 > U_c$. Hence, when E_- decreases due to dissipation, its absolute value increases. As a result, the wave amplitude ζ_0 increases.

Obviously the wave energy is frame-dependent. The wave mode is unstable in the presence of dissipation when its energy is negative in the reference frame where the fluid domain where the dissipation takes place is at rest. In the problem studied in this section this is just the reference frame where the fluid below the discontinuity is at rest.

When $U_0 = 0$ the wave mode with $\omega = \omega_+$ propagates in the positive x-direction, and the wave mode with $\omega = \omega_-$ propagates in the negative x-direction. These two modes are called forward and backward propagating respectively. We see that the backward propagating mode becomes a negative energy wave as soon as the shear velocity U_0 is so large that the mode phase velocity changes the sign and the mode propagates in the positive x-direction.

Waves with negative energy are called negative energy waves. We see that the dissipative instability of MHD tangential discontinuities is related to the existence of negative energy waves.

Ruderman et al. [6] considered another limiting case where $\tau_i\omega_i \gg 1$, which is satisfied in sufficiently rarefied plasmas in presence of strong magnetic field. In this case the first term in the expression for the viscosity tensor (6) strongly dominates four other terms, so that we can take $\mathbf{P} \approx \eta_0 \mathbf{P}_0$. Similarly the first term in the expression for the heat flux (15) strongly dominates two other terms, so that we can take $\boldsymbol{q} \approx -\kappa_{\parallel}\nabla_{\parallel}T$. Viscosity and thermal conduction of this form are called strongly anisotropic. Interestingly, in contrast to isotropic viscosity and thermal conduction, strongly anisotropic viscosity and thermal conduction do not prevent the existence of velocity and temperature discontinuities. Ruderman et al. [6] studied dissipative instabilities of MHD tangential discontinuities in fluids with strongly anisotropic viscosity and thermal conduction and arrived at the same conclusion as Ruderman and Goossens [5]: dissipation destabilizes MHD tangential discontinuities.

Equation (64) shows that the increment of dissipative instability is proportional to k^2. However this result was obtained in the long-wavelength approximation. Really, the condition

$R_e \gg 1$ used when deriving Eq. (64) is equivalent to $k \ll \rho_{01} V_{A1}/\eta_0$. Kikina [7] studied exactly the same problem as Ruderman and Goossens [5], however without magnetic field. She found that the instability increment tends to a limiting value when $k \to \infty$. Hence, the initial value problem for a tangential discontinuity in a fluid viscous at one side of the discontinuity and inviscid at the other side is well-posed. This fact enables Ruderman et al. [8] to study the absolute and convective instabilities of such a discontinuity. Although an analysis for MHD discontinuities similar to one made by Kikina [7] has not been carries out yet, there is no reason to expect that the result will be different. Hence, we can anticipate that the initial value problem for an MHD tangential discontinuity in the presence of dissipation is also well-posed.

5. Stability of shear flows with continuous velocity profile

5.1. Flow with continuous velocity and discontinuous magnetic field

When the velocity of a shear flow is continuous, it is, in general impossible to obtain the dispersion equation determining the instability threshold and increment. One exception is a flow with piece-wise linear continuous velocity profile in an incompressible fluid. The examples of studying stability of such flows can be found in Chandrasekhar [3]. Ruderman and Brevdo [9] generalized the problem to include the magnetic filed. They considered the following equilibrium flow. In Cartesian coordinates x, y, z there is constant magnetic field with the magnitude B_0 in the region $z > 0$, while the region $z < 0$ is magnetic-free. The magnetic field is parallel to the x-axis. The density is equal to ρ_{01} for $z < 0$ and ρ_{02} for $z > 0$. There is also a shear flow in the x-direction with the magnitude U_0 given by

$$U_0 = \begin{cases} U_m, & z \leq -h, \\ -U_m(z/h), & -h < z < 0, \\ 0, & z \geq 0. \end{cases} \tag{68}$$

It is assumed that the fluid is incompressible. Then the linearised hydrodynamic equations describing the flow perturbations in the region $z < 0$ can be reduced to one equation for the z-component of the velocity perturbation, $\nabla^2 w' = 0$. As a result, taking perturbations of all quantities proportional to $\exp(i\boldsymbol{k}\cdot\boldsymbol{r} - i\omega t)$ with $\boldsymbol{k}$ parallel to the xy-plane, we obtain the system of ordinary differential equations that can be solved analytically. Using this analytical solution and the boundary conditions at the discontinuity at $z = 0$ we obtain the dispersion equation

$$2\Omega^3 + 2(\Delta Q - \alpha)\Omega^2 + (Q - W)\Omega + QW = 0, \tag{69}$$

where

$$\alpha = kh, \quad s = \frac{\rho_{01}}{\rho_{02}}, \quad \Delta = \frac{s-1}{s+1}, \quad \Omega = \frac{\omega h}{V_m(1+\Delta)},$$
$$Q = \frac{2\alpha - 1 + e^{-2\alpha}}{2(1+\Delta)}, \quad W = \frac{\alpha^2(1-\Delta)}{M_A^2(1+\Delta)^2}, \quad M_A^2 = \frac{\mu_0\rho_2 U_m^2}{B_0^2}, \tag{70}$$

M_A being the Alfvén Mach number. The dispersion equation (69) was studied analytically and, for illustration, numerically. The main results are the following. The flow becomes unstable when $M_A > 1/(1+s)$. In the dimensional variables the instability criterion takes the form

$$U_m^2 > U_c^2 \equiv \frac{s}{(1+s)^2} V_{\rm HK}^2, \tag{71}$$

where $V_{\rm HK}$ is given by Eq. (39) with $V_{A1} = 0$. If we take $h = 0$ we obtain a particular case of MHD tangential discontinuity with the magnetic-free fluid at one side. Such a discontinuity is unstable when $U_m > V_{\rm HK}$. Since $U_c < V_{\rm HK}$, it follows that the presence of the transitional layer where the shear velocity linearly decreases from its maximum value to zero destabilizes the flow.

The second important result is that now only perturbations with $k < k_{\max}$ are unstable. As a result the instability increment is bounded and the initial value problem is well-posed.

5.2. Flow with continuous velocity and magnetic field

Let us now discuss the stability properties of MHD flows with continuous velocity and magnetic field. In this case a new phenomenon, called resonant absorption, appears. As it is well known, Alfvén waves can propagate in conducting fluids permeated by magnetic field [10]. When a conducting fluid is compressible, two additional wave modes exist: fast and slow magnetosonic waves (e.g. Goedbloed and Poedts [11]). Resonant absorption occurs when, in an inhomogeneous conducting fluid, global wave motion is locally in resonance with either Alfvén or slow magnetosonic waves at a particular magnetic surface. The resonance causes energy to build up in the vicinity of a resonance magnetic surface, at the expense of the energy in the global motion. As a result the energy of the global motion decreases. When the equilibrium is static, this energy decrease causes the global wave to decay. However when the equilibrium is steady, i.e. there is flow, negative energy waves can exist. As we have already seen in the previous section, the negative energy wave amplitude grows when its energy decreases, which can result in the instability. The negative energy wave instability related to resonant absorption is also called resonant instability.

In what follows we consider resonant instability of a shear flow of an incompressible fluid. In that case we can only have the Alfvén resonance. It occurs at the magnetic surface where the phase velocity of a global wave matches the local Alfvén velocity. This only can occur when there is the continuous variation of the Alfvén velocity. This is why we did not have the Alfvén resonance in the problem studied in the previous subsection.

The resonant instability of a shear flow of incompressible fluid was first studied by Hollweg et al. [12]. In what follows we give slightly modified version of their analysis. Let us consider a flow with the flow velocity in the x-direction of Cartesian coordinates x, y, z. The velocity magnitude is a function of z. The equilibrium magnetic field is also in the x-direction, and its magnitude as well as the fluid density are functions of z.

We assume that the equilibrium quantities vary only in the slab $|z| < a$, while they are constant in the regions $z < -a$ and $z > a$, where they are marked by the subscripts '1' and '2'. The total equilibrium pressure is constant,

$$p_0 + \frac{B_0^2}{2\mu_0} = 0. \tag{72}$$

In what follows we use the perturbation of the total pressure,

$$P' = p' + \frac{B_0 B_x'}{\mu_0}. \tag{73}$$

Now we use the system of linearised ideal MHD equations for an incompressible fluid, which consists of Eqs. (22), (23) and (30) with $\boldsymbol{F}_{\text{vis}} = 0$ and $\lambda = 0$. Eliminating all variables from this system in favour of P' and w' we obtain

$$(\omega - U_0 k_x)\frac{dP'}{dz} = i\rho_0 \left[(\omega - U_0 k_x)^2 - V_A^2 k_x^2\right] w', \tag{74}$$

$$\rho_0 \left[(\omega - U_0 k_x)^2 - V_A^2 k_x^2\right] \frac{d}{dz}\left(\frac{w'}{\omega - U_0 k_x}\right) = -ik^2 P'. \tag{75}$$

If there is z_0 such that $\omega = k_x U_0(z_0)$, then z_0 is a singular point of the system of Eqs. (74) and (75). However, this is only an apparent singularity. Really, introducing $\tilde{w} = w'/(\omega - U_0 k_x)$ and eliminating P' from Eqs. (74) and (75) we obtain the equation

$$\frac{d}{dz}\rho_0 \left[(\omega - U_0 k_x)^2 - V_A^2 k_x^2\right] \frac{d\tilde{w}}{dz} = k^2 \rho_0 \left[(\omega - U_0 k_x)^2 - V_A^2 k_x^2\right] \tilde{w}. \tag{76}$$

We see that z_0 is a regular point of this equation, so that $\tilde{w}$ is a regular function at z_0. Then it follows that w' and P' are also regular functions at z_0.

In what follows we only consider long-wavelength perturbations satisfying $ka \ll 1$. Then it follows from Eq. (74) that $dP'/dz = \mathcal{O}(ka)$ when $|z| < a$, and we can take $P' \approx$ const in this region. This approximation was first used by Hollweg and Young [13] to study resonant damping of surface waves propagating along a thin transitional layer. Using this assumption we immediately obtain from Eq. (75) that

$$w' = (\omega - U_0 k_x)\left(-ik^2 P' \int_{-a}^{z} \frac{d\tilde{w}}{\rho_0\left[(\omega - U_0 k_x)^2 - V_A^2 k_x^2\right]} + w_0'\right), \tag{77}$$

where w_0' is a constant. The integral in this expression is divergent if there is $z_A \in (-a, a)$ such that

$$|\omega - k_x U_0(z_A)| = k_x V_A(z_A), \tag{78}$$

and $z > z_A$. This condition means that, at $z = z_A$, the absolute value of the Doppler-shifted frequency of the global wave coincides with the local Alfvén frequency $V_A k_x$. As a result there is resonance between the global wave and the local Alfvén waves at $z = z_A$, which causes the global wave energy to be transferred to the local Alfvén waves in the vicinity of z_A. The point z_A is called the Alfvén resonant position. We assume that there is at most one Alfvén resonant position for any value of $\boldsymbol{k}$.

To evaluate the integral in Eq. (77) when the condition (78) is satisfied we notice that, when taking perturbations of all variables proportional to $e^{-i\omega t}$, we implicitly assumed that we made the Laplace transform. The Laplace transform is an analytic function of ω only in the half-plane $\omega_i = \Im(\omega) > \bar{\omega}_i$, where $\Im$ indicates the imaginary part of a quantity. When studying instabilities we have to take $\bar{\omega}_i$ larger than the instability increment. We will see in what follows that the condition $ak \ll 1$ implies that the instability increment is much smaller than $|\omega_r| = |\Re(\omega)|$, so that we can take $|\omega_i| \ll |\omega_r|$. Then, assuming that all the equilibrium quantities are analytic functions of z for $z \in (-a, a)$, we obtain that Eq. (78) is satisfied at $z = z_A + iz_i$, where

$$z_i = \frac{\omega_i}{k_x \Delta_A}, \qquad \Delta_A = \left.\frac{d(U_0 \pm V_A)}{dz}\right|_{z=z_A}. \tag{79}$$

Now we formally take $\omega_i \to 0$, so that $z_i \to 0$. To avoid singularity in the integral in Eq. (77) we deform the integration contour in such a way that the new contour consists of two intervals, $(-a, z_A - \epsilon)$ and $(z_A + \epsilon, z)$, and a semi-circle of radius ϵ centred at z_A, where $0 < \epsilon \ll a$. The semi-circle is below the z-axis if $z_i > 0$, and above the z-axis if $z_i < 0$. Then Eq. (77) reduces to the approximate expression

$$w' = (\omega_r - U_0 k_x)\left(\frac{-\pi k^2 P'}{2k_x \rho_A |\Delta_A| (\omega_r - U_A k_x)} - ik^2 P' \mathcal{P}\int_{-a}^{z} \frac{d\tilde{z}}{\rho_0\left[(\omega_r - U_0 k_x)^2 - V_A^2 k_x^2\right]} + w_0'\right), \tag{80}$$

where $\mathcal{P}$ indicates the principal Cauchy part of the integral and $\rho_A = \rho_0(z_A)$, $U_A = U_0(z_A)$. Recall that this expression is only valid for $z > z_A$. When $z < z_A$ the first term in the brackets is absent. In particular, $w'(-a) = w_0'(\omega_r - U_{01} k_x)$.

Since all the equilibrium quantities are constant in the regions $z < -a$ and $z > a$, the solution to Eqs. (74) and (75) in these regions is straightforward,

$$w' = w_1 e^{kz}, \quad P' = \frac{i\rho_{01}\left[(\omega - U_{01} k_x)^2 - V_{A1}^2 k_x^2\right] w_1 e^{kz}}{k(\omega - U_{01} k_x)}, \quad (z < -a), \tag{81}$$

$$w' = w_2 e^{-kz}, \quad P' = -\frac{i\rho_{02}\left[(\omega - U_{02}k_x)^2 - V_{A2}^2 k_x^2\right] w_2 e^{-kz}}{k(\omega - U_{02}k_x)}, \quad (z > a), \tag{82}$$

where w_1 and w_2 are constants. It follows from Eqs. (80)–(82) that

$$\frac{w_2}{\omega_r - U_{02}k_x} - \frac{w_1}{\omega_r - U_{01}k_x} = -k^2 P' e^{ak}\left(\frac{\pi}{2k_x\rho_A|\Delta_A|(\omega_r - U_A k_x)} + i\mathcal{P}\int_{-a}^{a}\frac{dz}{\rho_0\left[(\omega_r - U_0 k_x)^2 - V_A^2 k_x^2\right]}\right). \tag{83}$$

Recalling that P' is constant for $|z| < a$ we can obtain another expression for the left-hand side of this equation from Eqs. (81) and (82),

$$\frac{w_2}{\omega_r - U_{02}k_x} - \frac{w_1}{\omega_r - U_{01}k_x} = ikP' e^{ak}\left(\frac{1}{\rho_{02}\left[(\omega - U_{02}k_x)^2 - V_{A2}^2 k_x^2\right]} + \frac{1}{\rho_{01}\left[(\omega - U_{01}k_x)^2 - V_{A1}^2 k_x^2\right]}\right). \tag{84}$$

Comparing Eqs. (83) and (84) we obtain the dispersion equation

$$\frac{1}{\rho_{02}\left[(\omega - U_{02}k_x)^2 - V_{A2}^2 k_x^2\right]} + \frac{1}{\rho_{01}\left[(\omega - U_{01}k_x)^2 - V_{A1}^2 k_x^2\right]} = \frac{i\pi k}{2k_x\rho_A|\Delta_A|(\omega_r - U_A k_x)} - k\mathcal{P}\int_{-a}^{a}\frac{dz}{\rho_0\left[(\omega_r - U_0 k_x)^2 - V_A^2 k_x^2\right]}. \tag{85}$$

It is straightforward to estimate that the ratio of the right-hand side of this equation to its left-hand side is of the order of $ak \ll 1$. Hence we can use the regular perturbation method to calculate ω. In the first order approximation we obtain $\omega = k_x V_\pm$, where

$$V_\pm = \frac{\rho_{01}U_{01} + \rho_{02}U_{02} \pm \sqrt{\rho_{01}\rho_{02}[V_{\rm KH}^2 - (U_{02} - U_{01})^2]}}{\rho_{01} + \rho_{02}}, \tag{86}$$

and we have assumed that $|U_{02} - U_{01}| < V_{\rm KH}$, so that there is no KH instability. Proceeding to the next order approximation we eventually arrive at

$$\omega = k_x V_\pm + kk_x(i\Gamma_\pm + S_\pm), \tag{87}$$

where

$$\Gamma_\pm = \frac{1}{4\rho_A|\Delta_A|H_\pm(U_A - V_\pm)}, \quad S_\pm = \frac{1}{2H_\pm}\mathcal{P}\int_{-a}^{a}\frac{dz}{\rho_0\left[(V_\pm - U_0)^2 - V_A^2\right]}, \tag{88}$$

$$H_\pm = \frac{V_\pm - U_{02}}{\rho_{02}\left[(V_\pm - U_{02})^2 - V_{A2}^2\right]^2} + \frac{V_\pm - U_{01}}{\rho_{01}\left[(V_\pm - U_{01})^2 - V_{A1}^2\right]^2}. \tag{89}$$

The wave mode is unstable if $\Gamma_\pm > 0$. Using Eq. (85) in the first order approximation and (86) it is straightforward to show that $H_+ > 0$ and $H_- < 0$. Then the forward propagating mode is unstable when $V_+ < U_A$, i.e. when it propagates in the negative x-direction in the reference frame moving with the velocity U_A. The backward propagating mode is unstable when $V_- > U_A$, i.e. when it propagates in the positive x-direction in the reference frame moving with the velocity U_A. It can be shown that in both cases the unstable mode has negative energy in the reference frame moving with the velocity U_A. Since dissipation just takes place at $z = z_A$, we see that the instability condition is the same as in the previous section: a mode becomes unstable when it changes the propagation direction due to the effect of flow in the reference frame where the dissipation region is at rest. It turns out that this is a general condition for the negative energy instability (see, e.g., Ostrovskii et al. [14]).

Note that, in contrast to the dissipative instability, the condition that a mode has negative energy is not enough for the onset of the resonant instability. The second condition that also has to be satisfied is that there is a resonant position in the interval $(-a, a)$, i.e. that there is z_A satisfying Eq. (78). It is straightforward to see that this equation reduces to $V_+ = U_0(z_A) - V_A(z_A)$ for the forward propagating wave, and to $V_- = U_0(z_A) + V_A(z_A)$ for the backward propagating wave.

Let us consider one example. Suppose that both U_0 and V_A are linear functions of z, while ρ_0 is constant,

$$U_0 = \frac{U_{01} + U_{02}}{2} + \frac{z(U_{02} - U_{01})}{2a}, \qquad V_A = \frac{V_{A1} + V_{A2}}{2} + \frac{z(V_{A2} - V_{A1})}{2a}. \tag{90}$$

We also assume that $U_{01} < U_{02}$. The equation $V_+ = U_0(z_A) - V_A(z_A)$ reduces to

$$V_{A1} + V_{A2} + \sqrt{V_{\rm KH}^2 - (U_{02} - U_{01})^2} = (z_A/a)(U_{02} - U_{01} + V_{A1} - V_{A2}),$$

where now $V_{\rm KH}^2 = 2(V_{A1}^2 + V_{A2}^2)$. This equation has the solution $z_a \in (-a, a)$ only when

$$U_{02} - U_{01} + V_{A1} - V_{A2} > V_{A1} + V_{A2} + \sqrt{V_{\rm KH}^2 - (U_{02} - U_{01})^2}. \tag{91}$$

A necessary condition to satisfy this inequality is $U_{02} - U_{01} > 2V_{A2}$, which is compatible with $U_{02} - U_{01} < V_{\rm KH}$ only if $V_{A1} > V_{A2}$. Let us assume that this condition is satisfied. Then Eq. (91) reduces to $U_{02} - U_{01} > U_c$, where

$$U_c = V_{A1} + V_{A2}. \tag{92}$$

It is straightforward to see that $U_c < V_{\rm KH}$. It is also straightforward to show that $V_+ < U_A$ when $U_{02} - U_{01} > U_c$, so that the mode is unstable.

A similar analysis shown that the backward propagating mode can be unstable only if $V_{A1} < V_{A2}$. Then the instability condition is the same. Hence, summarizing, the condition for the onset of the resonant instability is

$$U_{02} - U_{01} > U_c. \tag{93}$$

The unstable wave mode is the forward propagating mode when $V_{A1} > V_{A2}$, and the backward propagating mode when $V_{A1} < V_{A2}$.

In accordance with Eq. (87) the increment of the resonant instability is proportional to kk_x. However, this result was obtained in the long wavelength approximation. If we extend the analysis to perturbations with arbitrary wavelength, we obtain the instability saturation. Moreover, this result is valid even when $U_{02} - U_{01} > V_{\rm KH}$. Hence, the initial value problem for a shear flow with continuous velocity profile is well-posed.

6. Heliopause stability

The solar system is moving with the supersonic speed of about 26 km/s with respect to the local interstellar medium. The sun produces a permanent supersonic flow of plasma with the speed of about 500 km/s called the solar wind. The collision of two supersonic flows results in a region of their interaction called the heliospheric interface. The heliospheric interface is bounded by the bow shock at the interstellar side, and by the termination shock at the solar wind side. Inside the heliospheric interface there is a surface called heliopause. The heliopause separates the flow of the interstellar plasma decelerated at the bow shock from the flow of the solar wind plasma decelerated at the termination shock. A schematic picture of the heliospheric interface is shown in Fig. 2. The model of the heliospheric interface with two shocks was first suggested and developed by Baranov et al. [15] in the approximation of infinitely thin interface. Then Baranov et al. [16, 17] improved this model by taking the finite thickness of the heliospheric interface into account. After that it was further developed by the effort of many researchers (see the review papers by Baranov [18, 19]).

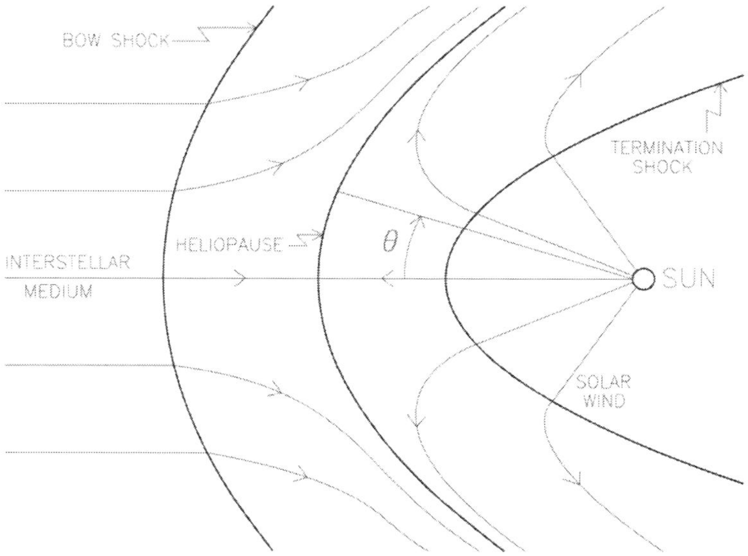

Figure 2. The schematic picture of the heliospheric interface. The lines with arrows are the streamlines.

The problem of the heliopause stability is very important for understanding physical processes in the heliospheric interface and proper interpretation of the observational data. The heliopause instability can result in turbulization of the plasma flow in the heliospheric interface, which can seriously affect the propagation of electromagnetic waves from the interstellar medium to the inner heliosphere. To our knowledge the first review of different forms of instabilities that can operate at the heliopause was given by Fahr et al. [20]. After that the problem of the heliopause stability remains an important problem in the heliospheric physics.

In the absence of magnetic filed the heliopause is unconditionally unstable. However, in accordance with the observational data, there is magnetic field in the interstellar medium with the magnitude of a few microgauss (μG). This magnetic field can stabilize the heliopause. Fahr et al. [20] applied the stability analysis for MHD tangential discontinuities to the near flanks of the heliopause ($\theta \lesssim 30°$, see Fig. 2), where the plasma flow can be considered as approximately incompressible. They carried out the local analysis assuming that the wavelengths of perturbations are much smaller than the characteristic size of the heliospheric interface. This assumption enabled them to consider the heliopause as a plane tangential discontinuity.

Ruderman and Fahr [21] generalized the analysis by Fahr et al. [20] and took the compressibility into account. As a result their analysis is not restricted to the near flans of the heliopause and is applicable to the whole heliopause. In their analysis Ruderman and

Fahr [21] used the dispersion equation (40) assuming that there is magnetic field only at one side of the discontinuity, i.e. $\boldsymbol{B}_{02} = 0$. To simplify the analysis they took into account that the ration of plasma densities at the two sides of the heliopause is small, $\rho_{02}/\rho_{01} \sim 0.01$, and used the regular perturbation method with this ratio as a small parameter. Their main results can be summarized as follows. The magnetic field can stabilize a part of the heliopause only if it is parallel to the flow velocity. The typical values of the magnetic filed magnitude and plasma number density in the interstellar medium are $B_{01} \approx 3$ μG and $n \approx 10^5$ m^{-3}. This gives $V_{A1} \approx 21$ km/s. Using these values Ruderman and Fahr [21] found $V_{\rm KH} \approx 210$ km/s. The speed of the flow of the solar wind compressed at the termination shock at the heliopause grows when θ increases, and it does not exceed 75 km/s for $\theta \lesssim 30°$ (e.g. Baranov et al. [17]). Thus, in accordance with the analysis by Ruderman and Fahr [21], the interstellar magnetic field can stabilize the near flanks of the heliopause corresponding to $\theta \lesssim 30°$.

These results were reconsidered by Ruderman and Brevdo [9]. In accordance with their analysis the transitional layer with the continuous velocity variation at the solar wind side of the heliopause would destabilize the flow. It follows from Eq. (71) that, for $s \approx 0.01$, $U_m \approx 0.1 V_{\rm KH} \approx 21$ km/s. Hence, if such a layer exists, then the heliopause is stable only when the solar wind speed near the heliopause is smaller then or of the order of 21 km/s, which corresponds to a small fraction of the heliopause.

7. Magnetopause stability

7.1. *Resonant instability of magnetopause*

The Earth magnetic field has the shape of a dipole. The pressure imposed by the solar wind distorts this magnetic field. However the Earth magnetic field prevents the direct penetration of the solar wind in the atmosphere. It stops the solar wind at the distance of about 15 Earth radii in the solar direction. As a result a magnetic cavity which is nearly free of the interplanetary plasma is formed. This cavity is called the magnetosphere, and its boundary is called the magnetopause. The magnetopause has the bullet shape with a long tail in the anti-solar direction. In simplest models of the magnetosphere the magnetopause is considered as an MHD tangential discontinuity. Studying the stability of the magnetopause is very important for understanding physical processes in the magnetosphere. In particular, the instability of the magnetopause can result in turbulization of the flow near the magnetopause, which can facilitate the indirect penetration of the solar wind in the magnetosphere.

The magnetic field in the magnetotail region is quasi-parallel to the solar wind flow, so that it can stabilise the magnetopause in this region. In the first studies of the magnetopause stability the threshold velocity of the solar wind needed for the onset of the KH instability was calculated. However later it was realised that the heliopause can be unstable at much smaller solar wind speed due to resonant instability. This possibility was studied by Ruderman and Write [22]. They studied the stability of the flow schematically shown in Fig. 3. In the equilibrium the magnetopause coincides with the xy-plane in Cartesian coordinates x, y, z. The flow velocity and magnetic field are in the x-direction. The flow speed is zero in the region $z < 0$ (i.e. in the magnetosphere), and is equal to constant quantity U in the region $z > 0$ (this region is called the magnetosheath). The magnetic field is equal to zero in the region $z > 0$, and to constant quantity B in the region $z < 0$. The pressure is equal to constant quantity p in the region $z > 0$. The magnetic pressure, $B^2/2\mu_0$, in the magnetosphere is much larger than the plasma pressure. This observation enables us to neglect the plasma pressure in the region $z < 0$. The density is equal to constant quantity ρ_2 in the magnetosheath. In the magnetosphere it is equal to constant quantity ρ_1 for $z < -a$, and it varies in the slab $-a < z < 0$. As a result the Alfvén speed V_A also varies in the slab $-a < z < 0$. It is assumed that $V_A(z)$ is a linear function.

If we take $a \to 0$, we obtain the MHD tangential discontinuity at $z = 0$. For the particular

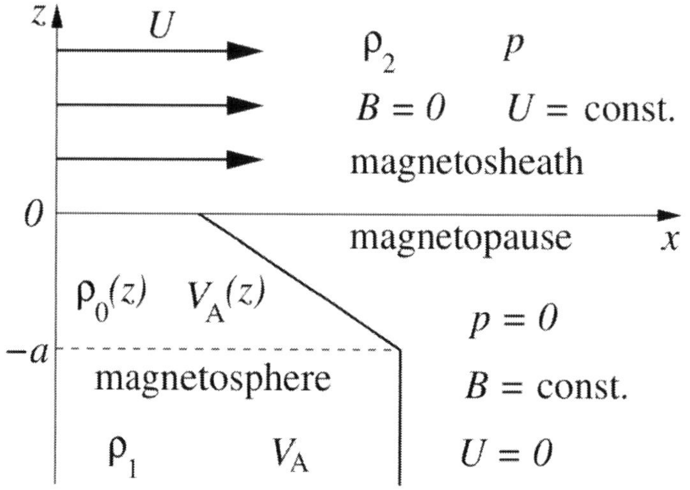

Figure 3. The schematic picture of the flow near the magnetopause. The solid piece-wise linear line shows the Alfvén speed profile.

equilibrium considered in this section the dispersion equation (40) reduces to

$$\frac{\rho_1(\omega^2 - V_A^2 k_x^2)}{m_1} + \frac{\rho_2(\omega - U k_x)^2}{m_2} = 0 \tag{94}$$

with m_1 and m_2 given by

$$m_1 = \sqrt{k^2 - \frac{\omega^2}{V_A^2}}, \qquad m_2 = \sqrt{k^2 - \frac{(\omega - U k_x)^2}{c_S^2}}, \tag{95}$$

where $\Re(m_{1,2}) > 0$ and $c_S^2 = \gamma p/\rho_2$. Introducing the dimensionless variables

$$\Omega = \frac{\omega}{kV_A}, \quad M = \frac{U}{V_A}, \quad \beta = \frac{c_S^2}{V_A^2},$$

rewriting Eq. (94) in the dimensionless form, squaring the obtained equation, and using the relation $\rho_1/\rho_2 = 2\beta/\gamma$ that follows from the total pressure balance at the magnetopause, we arrive at

$$\gamma^2(\Omega^2 - 1)(\Omega - M\cos\varphi)^4 = 4\beta(\Omega^2 - \cos^2\varphi)^2[(\Omega - M\cos\varphi)^2 - \beta], \tag{96}$$

where φ is the angle between the wavevector $\boldsymbol{k}$ and the x-direction. We have to be cautious when analysing this equation because it can contain spurious roots corresponding to the minus sign between the two terms on the left-hand side of Eq. (94). Equation (96) is a sixth-order polynomial equation containing three free parameters, M, β and φ (the ratio of specific heats γ can be considered as fixed). It is a difficult task to study this equation in the general case.

McKenzie [23] studied Eq. (96) in two particular cases. In both cases the left-hand side of Eq. (96) reduces to the product of quadratic and quartic polynomials. The first case corresponds to perturbations propagating along the equilibrium magnetic field ($\varphi = 0$), and the second to perturbations propagating at the angle $\varphi = \arcsin\frac{\gamma}{2}$. However studying particular cases does not provide us with the KH threshold velocity V_{KH}.

To make analytical progress Ruderman and Wright [22] assumed that $\rho_1/\rho_2 \ll 1$, which is equivalent to $\beta \ll 1$. This assumption is fairly well satisfied for at least a part of the tail's magnetopause. Their asymptotic analysis resulted in $M_{\rm KH} = 1$, or $V_{\rm KH} = V_A$.

When $M < 1$ and $\beta \ll 1$ Eq. (96) has only two non-spurious roots, Ω_+ and Ω_-, corresponding to the forward and backward propagating surface waves. If we now take the transitional layer into account but restrict our analysis to the long waves ($ak \ll 1$), then it can be shown that the backward propagating wave can be resonantly unstable. Ruderman and Wright [22] have shown that the two conditions for the onset of the resonant instability, namely, that the backward propagating wave is a negative energy wave, and that there is a resonant position $z_A \in (-a, 0)$, are satisfied when $M > M_c$ provided that $V_A(0)$ is sufficiently small, where

$$M_c = \frac{1}{\gamma}\sqrt{2\beta\left[\sqrt{\gamma^2 + \cos^4\varphi} - \cos\varphi^2\right]}. \tag{97}$$

We see that $M_c = \mathcal{O}(\beta^{1/2}) \ll 1 = M_{\rm KH}$, so that the resonant instability occurs for the flow velocity magnitude much smaller than that needed for the onset of the KH instability.

The analysis of the resonant instability of the magnetopause has been extended to include different additional effects like the finite plasma pressure in the magnetosphere, and to relax the assumption $\beta \ll 1$ (see, e.g., Erdélyi and Taroyan [24] and Taroyan and Erdélyi [25, 26, 27]). These studies confirmed that the resonant instability of the heliopause occurs for the flow velocity magnitude substantially smaller than the KH threshold.

7.2. Absolute and convective instabilities of magnetopause

The parts of the magnetopause that are situated at the solar side of the Earth are called the near flanks. While the magnetic field in the tail region of the magnetopause is almost parallel to the flow velocity, it is almost perpendicular to the flow velocity in the near flanks of the magnetopause. This implies that the near flans should be subject to the KH instability for any velocity magnitude in the magnetosheath, in particular, because the magnetic field does not affect perturbations propagating perpendicular to the magnetic field. Then we can expect that the near flanks of the magnetopause would be destroyed by the instability, and there would be a region of highly turbulent flow instead of this part of the magnetopause. However observations onboard of satellites clearly show that the near flanks of the heliopause do exist, and they are not perturbed very much.

A viable conjecture that can resolve this apparent contradiction between the theory and observations is that the near flanks of magnetopause are only convectively unstable. In that case the unstable perturbations are convected from the near flank regions to the far flanks. Then they can be either stabilized by the quasi-parallel magnetic filed in the magnetotail, or continue to grow and eventually destroy the magnetotail at large distances from the Earth. To verify this conjecture Wright et al. [28] studied the absolute and convective instabilities of the near flanks of the magnetopause. They considered the following equilibrium flow. In Cartesian coordinates x, y, z the flow velocity is in the y-direction, and the magnetic field is in the z-direction. The flow velocity in the magnetosheath ($x > 0$) is equal to the constant value U_0 except the boundary layer $0 < x < a$ where it monotonically decreases from U_0 to 0. The magnetosheath is magnetic-free ($B = 0$), and the density in the magnetosheath is equal to constant quantity ρ_2. The sound speed in the magnetosheath is c_S. The magnetopause coincides with the yz-plane. The density and magnetic field in the magnetosphere ($x < 0$) are constant and equal to ρ_1 and B respectively, so that the Alfvén speed is also constant and equal to V_A. There is also a reflecting boundary in the magnetosphere a distance $h \gg a$ from the magnetopause. This boundary mimics the region of sharp increase of magnetic field closer to the Earth that reflects magnetosonic waves. The sketch of the equilibrium state is shown in Fig. 4.

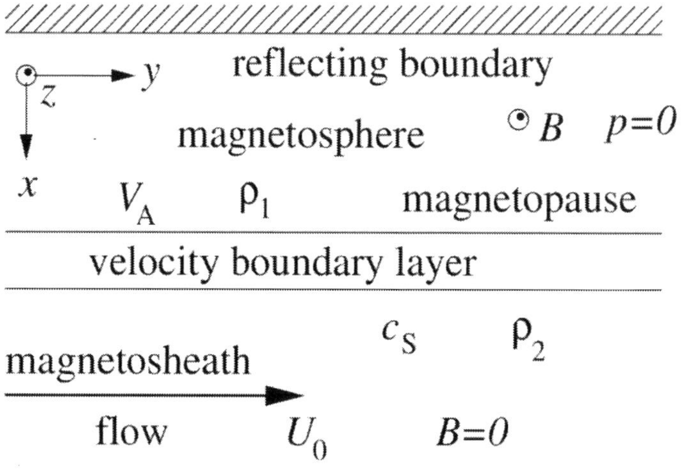

Figure 4. The schematic picture of the flow near the near flank of magnetopause.

Only perturbations independent of z were studied, so that the perturbations of all quantities were functions of x, y and t. The initial value problem was solved. Using the Fourier transform with respect to y and the Laplace transform with respect to t reduces the system of linearised ideal MHD equations to the system of ordinary differential equations with x as the independent variable. This system was solved numerically to determine the dispersion equation relating the frequency, ω, and the wave number, k, of the normal modes. This equation has the form $D(\omega, k) = 0$. After that perturbations of all variables can be written in the form of the inverse Laplace-Fourier transform. In particular the perturbed surface of the the magnetopause is determined by the equation $x = \zeta(y, t)$, where

$$\zeta(y,t) = \int_{i\sigma-\infty}^{i\sigma+\infty} \int_{-\infty}^{\infty} \frac{T(\omega,k)}{D(\omega,k)} e^{i(ky-\omega t)} dk\, d\omega. \tag{98}$$

Here $T(\omega, k)$ depends on the initial conditions, and $(i\sigma-\infty,\ i\sigma+\infty)$ is the Bromwich integration contour, and σ is chosen in such a way that this contour is above all singularities of function $D(\omega, k)$ considered as function of ω. This condition has to be satisfied for any fixed k. Since the instability increment is unbounded in the case of tangential discontinuity, the perturbation of such a discontinuity cannot be written in this form. However, due to the presence of the velocity boundary layer, the instability increment of the equilibrium flow considered in this subsection is bounded, $\Im(\omega) \le \omega_{im}$ for any k. Hence, it is enough to take $\sigma > \omega_{im}$ in Eq. (98).

The study of the absolute and convective instabilities is based of calculating the asymptotic behaviour of $\zeta(y, t)$ as $t \to \infty$. The instability is absolute when $\zeta(y, t)$ grows exponentially with time, and convective when $\zeta(y, t) \to 0$ as $t \to \infty$. To calculate this asymptotic behavior Wright et al. [28] used Briggs' [29] method first developed in plasma physics and then extended to hydrodynamics. This method is based on the extension of function $D(\omega, k)$ to the complex k-plane and the use of methods of complex analysis. The main result obtained by Wright et al. [28] is that, for any reasonable values of the flow parameters, the shear flow at the near flanks of the heliopause is only convectively unstable in the terrestrial reference frame.

Wright et al. [28] also studied the so-called signalling problem. This problem concerns propagation of perturbations excited by an external driver at a fixed spatial position. They found that the distance from the driver where the amplitude of a convectively unstable perturbation increases e-times (the so-called e-fold distance) is sufficiently large, so that we can expect to observe large-amplitude perturbations only in the magnetotail.

Wright et al. [28] investigated the absolute and convective instabilities for a very restricted rage of parameters. Mills et al. [30] extended the analysis to a much wider range of parameters. They confirmed the conclusion made by Wright et al. [28] that the near flunks of the magnetopause are only convectively unstable. Wright et al. [31] studied the excitation of magnetic field line resonances by growing and convected wave packets, and by spatially amplifying waves excited by an external driver. They concluded that excitation of field line resonances by spatially amplifying waves is more efficient than the excitation by wave packets.

8. Summary and conclusions

In this brief review we have discussed the linear theory of stability of MHD shear flows. We started our review from considering the stability of MHD tangential discontinuities. The main conclusions that we have made are the following. The magnetic field can stabilize a tangential discontinuity, however its ability to do this depends on the mutual direction of the field and the flow velocity. In particular, the magnetic field parallel to the flow velocity can stabilize the discontinuity. In general, the tangential discontinuity becomes unstable when the velocity jump across the discontinuity exceeds the Kelvin-Helholtz (KH) instability threshold $V_{\rm KH}$. This threshold depends on the magnetic field magnitude and direction as well as densities at the two sides of the discontinuity. For particular values of this parameters it can be zero. In particular, $V_{\rm KH} = 0$ when the magnetic field is perpendicular to the flow velocity. Another important conclusion is that the instability increment of an MHD tangential discontinuity is proportional to the wave vector, so that it is unbounded. This means that the initial value problem for the perturbed tangential discontinuity is ill-posed. This, in particular, makes studying the absolute and convective instabilities of MHD tangential discontinuities impossible.

There are two ways to make the initial value problem for a perturbed MHD shear flow well-posed: either to take dissipation into account, or to consider shear flows with continuous velocity profile. As an example of the first approach we considered the stability of an MHD tangential discontinuity in an incompressible fluid viscous at one side of the discontinuity and inviscid at the other side. The account of viscosity makes the instability increment bounded, so that the initial value problem for perturbations is now well-posed. In addition, contrary to our intuition, it destabilizes the discontinuity, so that it becomes unstable when the velocity jump is large than $U_c < V_{\rm KH}$. We have shown that this phenomenon is related to the existence of negative energy waves. This kind of instability is often called the negative energy wave instability. Since it is also related to dissipation, it has the second name: the dissipative instability.

As an example of the second approach we considered an MHD shear flow in an incompressible fluid with continuous velocity profile. In this case the instability increment is also bounded and the initial value problem is well-posed. We studied two different flows. In the first flow we still have an MHD tangential discontinuity, however only the density and magnetic field have jumps across it, while the velocity is continuous. The magnetic field is present only at one side of the discontinuity, while the fluid is magnetic-free at the other side. The velocity varies from zero to its maximum value in a layer bounded at one side by the discontinuity. Once again the existence of this transitional layer not only makes the instability increment bounded, but also destabilises the flow. But in this case this effect is not related to negative energy waves because in this problem there is no energy sink.

In the second flow there is no discontinuity, so that not only the velocity but also the density and magnetic field are continuous. The stability of this flow was studied in the long wavelength approximation, i.e. under the assumption that the perturbation wavelength is much larger than the thickness of the layer where the equilibrium quantities vary. In this approximation there are two surface waves propagating along the layer provided that the variation of the shear velocity is smalled than $V_{\rm KH}$. One of this two waves can become a negative energy wave. If an additional condition is satisfied, this wave is in resonance with local Alfvén waves at the resonant position inside the inhomogeneous layer. The resonance results in the decrease of the surface wave due to the energy transfer to the local Alfvén waves. Since the surface wave is a negative energy wave, this energy decrease causes its amplitude to grow exponentially. Since this negative energy wave instability is related to the existence of resonance, it also has the second name: the resonant instability.

In the last two sections we considered the application of theoretical results to the stability of the heliopause and magnetopause. We discussed the possibility to stabilise the heliopause by the interstellar magnetic field. The conclusion is that the typical interstellar magnetic field can only stabilise the near flanks of the heliopause. If the heliopause is modelled as an MHD tangential discontinuity, then the portion of the heliopause that can be stabilised corresponds to the polar angle $\theta \lesssim 30°$. However, if we add a transitional layer from the solar wind side of the heliopause where the velocity varies continuously, than the stable part of the heliopause shrinks dramatically.

We discussed the resonant instability of the magnetopause tail regions and showed that they can become unstable for the values of the plasma velocity in the magnetosheath much smaller than $V_{\rm KH}$. In the near flanks of the magnetopause the magnetic field is quasi-perpendicular to the velocity, so that it cannot stabilise this portion of the magnetopause. However observations show that the near flanks of the heliopause exist and remain almost unperturbed. We showed that this apparent contradiction between theory and observations can be resolved if we study the absolute and convective instabilities of the magnetopause. It turns out that the near flanks of the magnetopause are only convectively unstable. As a result, growing wave packets are convected away of this region before their amplitudes become sufficiently large. Even if perturbations are excited by an external driver at a fixed spatial position, the spatial growth rate of these perturbations is sufficiently small, so that we can expect to observe large-amplitude perturbations only in the magnetotail.

Acknowledgement

This work was performed in the framework of FCP "Kadry" (1.5) project - goskontrakt No. 02.740.11.5025.

References

[1] Braginskii S I 1965 *Review of Plasma Physics* **1** 205
[2] Syrovatskii S I 1957 *Usp. Fiz. Nauk* **62** 247
[3] Chandrasekhar S 1961 *Hydrodynamic and Hydromagnetic Stability* (Oxford: Clarendon Press)
[4] Fejer J A 1964 *Phys. Fluids* **7** 499
[5] Ruderman M S and Goossens M 1995 *J. Plasma Phys.* **54** 149
[6] Ruderman M S, Verwichte E, Erdélyi R and Goossens M 1996 *J. Plasma Phys.* **56** 285
[7] Kikina N G 1967 *Sov. Phys. Acoust.* **13** 184
[8] Ruderman M S, Brevdo L and Erdélyi R 2004 *Proc. Roy. Soc. Lond. A* **460** 847
[9] Ruderman M S and Brevdo L 2006 *Astron. Astrophys.* **448** 1177
[10] Alfvén H 1942 *Nature* **3805** 405
[11] Goedbloed J P and Poedts S 2004 *Principles of Magnetohydrodynamics* (Cambridge: University Press)
[12] Hollweg J V, Yang G, Cadez V M and Gakovic B 1990 *Astrophys. J.* **349** 335
[13] Hollweg J V and Yang G 1988 *J. Geophys. Res.* **93** 5423
[14] Ostrovskii L A, Rybak S A and Tsimring L Sh 1986 *Sov. Phys. Usp.* **29** 1040
[15] Baranov V B, Krasnobaev K V and Kulikovskii A G 1970 *Sov. Phys. Dokl.* **15** 791
[16] Baranov V B, Krasnobaev K V and Ruderman M S 1976 *Astrophys. Space Sci.* **41** 481
[17] Baranov V B, Lebedev M G and Ruderman M S 1979 *Astrophys. Space Sci.* **66** 441
[18] Baranov V B 1990 *Space Sci. Rev.* **52** 89
[19] Baranov V B 2009 *Space Sci. Rev.* **142** 23
[20] Fahr H J, Neutch W, Grzedzielski S, Macek W and Ratkiewicz R 1986 *Space Sci. Rev.* **43** 329
[21] Ruderman M S and Fahr H J 1993 *Astron. Astrophys.* **275** 635
[22] Ruderman M S and Wright A N 1998 *J. Geophys. Res.* **103** 26,573

[23] McKenzie 1970 *Planet. Space. Sci.* **18** 1
[24] Erdélyi R and Taroyan Y 2003 *J. Geophys. Res.* **108** Article No. 1043
[25] Taroyan Y and Erdélyi R 2002 *Phys. Plasmas* **9** 3121
[26] Taroyan Y and Erdélyi R 2003 *Phys. Plasmas* **10** 266
[27] Taroyan Y and Erdélyi R 2003 *J. Geophys. Res.* **108** Article No. 1301
[28] Wright A N, Mills K J, Ruderman M S and Brevdo L 2000 *J. Geophys. Res.* **105** 385
[29] Briggs R J 1964 *Electron-stream interaction with plasmas* (Cambridge, MA: MIT Press)
[30] Mills K J, Longbottom A W, Wright A N and Ruderman, M S 2000 *J. Geophys. Res.* **105** 27,685
[31] Wright A N, Mills K J, Longbottom A W and Ruderman M S 2002 *J. Geophys. Res.*, **107** Article No. 1242

Journal of Physics: Conference Series **216** (2010) 012017 doi:10.1088/1742-6596/216/1/012017

Influence of spatial heterogeneities on spreading dynamics

Nikos Savva and Serafim Kalliadasis
Department of Chemical Engineering, Imperial College London, London SW7 2AZ, UK

Abstract. Consider the motion of a two-dimensional droplet of a partially wetting fluid over topographical substrates. Assuming that spreading occurs in the Stokes regime, a singular perturbation procedure yields a set of integro-differential equations for the locations of the two droplet fronts. By analyzing these equations, we illustrate a number of intriguing features that are not present when the substrate is flat, such as the existence of multiple equilibria which allows for a hysteresis-like effect on the apparent contact angle. Additionally, we demonstrate a stick-slip behavior of the contact line as it moves along the local variations and the interesting possibility of a relatively brief recession of one of the contact lines.

1. Introduction

The influence of spatial heterogeneities on droplet spreading is arguably a much less theoretically studied subject compared to spreading over ideally flat substrates (e.g. [1, 2, 3, 4] and the references therein) even though they are known to impact the dynamics significantly regardless of their size. Specific spatial substrate variations may be desired in some manufacturing processes, but more often than not small-scale spatial defects are unavoidable and manifest themselves as 'noise' in the system. The corresponding substrates are commonly called 'rough' and have been the subject of the majority of experimental studies on the influence of spatial heterogeneities on spreading dynamics (e.g. Schwartz & Tejada [5] and Semal *et al.* [6]).

The few theoretical works on droplets over topographical substrates have been devoted primarily to equilibrium configurations. Wenzel was the first to deduce an expression for the apparent contact angle on rough substrates based on simplistic energy arguments, that could be verified in experiments with relatively large contact angles [7, 8, 9], but is nevertheless violated for the grooved substrates considered in the present study as demonstrated in the recent experiments of Chung *et al.* [10]. Axisymmetric droplet equilibria for substrates with concentric grooves were studied by Johnson & Dettre [11] and Huh & Mason [12] while Cox [13] generalized their work to include arbitrary, three-dimensional substrates. The last author in particular investigated the case of an infinite fluid wedge and was primarily concerned with the existence of multiple droplet equilibria. His work suggested that the hysteresis that can be induced by the substrate features would be higher for parallel-grooved, two-dimensional substrates. He also conjectured that contact line spreading normal to the grooves will be harder than spreading in all other directions and that it will exhibit a stick-slip behavior. Evidence for such behavior was reported in the exploratory experiments by Oliver *et al.* [14] and more recently by Chung *et al.* [10], mentioned earlier. The methodologies applied in the above studies were limited in scope however, principally because only equilibrium configurations were considered and quantitative statements about the dynamics were not made. As a matter of fact, dynamics has received much less attention compared to equilibrium, with a few notable exceptions which examined time-dependent problems numerically. We note, for example, the work of Gaskell *et al.* [15], who used the computationally advantageous precursor-film model to study droplet spreading over a three-dimensional topography by solving the long-wave approximation and the work of Gramlich *et al.* [16] who examined the motion of a single contact line without a precursor film over a two-dimensional topography by solving the Stokes equations.

Here we undertake a systematic investigation of both the equilibrium shapes and the motion of two-dimensional droplets over topographical substrates. The restriction to two dimensions implies that there are no transverse variations and the contact line is essentially treated as a set of two points. The inclusion of three-dimensional effects is much more involved, but we believe that our model is able to capture many of the qualitative features of the spreading behavior of a three-dimensional droplet. We shall hereinafter refer to a two-dimensional droplet simply as droplet.

Our starting point is the generalization of the work of Hocking on flat substrates [17] to topographical ones. We assume that substrate variations occur at lengthscales that are much longer than the slip length. We subsequently assess analytically the effects of the substrate topography on the motion of a droplet in the limit of $Ca \ll 1$, where Ca is the capillary number, defined as

$$Ca = \frac{\mu U}{\sigma}, \tag{1}$$

where μ is the fluid viscosity, σ is the surface tension and U is the (time-dependent) rate of spreading. Indeed, typical experimental configurations fall within this regime, but more importantly this assumption allows us to regard the droplet motion as quasi-static.

Assuming small angles everywhere, a single equation that describes the evolution of the droplet thickness is used, which is obtained via a long-wave expansion of the Stokes equations. We expect sharp variations in the droplet free-surface close to the contact points which suggest the presence of boundary layers there. We thus identify in §3 the bulk of the droplet as the outer region and the region near the contact lines as the inner regions. By asymptotically matching the two, we deduce the rate of spreading of the two moving fronts.

The equations obtained by matching are solved numerically for a number of test cases in §4. The dynamics of spreading, explored in more detail in §4.1, reveals that it is possible for the contact line to exhibit a stick-slip movement, especially during spreading on low-amplitude, short-wavelength substrates. On such substrates, a hysteresis-like effect is also observed and discussed briefly. By examining the phase-plane portrait of the two contact lines in § 4.2, we analyze the existence of multiple equilibrium configurations and their nature as well as how these are affected by varying the substrate properties. A summary of our results is offered in § 5.

2. Problem setup

We consider a droplet of a partially wetting fluid of density ρ, surface tension σ and viscosity μ spreading on an arbitrary substrate. The cross-section of the droplet lies in the $x - z$ plane and is of infinite extent in the $y-$direction. The substrate is given by $z = \eta(x)$, and at time t the profile of the free surface is located at $z = H(x,t) + \eta(x)$, where $H(x,t)$ corresponds to its thickness. The substrate is assumed to be a smooth function of x, so that there are no sharp edges that would prevent a clear definition of a contact line there [18]. Under the long-wave approximation [19] in which the slope of the droplet profile is assumed to be everywhere small we deduce a single equation for the droplet thickness:

$$\frac{3\mu}{\sigma}\partial_t H + \partial_x\left[H^2\left(H + 3\bar{\lambda}\right)\partial_x^3\left(H + \eta\right)\right] = 0, \tag{2}$$

where $\bar{\lambda}$ corresponds to the slip coefficient, introduced to alleviate the non-integrable stress singularity that would otherwise appear at the contact points [20]. Equation (2) is made non-dimensional by writing

$$x = Lx_*, \quad t = \frac{3\mu L}{\sigma \tan^3 \alpha_s} t_*, \quad \bar{\lambda} = \tfrac{1}{3} L\lambda \tan\alpha_s, \; \eta = L\eta_* \tan\alpha_s \text{ and } H = LH_* \tan\alpha_s, \tag{3a}$$

where α_s is the static contact angle and L is a length scale defined by the droplet cross-sectional area, A:

$$L = \sqrt{\frac{A}{2\tan\alpha_s}}. \tag{4}$$

Equation (2), written in non-dimensional form, becomes:

$$\partial_{t_*} H_* + \partial_{x_*} \left[H_*^2 \left(H_* + \lambda \right) \partial_{x_*}^3 \left(H_* + \eta_* \right) \right] = 0. \tag{5}$$

For convenience we shall henceforth drop the stars in (5) and proceed using the non-dimensional variables. For a droplet with contact lines at $x = a_\pm(t)$ such that $a_-(t) \leq x \leq a_+(t)$, the boundary conditions at $x = a_\pm(t)$ are

$$H = 0 \text{ and } \partial_x H = \mp \tan\theta_\pm, \tag{6}$$

where the $\theta_\pm$ are given by

$$\tan\theta_\pm = \frac{1 + \tan\alpha_s^2 \eta_\pm'^2}{1 \pm \tan\alpha_s^2 \eta_\pm'}, \tag{7}$$

where we set $\eta'_\pm = \partial_x \eta|_{x=a_\pm}$. The last geometric conditions essentially fix the angles the free-surface makes with the substrate to their static value. Finally, the cross-sectional area of the droplet remains constant, leading to the global condition

$$\int_{a_-}^{a_+} H \, \mathrm{d}x = 2, \tag{8}$$

with the chosen lengthscales. The solution to (5) depends on two parameters, the non-dimensional slip length, λ, which is always small, and the contact angle, α_s. We thus expect to have a boundary layer in the vicinity of the droplet contact line, where an abrupt change in $\partial_x H$ takes place to match the microscopic, static contact angle to the apparent contact angle in the bulk.

In what follows, we will examine the effects of the substrate profile $\eta(x)$ on the spreading dynamics. In principle, $\eta(x)$ can be arbitrary as long as it satisfies $\partial_x \eta \lesssim \mathcal{O}(1)$. In addition, we require that the contact line is not trapped within a feature of the substrate. This can be avoided by appropriately choosing the substrate amplitude and by requiring the substrate variations to occur at a lengthscale that is much longer than the slip length.

3. Singular Perturbation Expansion

The problem is analytically tractable if we restrict our discussion to the limiting case $Ca \ll 1$ as noted in § 1 and as done in the earlier studies on flat substrates for both the slip [17] and constant-thickness precursor film [21] models. In this limit, surface tension forces dominate viscous ones. This implies that in a region sufficiently far from the contact line, the droplet shape is primarily affected by surface tension, allowing a different treatment of the governing equation in the bulk of the fluid (outer region) from the vicinity of the contact lines (inner regions). By properly matching the outer solution with the inner ones, we can obtain the velocities of the two moving fronts. Matching with a precursor-film inner region is discussed in [22, 23].

3.1. Outer region

For convenience, we transform the domain $a_- \le x \le a_+$ to $-1 \le y \le 1$, via the mapping:

$$x = \tfrac{1}{2}(a_+ - a_-)\,y + \tfrac{1}{2}(a + a_-), \tag{9}$$

which allows us to write (5) for the outer region in the (non-moving) coordinate y as:

$$\partial_t H - \frac{\dot a_+ (1+y) + \dot a_- (1-y)}{a_+ - a_-}\partial_y H + \frac{16}{(a_+ - a_-)^4}\partial_y \left[H^3 \partial_y^3 (H + \eta)\right] = 0, \tag{10}$$

where $\dot a_\pm = \mathrm{d}a_\pm/\mathrm{d}t$ correspond to the spreading rates of the two contact lines at $x = a_\pm(t)$. In (10), we dropped λ because slip is not significant in the outer region. The limit $Ca \ll 1$ is equivalent to $|\dot a_\pm| \ll 1$, which prompts us to introduce a quasi-static expansion for $H(x,t)$ in the form:

$$H(x,t) = H_0(y, a_\pm) + H_1(y, \dot a_\pm, a_\pm) + \cdots, \tag{11}$$

where the time dependence of $H(x,t)$ enters through $a_\pm(t)$ and their time derivatives. The task here is to approximate $H(x,t)$ as we approach the contact lines with the aim of asymptotically matching with the solution coming from the inner region. To leading order in $\dot a_\pm$ we have:

$$\partial_y^3 (H_0 + \eta) = 0. \tag{12}$$

This equation is third order in y and cannot, in general, satisfy all conditions (6) and (8). Since the contact angle boundary conditions apply at the contact lines, which are outside the domain of the outer solution, we use (6) and (8) to obtain $H_0(y,t)$:

$$\begin{aligned} H_0 = {} & \frac{3}{2}\left[\frac{2}{a_+ - a_-} + \bar\eta - \tfrac{1}{2}(\eta_+ + \eta_-)\right](1 - y^2) + \tfrac{1}{2}\left[\eta_+ (1+y) + \eta_- (1-y)\right] \\ & - \eta\left(\tfrac{1}{2}(a_+ - a_-)\,y + \tfrac{1}{2}(a_+ + a_-)\right), \end{aligned} \tag{13}$$

where $\eta_\pm = \eta(a_\pm)$ and $\bar\eta = \int_{a_-}^{a_+} \eta(\xi)\,\mathrm{d}\xi/(a_+ - a_-)$. The term $\partial H_0/\partial t$ is approximated as $\partial H_0/\partial t = \dot a_+ \partial H_0/\partial a_+ + \dot a_- \partial H_0/\partial a_-$ and appears in the equation for the next term, $H_1(y,t)$, which is solved subject to homogeneous boundary conditions, namely

$$H_1(\pm 1) = 0 \quad \text{and} \quad \int_{-1}^{+1} H_1 \,\mathrm{d}y = 0. \tag{14}$$

Finding H_1 cannot be done analytically for arbitrary $\eta(x)$, but what we are interested in is the leading-order behavior of H_1 as $y \to \pm 1$. The third derivative of H_1 is singular at $y = \pm 1$ and behaves like

$$\partial_y^3 H_1 = \frac{a_+ - a_-}{2\phi_\pm^2 (1 \mp y)^2}\dot a_\pm + \mathcal{O}\left((1 \mp y)^{-1}\right),$$

where

$$\phi_\pm = \frac{2}{a_+ - a_-}\left[\frac{6}{a_+ - a_-} + 3\bar\eta - 2\eta_\pm - \eta_\mp \pm \tfrac{1}{2}\eta'_\pm (a_+ - a_-)\right]. \tag{15}$$

Here $\phi_\pm$ correspond to the apparent contact angles obtained by evaluating $\partial_x H_0$ at $x = a_\pm$. In the limit of $\eta \ll H$, $\phi_\pm$ can be viewed as mesoscopic angles and the droplet spreads with an apparent angle, $\psi = 12/(a_+ - a_-)^2$. Successive integrations yield the leading order behavior for $\partial_y H_1$ as the contact lines are approached:

$$\partial_y H_1 \sim -\dot a_\pm \frac{a_+ - a_-}{2\phi_\pm^2}\left[\ln \frac{e}{2}(1 \mp y) + \beta_\pm\right] \quad \text{as } y \to \pm 1, \tag{16}$$

where the constants $\beta_\pm$ are found by appropriately integrating the equation for H_1 and making use of the conditions (14) to find:

$$\beta_\pm = \int_{-1}^{1} \frac{1}{1 \mp y} \left\{ 1 - \frac{(a_+ - a_-)^3 \phi_\pm^2 \left(1-y^2\right)^3 \left[(\dot{a}_+\phi_+ - \dot{a}_-\phi_-)\, y + \dot{a}_+\phi_+ + \dot{a}_-\phi_-\right]}{128 \dot{a}_\pm H_0^3} \right\} \mathrm{d}y. \quad (17)$$

By writing (13) and (16) in terms of x and combining them, we find the slope of H as $x \to a_\pm(t)$:

$$\mp \partial_x H \sim \phi_\pm \pm \frac{\dot{a}_\pm}{\phi_\pm^2} \left[\ln \left(e \frac{\pm a_\pm \mp x}{a_+ - a_-} \right) + \beta_\pm \right]. \quad (18)$$

These asymptotic expansions will ultimately be matched asymptotically with the inner solutions to be derived in the vicinity of the contact lines. For the expansions (18) to be valid, we require that matching occurs sufficiently close to the contact lines, so that the linear term of $\partial_x H$ is of higher order than the logarithmic term, which in turn is of higher order than the constant term in the expansion.

3.2. *Inner region*

The inability of the outer solution to satisfy the constant contact-angle boundary condition near the contact lines, necessitates the presence of inner regions where slip is significant and the above condition is exactly satisfied. To treat the inner region near $x = a_\pm$ for which $x - a_\pm = \mathcal{O}(\lambda)$, we make the change of coordinates

$$H = \lambda \Phi \quad \text{and} \quad x = a_\pm \mp \lambda \xi / \theta_\pm, \quad (19)$$

which transforms (5) to

$$\lambda \partial_t \Phi \pm \dot{a}_\pm \theta_\pm \partial_\xi \Phi + \theta_\pm^4 \partial_\xi \left[\Phi^2 (\Phi + 1) \partial_\xi^3 \Phi \right] = 0, \quad (20)$$

where it is assumed that $\lambda \left| \xi \partial \theta_\pm / \partial a_\pm \right| \ll 1$. Note also that the contribution $\partial_x [H^2 (H + \lambda) \partial_x^3 \eta]$ in (5) originating from the substrate is neglected, as we have made the assumption that the variations in $\eta(x)$ take place at lengthscales that are much longer than the slip length, and hence the substrate is locally treated as being flat. However, the substrate topography still affects the inner region dynamics through $\theta_\pm$ in equation (7), which depends on $\partial_x \eta$. These contributions are generally small, of $\mathcal{O}\left(\tan^2 \alpha_s\right)$, but are included in our model to exactly force the contact angle boundary condition.

Since we have assumed that the droplet spreads slowly, we can write

$$\Phi = \Phi_0 + \Phi_1 + \cdots \quad (21)$$

with $\Phi_0 = \xi$ and $\Phi_0 \gg \Phi_1$. The equation for Φ_1 thus becomes:

$$\partial_\xi^3 \Phi_1 = \mp \frac{\dot{a}_\pm / \theta_\pm^3}{\xi (\xi + 1)}, \quad (22)$$

which is solved subject to the conditions that $\Phi_1 = \partial_\xi \Phi_1 = 0$ at $\xi = 0$ and $\Phi_1 / \xi^2 \to 0$ as $\xi \to \infty$ to get:

$$\partial_\xi \Phi_1 = \pm \frac{\dot{a}_\pm}{\theta_\pm^3} \left[(\xi + 1) \ln (\xi + 1) - \xi \ln \xi \right] \quad (23)$$

The asymptotic behavior of $\partial_\xi \Phi$ as $\xi \to +\infty$ is readily obtained as:

$$\partial_\xi \Phi \sim 1 \pm \frac{\dot{a}_\pm}{\theta_\pm^3} (\ln \xi + 1) \quad \text{as } \xi \to +\infty. \tag{24}$$

which, in terms of the outer variables, is expressed as:

$$\mp \partial_x H \sim \theta_\pm \pm \frac{\dot{a}_\pm}{\theta_\pm^2} \left[\ln \left(\theta_\pm \frac{\pm a_\pm \mp x}{\lambda} \right) + 1 \right] \quad \text{as } \frac{\pm a_\pm \mp x}{\lambda} \to \infty. \tag{25}$$

3.3. Asymptotic matching

Apparently the logarithmic terms in the inner (25) and outer (18) regions do not generally match and therefore an intermediate region would be required to properly continue the matching procedure, as in the flat substrate case [17]. We notice, however, that matching can actually be successful if it is done for $(\partial_x H)^3$ instead of $\partial_x H$. Thus the sole purpose of an intermediate region is precisely to justify matching $\partial_x H$ and for the sake of brevity it is omitted in the present treatment. Hence, considering the cube of (18) and (25), the logarithmic terms match provided that the matching is carried within the overlap region of validity of the two asymptotic expansions. Matching the constant terms then in the asymptotic expansions yields:

$$\phi_\pm^3 - \theta_\pm^3 = \pm 3 \dot{a}_\pm \left[\ln \left(\theta_\pm \frac{a_+ - a_-}{\lambda} \right) - \beta_\pm \right]. \tag{26}$$

Hence, the order of magnitude of the speed of the moving front is $|\dot{a}_\pm| \sim \mathcal{O}(1/|\ln \lambda|)$. Equations (26) constitute a system of integro-differential equations for $a_\pm(t)$, which are coupled through $\beta_\pm$. Explicit expressions for $\dot{a}_\pm$ can be found after term re-arrangement in (26) and solving the resulting system:

$$\dot{a}_\pm = \pm \frac{\delta_\pm I_\mp + \delta_\mp \phi_\pm I_0}{I_+ I_- - \phi_+ \phi_- I_0^2}, \tag{27}$$

where

$$\delta_\pm = \tfrac{1}{3} \left(\phi_\pm^3 - \theta_\pm^3 \right),$$

$$I_\pm = \ln \left[\theta_\pm \frac{a_+ - a_-}{\lambda} \right] + \int_{-1}^{+1} \frac{1}{1 \mp y} \left[\frac{\phi_\pm^3 (a_+ - a_-)^3 (1 - y^2)^4}{128 H_0^3 (1 \mp y)} - 1 \right] \mathrm{d}y,$$

$$I_0 = \frac{\phi_+ \phi_- (a_+ - a_-)^3}{128} \int_{-1}^{1} \frac{(1 - y^2)^3}{H_0^3} \mathrm{d}y.$$

Equations (27) reveal an interesting interplay between the dynamics of each contact line. Equilibrium not only depends on whether the apparent contact angle becomes equal to the static one ($\delta_\pm = 0$), but also on how fast the other contact line moves. In fact, as our results indicate asymmetric spreading is almost always accompanied by a recession of one of the contact lines before the droplet reaches equilibrium.

4. Results

4.1. Spreading dynamics

We now proceed with presenting a few typical solutions to (27), aiming to illustrate the essential features of the spreading dynamics and to gain further insight into the effects of substrate heterogeneities. Since the overall picture is not drastically altered by variations in the slip length and the static contact angle, in all cases presented hereinafter, we fix $\lambda = 10^{-4}$ and $\alpha_s = 5\,^\circ$ using $a_\pm(0) = \pm 1$ as initial conditions for time integration.

Journal of Physics: Conference Series **216** (2010) 012017 doi:10.1088/1742-6596/216/1/012017

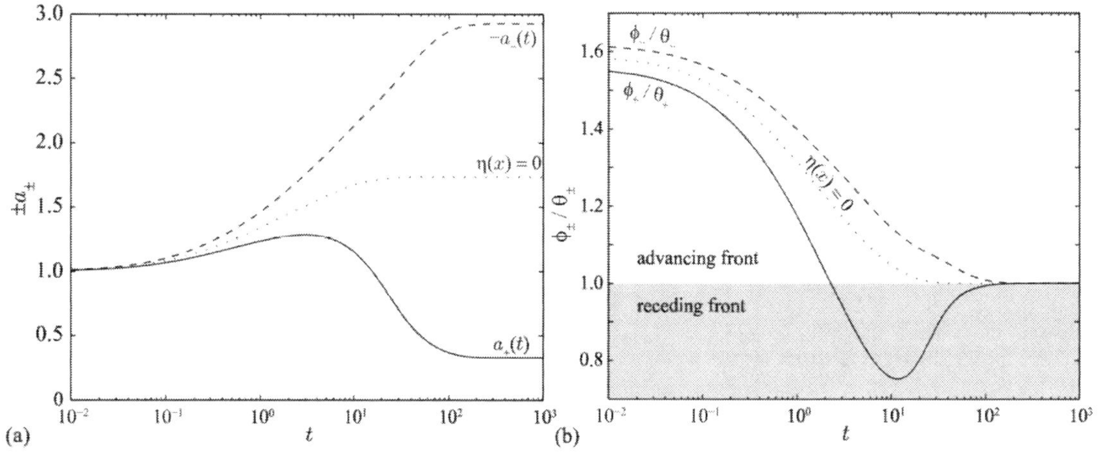

Figure 1. Spreading on a substrate $\eta(x) = (2\sin x + \sin 2x + 2\cos x)/10$ with $a_{\pm}(0) = \pm 1$, $\lambda = 10^{-4}$ and $\alpha_s = 5\,^{\circ}$. (a) Evolution of the moving fronts (solid and dashed lines); the dotted curve corresponds to the speed of the contact line for $\eta(x) = 0$. (b) The corresponding ratios $\phi_{\pm}/\theta_{\pm}$. When these ratios become less than 1 (shaded region), recession of the contact line occurs.

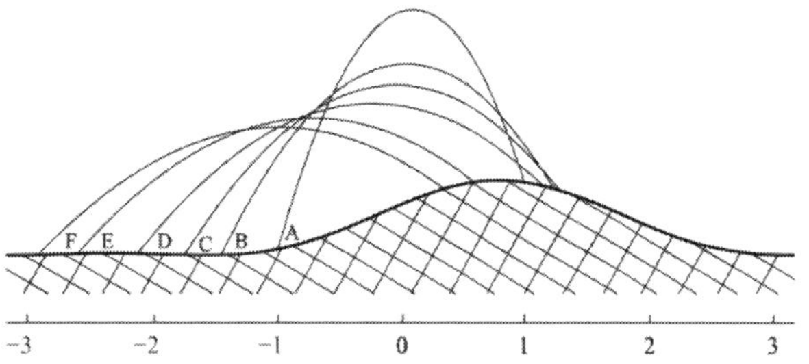

Figure 2. Evolution of the leading-order outer solution, (13) for the computations of Fig. 1. Curves A–F correspond to times $t = 0$, 1, 3, 10, 40 and 200 respectively.

We begin by considering the substrate $\eta(x) = (2\sin x + \sin 2x + 2\cos x)/10$, which corresponds to a 2π-periodic series of bumps. In Fig. 1(a), we present the evolution of the moving fronts as a function of time. At first, both fronts advance, but at a later stage the right front begins to recede as the droplet slides downhill towards equilibrium (see Fig. 2, where we show the outer solution, (13), at different times). As it turns out, this brief recession of the contact line is always present unless spreading is perfectly symmetric. This behavior can be alternatively observed by looking at the evolution of the ratio of $\phi_{\pm}/\theta_{\pm}$ evaluated at the two contact lines, as shown in Fig. 1(b). As our analysis suggests, contact line advancement occurs provided that $\phi_{\pm} > \theta_{\pm}$. Even though the right front recedes for a relatively long period of time, it does so very slowly.

The moving contact lines may exhibit a stick-slip-type behavior while they move along the substrate features as Cox conjectured [13]. To illustrate this effect, we consider the small-amplitude, periodic substrate given by $\eta(x) = 0.003\sin 80x + 0.005\cos 40x$. Figure 3(a) shows the location of the two moving fronts as a function of time. For comparison, we also plot the contact line location for ideally flat substrates. Even though at early times the droplet tends to spread nearly symmetrically following the overall trend of an ideally flat substrate, the contact line velocities are highly oscillatory as can be seen in Fig. 3(b), where we plot their time evolution. Sticking and slipping is particularly evident for the left contact line, as the staircase-like curve for $a_{-}(t)$ suggests. We can therefore conclude that substrate heterogeneities tend to have a greater impact on the dynamics as the droplet approaches equilibrium (see Fig. 4(a)). What

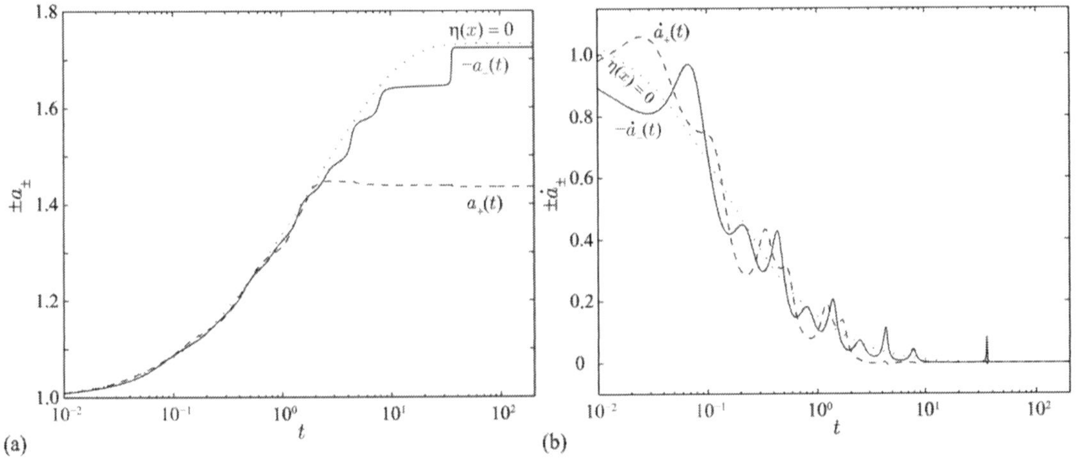

Figure 3. Spreading on a substrate $\eta(x) = 0.003\sin 80x + 0.005\cos 40x$ with $a_{\pm}(0) = \pm 1$, $\lambda = 10^{-4}$ and $\alpha_s = 5\,°$. (a) Evolution of moving fronts' locations (solid and dashed lines) demonstrating stick-slip-type behavior; the dotted curve corresponds to the contact line location for $\eta(x) = 0$. (b) The corresponding velocities of the contact points.

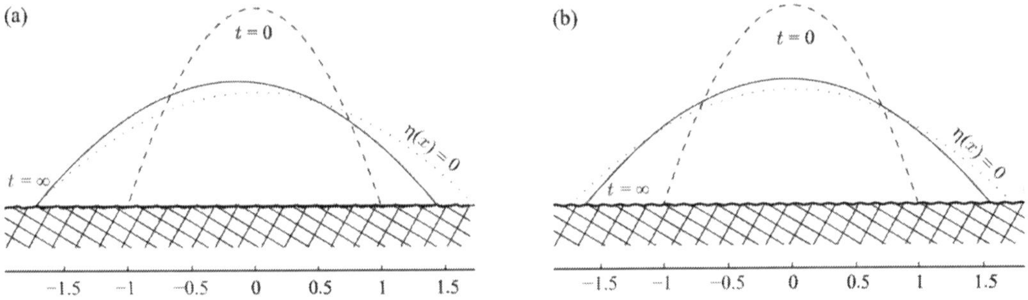

Figure 4. Droplet profiles for spreading on a substrate (a) $\eta(x) = 0.003\sin 80x + 0.005\cos 40x$ and (b) $\eta(x) = 0.005\cos 40x$, at times $t = 0$ (dashed curve) and $t = \infty$ (solid curve). The dotted curve corresponds to the equivalent droplet equilibrium profile had the substrate been flat illustrating the substrate-induced hysteresis effect upon comparison with the solid curve.

is also surprising is the fact that the right contact line is nearly stationary while sticking and slipping occurs. This example suggests that the temporary pinning of the contact line, reported in experimental studies [2], may be attributed at least in part to the details of the substrate topography. Of course other surface features, such as chemical defects/heterogeneities can also be responsible.

As the preceding example suggests, even small-amplitude spatial heterogeneities can dramatically alter the spreading dynamics. In Fig. 4(a) we show the equilibrium attained for the example in the previous paragraph, where we can see an appreciable shift of the droplet to the left. Contrary to the flat-substrate case, we also notice that it is possible for the apparent contact angle at equilibrium to be quite different from the static contact angle. Further evidence for substrate-induced hysteresis can be found in Fig. 4(b), where we plot the equilibrium attained while spreading symmetrically on the substrate $\eta(x) = 0.005\cos 40x$. This observation prompts us to investigate under what conditions the hysteresis-like effect induced by topographical substrate variations is more prominent. As noted earlier, when $|\eta(x)| \ll 1$

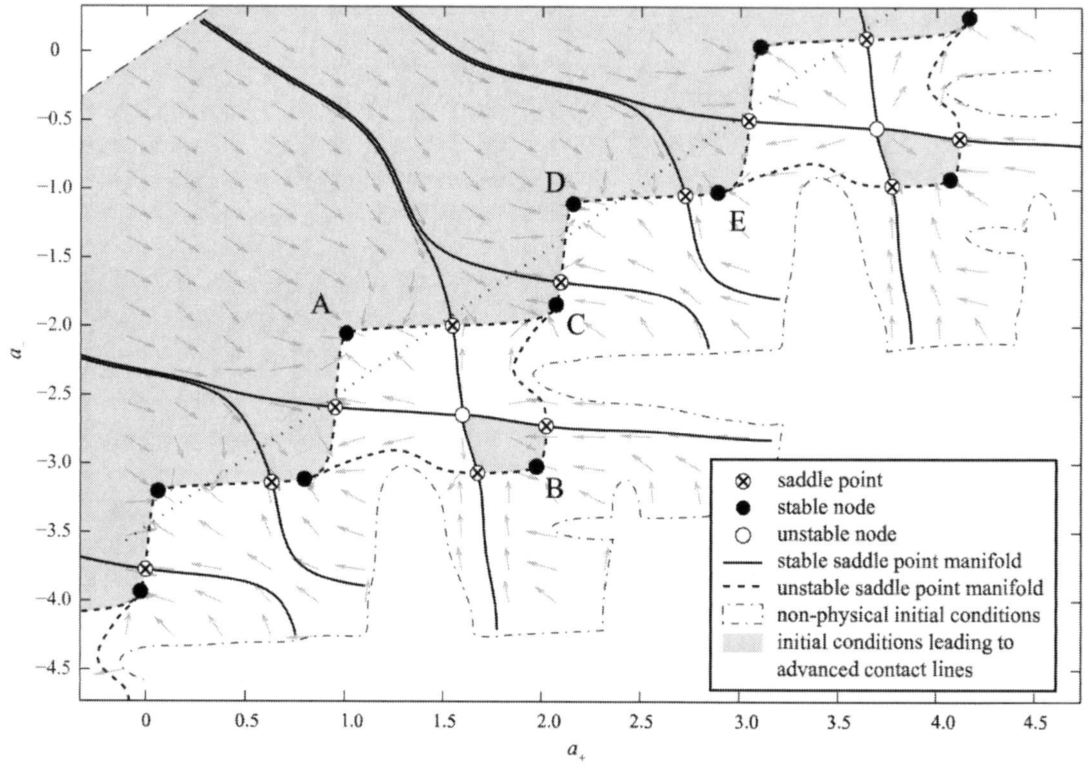

Figure 5. The $(a_+, -a_-)$ phase plane for $\eta(x) = (\cos 6x + \sin 3x)/12$. There are multiple droplet equilibria, that may correspond to stable nodes (solid circles), unstable nodes (open circles) or saddle points (crossed circles). For comparison the dotted line shows the line equilibria for ideally flat substrates. Solid and dashed curves correspond to stable and unstable saddle point manifolds, respectively, which delimit the basins of attraction of the stable fixed points. Since not all initial conditions are permissible, the dash-dotted line is used to separate the physical and non-physical (free surface touches the substrate at a point different to a contact one) parts of the $a_+ - a_-$ phase plane. The shaded region corresponds to the region of the phase plane that is accessible when both contact lines advance. The translational invariance due to the substrate periodicity implies a finite number of equilibria, namely 5 stable nodes (denoted by A–E), 6 saddle points and 1 unstable node.

the apparent contact angle for both contact lines is $\psi = 12/(a_+ - a_-)^2$, which at equilibrium will be different from the actual angle the substrate makes with the droplet free surface by an amount:

$$\Delta\phi_\pm = \phi_\pm - \psi = \frac{2}{a_+ - a_-}\left[3\bar{\eta} - (2\eta_\pm + \eta_\mp) \pm \tfrac{1}{2}\eta'_\pm (a_+ - a_-)\right]. \tag{28}$$

Given that the amplitude of the features is small, this substrate-induced hysteresis in the apparent contact angle can be appreciable when variations in $\eta(x)$ are relatively 'fast' but $\partial_x \eta \lesssim 1$, to be in accord with the assumptions of our long-wave expansion.

4.2. Phase-plane analysis

Our expectation that multiple equilibria exist, motivates the investigation of the droplet spreading dynamics on the phase plane. When the droplet reaches equilibrium, $\dot{a}_\pm = 0$, so that (18) becomes $\phi_{\pm\mathrm{eq}} = \theta_{\pm\mathrm{eq}}$, where we use $a_{\pm\mathrm{eq}}$ to denote the equilibrium positions of the

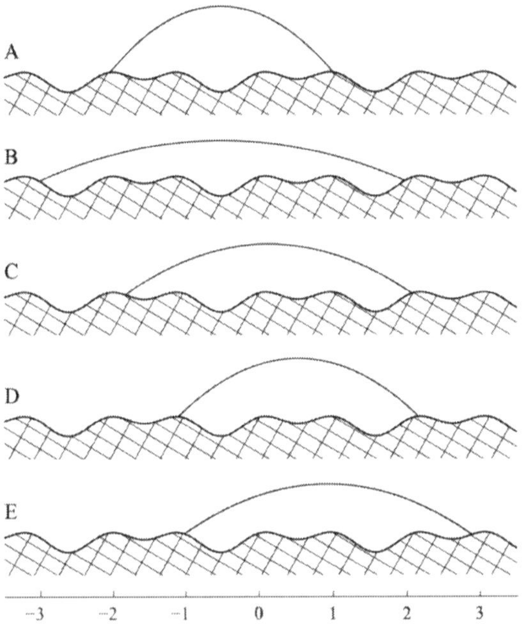

Figure 6. Equilibrium droplet profiles corresponding to the stable equilibria A–E of Fig. 5. Equilibrium shapes B,C and E enhance wetting, but their basins of attraction are significantly smaller than shapes A and D.

contact lines at $x = a_{\pm}(t)$. Spreading on a flat substrate is translationally invariant, with $a_{+\text{eq}}$ and $a_{-\text{eq}}$ related through

$$a_{-\text{eq}} = a_{+\text{eq}} - 2\sqrt{3}, \tag{29}$$

so that there exists a continuum of equilibrium configurations. The introduction of spatial heterogeneities drastically changes the attainable steady states. To demonstrate this, consider the substrate $\eta(x) = (\cos 6x + \sin 3x)/12$. Now the continuum of equilibria given by (29) is destroyed and countably infinite new equilibrium positions emerge. In Fig. 5 we show the equilibrium points on the $a_+ - a_-$ phase plane along with the direction field that corresponds to solutions to (27). Our analysis reveals that these equilibria can be either stable or unstable nodes or saddle points, but they do not necessarily correspond to different states. The periodicity of the substrate considered has an inherent translational invariance with respect to the wavelength of the topographical feature. Accounting for the substrate periodicity, we have in fact five distinct stable equilibria, shown as points A–E on the phase plane. Three of these are symmetric about the topographical features (points A, B and D). All equilibrium shapes corresponding to these contact line locations are given in Fig. 6. It is important to note that not all $a_{\pm}(0)$ would yield physically meaningful solutions to (27). In fact, a rather large part of the phase plane is excluded, since we require that $a_+ > a_-$ and that the droplet free surface does not touch or cross the topographical features (irregularly-shaped domain in the lower right of Fig. 5). On the remaining phase-plane, we see that the equilibrium shapes A and D have significantly larger basins of attraction than the other three equilibria. For these two shapes it is readily seen that the droplet wets the droplet less than the flat-case equilibrium. Wetting-enhancing equilibria, B, C and E, are more accessible through contact line recession than spreading. In principle however for the majority of permissible initial conditions both fronts are advancing at early times (shaded region in Fig. 5).

5. Conclusions

We have developed a model for describing the surface-tension dominated motion of a partially wetting two-dimensional droplet over spatially heterogeneous substrates in the Stokes-flow regime. Assuming small contact angles that remain constant and equal to their static value along the contact lines and that substrate variations occur over lengthscales that are much larger than the slip length, we obtained a single equation for the evolution of the droplet thickness and its contact lines obtained via a long-wave approximation of the Stokes equations. The force singularity at the contact lines was alleviated through the Navier slip condition. For small capillary numbers, the droplet approaches equilibrium quasi-statically, with the time-dependence entering the problem through the location of the contact lines. This allowed us to regard the dynamics as a perturbation problem in terms of the spreading rates of the two droplet fronts. By matching the flow in the vicinity of the contact lines with the flow in the bulk of the droplet, we were able to obtain a set of two coupled integro-differential equations for the location of the droplet fronts.

As the droplet moves along the topographical features, the contact lines may exhibit a stick-slip-type behavior, especially if the characteristic wavelength of these features is small. Such behavior is more readily observed as the droplet approaches equilibrium, where the forcing of the substrate can more easily influence the slowly moving droplet. This effect should not be confused with the sticking and slipping that may occur when chemical defects are present. What this calculation suggests, however, is that the substrate features may amplify this effect. We have also demonstrated via a phase-plane analysis that substrate heterogeneities allow for the existence of multiple equilibrium droplet configurations. The complexity of the phase plane, the number of equilibria and their nature are highly dependent on the density and height of the topographical features. As a matter of fact, we observe that a spatially heterogeneous substrate can induce a hysteresis-like effect, which can become appreciable for sufficiently 'rough' substrates.

Even though the inclusion of more realistic, three-dimensional substrates would yield results that can be directly compared with existing experiments, we believe that many of the essential features can be adequately studied with our model. It is thus hoped, that through the formalism developed in the present study, other related problems pertaining to the dynamics of droplet spreading can be addressed, such as the effects of random substrates on the statistics of the location of the contact lines. We shall examine this and related issues in future studies.

References

[1] Dussan V E B 1979 *Ann. Rev. Fluid Mech.* **11** 371–400
[2] deGennes P G 1985 *Rev. Mod. Phys.* **27** 827–863
[3] Blake T D 1993 Dynamic contact angles and wetting kinetics *Wettability* ed Berg J C (New York: Marcel Dekker Inc.)
[4] Bonn D, Eggers J, Indekeu J, Meunier J and Rolley E 2009 *Rev. Mod. Phys.* **81** 739–805
[5] Schwartz A M and Tejada S B 1972 *J. Colloid and Interface Sci.* **38** 359–375
[6] Semal S, Blake T D, Geskin V, de Ruijter M J, Castelein G and de Coninck J 1999 *Langmuir* **15** 8765–8770
[7] Hitchcock S J, Carroll N T and Nicholas M G 1981 *J. Mater. Sci.* **16** 714–732
[8] Shibuichi S, Onda T, Satoh N and Tsujii K 1996 *J. Phys. Chem.* **100** 19512–19517
[9] Bico J, Tordeux C and Quéré D 2001 *Europhys. Lett.* **55** 214–220
[10] Chung J Y, Youngblood J P and Stafford C M 2007 *Soft Matter* **3** 1163–1169
[11] Johnson R E and Dettre R H 1964 *Adv. Chem. Ser.* **43** 112–135
[12] Huh C and Mason S G 1977 *J. Coll. Interface Sci.* **60** 11–38
[13] Cox R G 1983 *J. Fluid Mech.* **131** 1–26
[14] Oliver J F, Huh C and Mason S G 1977 *J. Adhesion* **8** 223–234
[15] Gaskell P H, Jimack P K, Sellier M and Thompson H M 2004 *Int. J. Numer. Meth. Fluids* **45** 1161–1186
[16] Gramlich C M, Mazouchi A and Homsy G M 2004 *Phys. Fluids* **16** 1660–1667
[17] Hocking L M 1983 *Q. J. Mech. Appl. Math.* **36** 55–69
[18] Oliver J F, Huh C and Mason S G 1977 *J. Coll. Interface Sci.* **59** 568–581
[19] Greenspan H P 1978 *J. Fluid Mech.* **84** 125–143
[20] Huh C and Scriven L E 1971 *J. Coll. Interface Sci.* **35** 85–101
[21] Pismen L M and Eggers J 2008 *Phys. Rev. E* **78** 056304
[22] Kalliadasis S and Chang H C 1994 *Phys. Fluids* **6** 12–23
[23] Kalliadasis S and Chang H C 1996 *Ind. Eng. Chem. Res.* **35** 2860–2874

Journal of Physics: Conference Series **216** (2010) 012018 doi:10.1088/1742-6596/216/1/012018

Coherent structures theory for the generalized Kuramoto–Sivashinsky equation

D. Tseluiko, S. Saprykin, S. Kalliadasis

Department of Chemical Engineering, Imperial College London, London SW7 2AZ, United Kingdom

Abstract. We examine coherent structures interaction and formation of bound states in active–dispersive–dissipative nonlinear media. A prototype for such media is a simple weakly nonlinear model, the generalized Kuramoto–Sivashinsky (gKS) equation, that retains the fundamental mechanisms of any nonlinear process involving wave evolution, namely, a dominant nonlinearity, instability, stability and dispersion. We develop a weak interaction theory for the solitary pulses of the gKS equation by representing the solution as a superposition of the pulses and an overlap function. We derive a linearized equation for the overlap function in the vicinity of each pulse and project the dynamics of this function onto the discrete part of the spectrum of the linearized interaction operator. This leads to a coupled system of ordinary differential equations describing the evolution of the locations of the pulses. By analyzing this system, we prove a criterion for the existence of a countable infinite or finite number of bound states, depending on the strength of the dispersive term in the equation. The theoretical findings are corroborated by computations of the full equation.

1. Introduction

Consider the generalized Kuramoto–Sivashinsky (gKS) equation,

$$h_t + hh_x + h_{xx} + \delta h_{xxx} + h_{xxxx} = 0, \qquad (x,\, t) \in (-\infty,\, \infty) \times [0,\, \infty). \tag{1}$$

This equation is the simplest possible nonlinear evolution prototype that retains the fundamental elements of any nonlinear process that involves wave evolution: the dominant nonlinear term hh_x, instability and energy production h_{xx}, stability and energy dissipation h_{xxxx}, and dispersion δh_{xxx}, with δ a parameter that measures the relative importance of dispersion. As far as the nonlinearity is concerned, its functional form can be easily obtained from symmetry considerations: indeed, the only other dominant quadratic nonlinearity is h^2, which is obviously ruled out for systems whose spatial average does not drift, i.e. $d\langle h\rangle_x/dt = 0$.

In the limit $\delta \to 0$ the equation reduces to the usual Kuramoto–Sivashinsky (KS) equation, first derived in [1], [2] and [3] independently with a weakly nonlinear expansion for small-amplitude falling-film waves in the limiting case of large surface tension and away from criticality (we shall define these terms shortly). The same equation was also obtained later in a wide variety of applications such as chemical physics/reaction-diffusion systems, e.g. propagation of concentration waves [4, 5] and combustion, e.g. flame-front instabilities [6].

The gKS equation has been derived in many physical contexts, including plasma waves with dispersion due to finite ion banana width [7] and liquid films in different settings, i.e. from a film flowing down a planar substrate [8] or a vertical fiber [9] to films in the presence of various additional effects and complexities, e.g. films sheared by a turbulent gas [10], falling liquid films in the presence of a viscous stress at the free surface [11], liquid films flowing down a uniformly heated wall [12] and reactive falling films [13].

For the problem of a film falling down a planar substrate, in particular, the gKS equation is obtained from a weakly nonlinear expansion for small-amplitude waves and under the assumptions of strong surface tension and near-critical conditions [8],

$$We = O(\epsilon^{-2}), \qquad Re > Re_c, \qquad Re - Re_c = O(\epsilon^2), \tag{2}$$

where $Re = U_0 h_0/\nu$ is the Reynolds number assumed to be at most of $O(1)$ and based on the average velocity $U_0 = g \sin\theta h_0^2/(3\nu)$ of the Nusselt flat film flow, with h_0, g, ν and θ denoting the Nusselt flat film thickness, gravitational acceleration, liquid kinematic viscosity and inclination angle with respect to the horizontal direction, respectively. $Re_c = 5\cot\theta/6$ is the critical Reynolds number for the instability onset and $We = \gamma/(3\mu U_0)$ is the Weber number with γ and μ denoting the surface tension and dynamic viscosity of the fluid, respectively. x denotes a streamwise coordinate, t is time and h is a measure of the (small) deviation amplitude from the dimensionless Nusselt flat film of unity (non-dimensionalized with h_0). $\epsilon \ll 1$ is the so-called long-wave or film parameter, and is typically defined as the ratio of h_0 to a (long) lengthscale over which streamwise variations occur. For film flow down a vertical plane ($Re_c = 0$) we still require strong-surface tension and near-critical conditions but with different orders-of-magnitude assignments to (2) [8]:

$$We = O(\epsilon^{-1}), \qquad Re = O(\epsilon). \tag{3}$$

In both cases, δ is an $O(1)$ parameter defined as:

$$\delta = \sqrt{\frac{15}{2}\frac{1}{(Re - Re_c)We}}. \tag{4}$$

On the other hand, for an inclined plane with $We = O(\epsilon^{-2})$ and $Re = O(1)$, $Re - Re_c = O(1)$ or a vertical plane with $We = O(\epsilon^{-2})$ and $Re = O(1)$ – away from criticality and strong surface tension effects as noted earlier – the KS equation is the pertinent weakly nonlinear model.

As noted earlier, in the limit $\delta \to 0$, the gKS equation reduces to the KS one. On the other hand, in the limit $\delta \to \infty$, the gKS equation reduces to the Korteweg–de Vries (KdV) one: introduce the change of variables in equation (1)

$$\tau = \delta t, \qquad H = \frac{1}{\delta}h, \tag{5}$$

to write it as

$$H_\tau + HH_x + H_{xxx} + \varepsilon(H_{xx} + H_{xxxx}) = 0, \tag{6}$$

where $\varepsilon = 1/\delta$. In the limit $\varepsilon \to 0$ (or, equivalently, $\delta \to \infty$), equation (6) reduces to the KdV equation.

For small δ the gKS equation is a perturbed KS one and hence exhibits complicated chaotic dynamics in both space and time. However, a sufficiently large δ arrests the spatio-temporal chaos such that the solution evolves into a regular array of quasi-stationary pulses with roughly the same shape and speed and which interact indefinitely with each other through attractions or repulsions [14]. On the other hand, for moderate values of δ the spatio-temporal evolution of the gKS equation is dominated by quasi-stationary solitary pulses which continuously interact with each other through coalescence events or attractions/repulsions. Such pulses then become elementary processes representing the behavior of the full system and therefore it is feasible to consider the solution as their superposition. We focus on the development of a weak interaction theory by assuming that the pulses are sufficiently separated from each other, i.e. we ignore coalescence events between pulses which just repel or attract each other by interacting through their tails only.

The idea of weak interaction theory for solitary pulses in other systems has been implemented, for example, by [15] and [16]. As far as the gKS equation is concerned, previous efforts to develop weak interaction approaches, include [17], [18] and [19]. However, all previous studies for the gKS equation appear to be either incomplete or sometimes overlook important details and subtleties. Our aim is to obtain a clear and complete understanding of the pulse interaction problem for the gKS equation and to scrutinize our results by detailed comparisons with computations.

The rest of the paper is organized as follows. In Section 2 we develop a weak interaction theory of solitary pulses for the gKS equation. In Section 3 we analyze bound states of the pulses. Section 4 is devoted to discussion of theoretical results and conclusions.

2. Pulse-interaction theory for the gKS equation

From now on, we consider the gKS equation (1) in a frame moving with the velocity c_δ of a solitary pulse,

$$h_t - c_\delta h_x + hh_x + h_{xx} + \delta h_{xxx} + h_{xxxx} = 0. \tag{7}$$

Let $h_0 = h_0(x)$ be a stationary pulse. It satisfies the steady version of (7). We assume that the solution, h, is described as a superposition of n quasi-stationary pulses $h_1, \ldots, h_n$ located at $x_1(t), \ldots, x_n(t)$, respectively, namely,

$$h_i(x, t) = h_0(x - x_i(t)), \qquad i = 1, \ldots, n, \tag{8}$$

and a small overlap (or correction) function $\hat{h}$. Thus, we use the ansatz $h = \sum_{i=1}^{n} h_i + \hat{h}$ for the solution and aim to derive a system of equations governing the locations of the pulses. We consider weak interaction assuming that the pulses are sufficiently separated and, therefore, that for each pulse it is sufficient to take into account its interaction with only immediate neighbors. The idea of weak interaction theory for solitary pulses in other systems was analyzed by, for example, Ei [15] and Sandstede [16]. More precisely, we assume that $l_i \equiv x_{i+1} - x_i = \log \varepsilon + O(1)$ for $i = 1, \ldots, n-1$, where $\varepsilon \ll 1$. We additionally assume that the velocities of the pulses, x_i', $i = 1, \ldots, n$, and the overlap function, $\hat{h}$, are $O(\varepsilon)$. The linearized equation for the overlap function, $\hat{h}$, in the vicinity of the ith pulse takes the form

$$\hat{h}_t - x_1' h_{1x} = \mathcal{L}_1 \hat{h} - (h_1 h_2)_x \tag{9}$$

for $i = 1$,

$$\hat{h}_t - x_i' h_{ix} = \mathcal{L}_i \hat{h} - (h_{i-1} h_i)_x - (h_i h_{i+1})_x \tag{10}$$

for $2 \le i \le n-1$ and

$$\hat{h}_t - x_n' h_{nx} = \mathcal{L}_n \hat{h} - (h_{n-1} h_n)_x \tag{11}$$

for $i = n$, where $\mathcal{L}_i$'s are linear operators defined by

$$\mathcal{L}_i f = c_\delta f_x - f_{xx} - \delta f_{xxx} - f_{xxxx} - (h_i f)_x \tag{12}$$

for $i = 1, \ldots, n$. The formal adjoint operators, $\mathcal{L}_i^*$'s, with respect to the usual inner product, $\langle f, g\rangle = \int_{\infty}^{\infty} f\bar{g}\, dx$ (where bar denotes complex conjugate), in the space $L^2_{\mathbb{C}}$ of complex-valued measurable functions whose absolute values squared are Lebesgue integrable, are given by:

$$\mathcal{L}_i^* f = -c_\delta f_x - f_{xx} + \delta f_{xxx} - f_{xxxx} + h_i f_x. \tag{13}$$

We aim to project the dynamics in the vicinity of the ith pulse onto the null space of $\mathcal{L}_i$ spanned by the translational mode h_{ix}. Our analysis showed that the null space of $\mathcal{L}_i^*$ is spanned by a constant and a function, which we denote by Ψ^i, tending exponentially to different constants as $x \to \pm\infty$. Therefore, zero is not in the point spectrum of $\mathcal{L}_i^*$ on an infinite interval, and the projection onto the null space of $\mathcal{L}_i$ cannot be made in a straightforward way. Nevertheless, projections can be made rigorous by choosing an appropriate weighted space, namely,

$$L_a^2 = \{f : \mathrm{e}^{ax} f \in L_{\mathbb{C}}^2\}, \tag{14}$$

where a is a positive sufficiently small number, with the inner product $\langle f, g\rangle_a = \langle \mathrm{e}^{ax} f, \mathrm{e}^{ax} g\rangle$, where $\langle \cdot, \cdot\rangle$ denotes the usual inner product in $L_{\mathbb{C}}^2$. As noticed by Pego & Weinstein [20], studying the spectrum of $\mathcal{L}_i$ in L_a^2 is equivalent to studying the spectrum of the operator defined by $\mathcal{L}_i^a f = \mathrm{e}^{ax}\mathcal{L}_i(\mathrm{e}^{-ax} f)$ on $L_{\mathbb{C}}^2$. With such a construction, zero is an isolated eigenvalue of both $\mathcal{L}_i^a$ and $\mathcal{L}_i^{a*}$ of both algebraic and geometric multiplicity unity with the eigenfunction and the adjoint eigenfunction given by $\mathrm{e}^{ax} h_{ix}$ and $\mathrm{e}^{-ax}(\Psi^i - \lim_{x\to\infty}\Psi^i)$, respectively, and projections, P_i^a, onto the null spaces of $\mathcal{L}_i^a$'s can be made in a straightforward way, $P_i^a(f) = \langle f, \mathrm{e}^{-ax}(\Psi^i - \lim_{x\to\infty}\Psi^i)\rangle \mathrm{e}^{ax} h_{ix}$. The essential spectrum is shifted in the complex plane. For certain values of δ, the essential spectrum can be moved entirely to the left half-plane, meaning that for these values of delta the pulses tolerate sufficiently small disturbances that decay sufficiently fast as $x \to \infty$. For example, for $\delta = 0.4$ the essential spectrum belongs to the left half-plane for $a_1 < a < a_2$, where $a_1 \approx 0.13$ and $a_2 \approx 0.57$, and therefore pulses tolerate any small disturbance that is $o(\mathrm{e}^{-a_1 x})$ as $x \to \infty$. Assuming that "$\hat{h}$ is free of translational modes", i.e. that it is in the null spaces of the projections, we arrive at the following system describing the dynamics of the locations of the pulses:

$$x_1' = S_1(x_2 - x_1), \tag{15}$$
$$x_i' = S_2(x_i - x_{i-1}) + S_1(x_{i+1} - x_i), \quad 1 < i < n, \tag{16}$$
$$x_n' = S_2(x_n - x_{n-1}), \tag{17}$$

where

$$S_1(l) \equiv -\int_{-\infty}^{\infty} h_0(x + l/2) h_0(x - l/2) \Psi_x^0(x + l/2)\,\mathrm{d}x, \tag{18}$$
$$S_2(l) \equiv -\int_{-\infty}^{\infty} h_0(x + l/2) h_0(x - l/2) \Psi_x^0(x - l/2)\,\mathrm{d}x, \tag{19}$$

We have solved system (15)–(17) for $\delta = 0.4$ numerically for two pulses and compared results with numerical solutions of the full equation (7) when the initial condition was a superposition of two pulses with the same separation distance as that for (15)–(17). We found very good agreement between these results. In Fig. 1(a) we present a typical solution of system (15)–(17) where the world lines of 24 pulses are shown. We note that the results are shown in the frame moving with the velocity of a solitary pulse. Here $\delta = 0.4$. We can observe both attractions and repulsions, formation of two- and three-pulse bound states. Fig. 1(b) shows a histogram obtained on the statistics on 3000 pulse separation distances at $t = 1000$. The initial distribution of the separation distances was taken to be normal with mean 18 and standard deviation 3. We can observe three clear peaks. We note that the peaks are formed at around 9.5, 14 and 18.5, which are in very good agreement with the stable two-pulse bound state distances discussed in the next section and shown in Fig. 2.

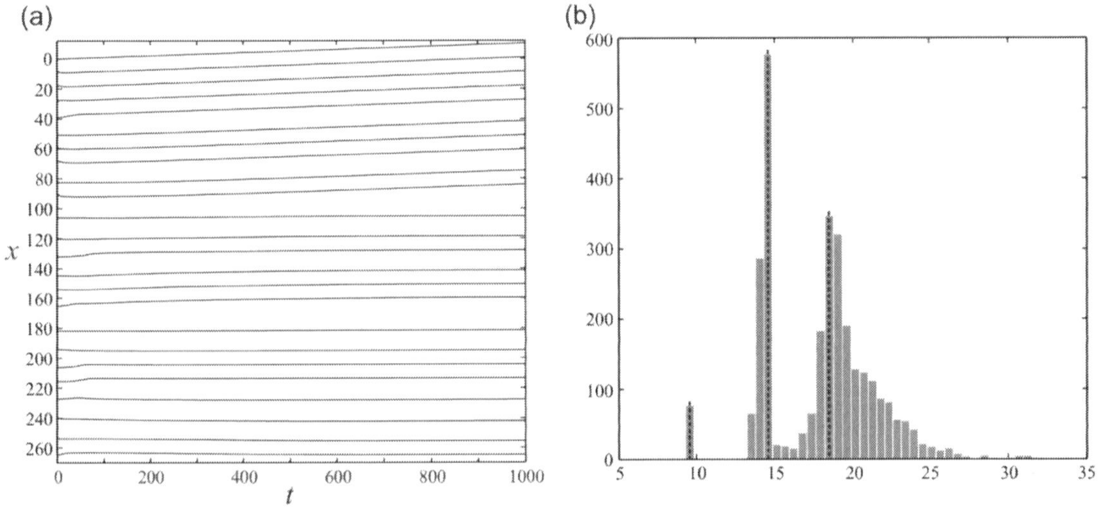

Figure 1. (a) Evolution of pulses of the gKS equation for $\delta = 0.4$ obtained by solving system (15)–(17). Attractions and repulsions can be observed as well as formation of bound states. (b) The histogram obtained on the statistics on 3000 pulse separation distances at $t = 1000$.

3. Bound-states theory

We note that system (15)–(17) can transformed to the following system for the separation distances l_i's:

$$l_1' = S_2(l_1) + S_1(l_2) - S_1(l_1), \tag{20}$$
$$l_i' = S_2(l_i) + S_1(l_{i+1}) - S_2(l_{i-1}) - S_1(l_i), \qquad 1 < i < n-1,$$
$$l_{n-1}' = S_2(l_{n-1}) - S_2(l_{n-2}) - S_1(l_{n-1}). \tag{21}$$

Studying the fixed points of this system, we obtain the bound states of the pulses, i.e. the states when the pulses travel together as a whole.

For instance, for a bound state of two pulses we must have:

$$S_1(l_1) = S_2(l_1). \tag{22}$$

The graphs of S_1 and S_2 are shown in Fig. 2 for $\delta = 0.4$. The abscissas of the intersection points indicate the separation distances for which bound states can be formed, and the ordinates indicated the corresponding velocities of the bound states relative to c_δ. Black circles and crosses correspond to stable and unstable bound states, respectively. It is also interesting to note that the ordinates of the intersection points are always negative, i.e. the velocity of a two-pulse bound state is always less than that of an individual pulse. Another interesting observation is that for $\delta = 0.4$ we, apparently, get a countable infinite number of bound states. This observation can be proved analytically by analyzing the behavior of S_1 and S_2 as $L \to \infty$, namely, it can be shown that if δ is less than a threshold value $\delta^* \approx 0.85$, then there is countable infinite number of intersections of S_1 and S_2. Otherwise, there is a finite number of intersections of S_1 and S_2.

4. Discussion

We developed a weak interaction theory of the pulses of the gKS equation by representing its solution as a superposition of such pulses and an overlap function and deriving a system

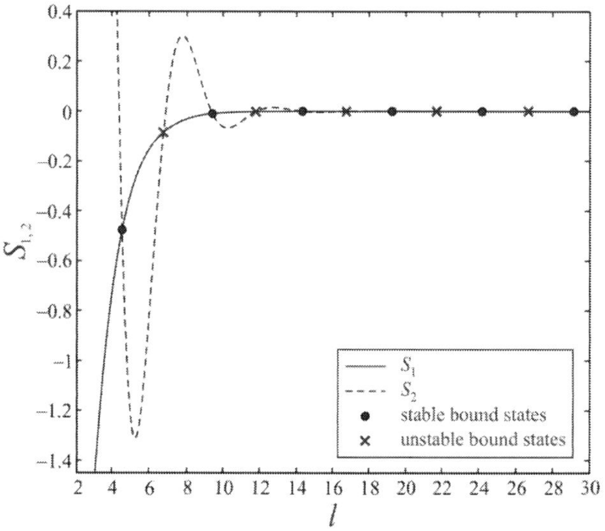

Figure 2. Dependence of S_1 and S_2 on the separation distance between two pulses (solid and dashed lines, respectively) for $\delta = 0.4$. Circles and crosses correspond to stable and unstable two-pulse bound states, respectively.

of linearized equations for the overlap function in the vicinity of each pulse. We found that zero is an eigenvalue of the linearized operators of geometric and algebraic multiplicity unity which is embedded into the essential spectrum. This eigenvalue is associated with translational invariance of the equation. However, zero is not an eigenvalue for the corresponding adjoint operators. The null spaces of the adjoint operators are spanned by a constant and a function having a jump at infinity. Despite this fact, we showed that projections onto the null spaces, spanned by translational modes, can be made rigorous in an appropriate weighted space and the derivation of a dynamical system describing interaction of the pulses due to translational modes can be rigorously justified. This system can be written in terms of the separation distances between consecutive pulses. By analyzing its fixed points, we obtained bound states of any number of pulses. In particular, we analyzed in detail bound states of two pulses and proved a criterion for the existence of a countable infinite or finite number of bound states, depending on the strength of the dispersive term in the equation. Interestingly, this criterion exactly coincides with Shilnikov's criterion on the existence of subsidiary homoclinic orbits, see [21]. Our approach, however, in addition to providing an existence result for the bound states, also gives the description of the dynamics of the pulses. Besides, it can be extended to higher dimensions.

Acknowledgments

We acknowledge financial support from the Engineering and Physical Sciences Research Council of the UK (EPSRC) through grants no. EP/F009194 and EP/F016492.

References

[1] Homsy G M 1974 *Lectures in Applied Mathematics: Nonlinear wave motion* **15** 191–194
[2] Lin S P 1974 *J. Fluid Mech.* **63** 417–429
[3] Nepomnyashchy A A 1974 *Fluid* Dyn. **9** 354–359
[4] Kuramoto Y 1978 *Prog. Theor. Phys. Suppl.* **64** 346–367
[5] Kuramoto Y and Tsuzuki T 1975 *Prog. Theor. Phys.* **54** 687–699

[6] Sivashinsky G I 1977 *Acta Astronaut.* **4** 1176–1206
[7] Cohen B I, Krommes J A, Tang W M and Rosenbluth M N 1976 *Nucl. Fusion* **16** 971–92
[8] Kalliadasis S, Ruyer-Quil C, Scheid B and Velarde M G 2009 *Film Flows, Wave Instabilities and Thermocapillarity* (Springer Series on Applied Mathematical Sciences, to appear)
[9] Tseluiko D, Saprykin S, Duprat C, Giorgiutti-Dauphiné F and Kalliadasis S 2009 *Phys. Rev. Lett.* submitted
[10] Jurman L A and McCready M J 1989 *Phys. Fluids A* **1** 522–36
[11] Oron A and Edwards D A 1993*Phys. Fluids A* **5** 506–8
[12] Kalliadasis S, Demekhin E A, Ruyer-Quil C and Velarde M G 2003 *J. Fluid Mech.* **492** 303–38
[13] Trevelyan P M J and Kalliadasis S 2004 *Phys. Fluids* **16** 3191–208
[14] Kawahara T 1983 *Phys. Rev Lett.* **51** 381–383
[15] Ei S-I 2002*J. Dyn. Differ. Equ.* **14** 85–137
[16] Sandstede B 2002 *Handbook of Dynamical Systems II*, ed B. Fiedler (Amsterdam: North-Holland) chapter 18 pp 983–1055
[17] Elphick C, Ierley G R, Regev O and Spiegel E A 1991 *Phys. Rev. A* **44** 1110–22
[18] Ei S-I and Ohta T 1994*Phys. Rev. E* **50** 4672–78
[19] Chang H-C and Demekhin E A 2002 *Complex Wave Dynamics on Thin Films* (Amsterdam: Elsevier)
[20] Pego R L and Weinstein M I 1992 *Phil. Trans. R. Soc. Lond. A* **340** 47–94
[21] Glendinning P and Sparrow C 1984 *J. Stat. Phys.* **35** 645–96

Journal of Physics: Conference Series **216** (2010) 012019 doi:10.1088/1742-6596/216/1/012019

Bulging and bending of Kelvin-Helmholtz billows controlled by symmetry and phase of initial perturbation

E Kit, D Gerstenfeld, A Y Gelfgat,
School of Mechanical Engineering, Faculty of Engineering, Tel-Aviv University, 69978 Tel-Aviv, Israel

N V Nikitin
Institute of Mechanics, Moscow State University, 119899 Moscow, Russia

E-mail: gelfgat@eng.tau.ac.il, kit@eng.tau.ac.il

Abstract
A computational study of two possible modes of instability of developed Kelvin-Helmholtz (KH) billows is carried out. It is shown that depending on one of three factors, (i) symmetry of initial perturbation in the cross-flow direction, (ii) phase shift of initial perturbation in the streamwise direction, or (iii) excitation of either spanwise or oblique initial wave, the resulting instability can set in as a spanwise bending or bulging of a two-dimensional KH roller. To ensure that the observed secondary bifurcations result from the linear instability the study is carried out considering a two–dimensional flow combined of a distorted mean flow and developed "frozen" KH billow in under and/or over-saturated state as a base flow subject to three-dimensional instabilities.

1. Introduction

The present paper is devoted to study of secondary three-dimensional instabilities of Kelvin-Helmholtz (KH) billows that take place qualitatively similar to instabilities of Stuart vortices studied by Pierrehumbert and Widnall [1] and later by Schoppa et al. [2]. Pierrehumbert and Widnall [1] found three different modes of the Stuart vortex instability. The first one corresponds to subharmonic instabilities caused by mutual rollup and pairing of two-dimensional KH billows while "pairing" was attributed to development of two-dimensional billows at subharmonic frequency and "helical pairing" mode corresponds to the three-dimensional billows at subharmonic frequency too. The second, three-dimensional mode, called "translative", corresponds to a bending of primary two-dimensional billows at fundamental frequency. Later it was studied by Rogers & Moser [3] who called it "bending mode". The third mode, also three-dimensional and occurring at fundamental frequency, had a smaller growth rate and was not studied in detail in [1]. Later, Schoppa et al. [2,4] showed that this mode can be triggered by a certain initial disturbance and called it "core dynamics instability". They showed that time evolution of this mode causes bulging of initially two-dimensional spanwise vorticity surfaces, so that in the following text we call it "bulging mode".

The spanwise undulations of KH rollers, possibly corresponding to the bending instability, were observed in many experiments, e.g., [5-12]. In several three-dimensional numerical studies, e.g., [14-16], initially two-dimensional KH billows arrived to a wavy spanwise pattern, also indicating that the bending instability is probably the fastest growing 3D mode.

Rogers & Moser [3] concluded that secondary three-dimensional instabilities of KH billows develop either after oversaturation of two-dimensional rollers, or after their pairing. They stated in the summary: "Long-term growth of three-dimensionality requires long-term presence of spanwise vorticity in the braid region. In a developing layer, this occurs only at late time, after oversaturation of the two-dimensional rollers. The translative instability is thus an instability of the late-time oversaturated flow. Oversaturation can be prevented when the mixing layer is pairing". Schoppa et al. [4] observed that initial growth of the bulging mode, called also core dynamic instability (CDI), is interrupted by pairing after which the bulging mode is stabilized. Then growth of this mode was observed again after the pairing. Schoppa et al. [4] concluded that "... for linear 3D perturbation evolution in the presence of pairing, excitation of CDI by subharmonic oblique modes generates three-dimensionality which is comparable to that of nearly the most unstable translative mode". Seemingly, these conclusions were never questioned. In their detailed study of bulging, or CDI, instabilities Schoppa et al. [2] considered the 2D mixing layer flow during or after KH billows pairing. At the same time in the above cited studies the bending and bulging instabilities were addressed to similar instabilities of Stuart vortex [1,2]. Our recent experiments [12] and computations, which we describe below, show a clear evidence of both bending and bulging instabilities (see Fig. 1) that develop well before pairing. This motivates our revisiting of this problem. In the following we show that by a controlled initial perturbation it is possible to trigger either bending or bulging mode. Selection of either a bulging or bending mode depends on a combination of three following factors: perturbation symmetry in the cross-flow direction, location of a maximum of the perturbation amplitude inside the roller or inside the braid, i.e., on the streamwise phase, and on excitation of either a spanwise or an oblique wave. An alteration of one of these three factors leads to a replacement of bending by bulging or vice versa.

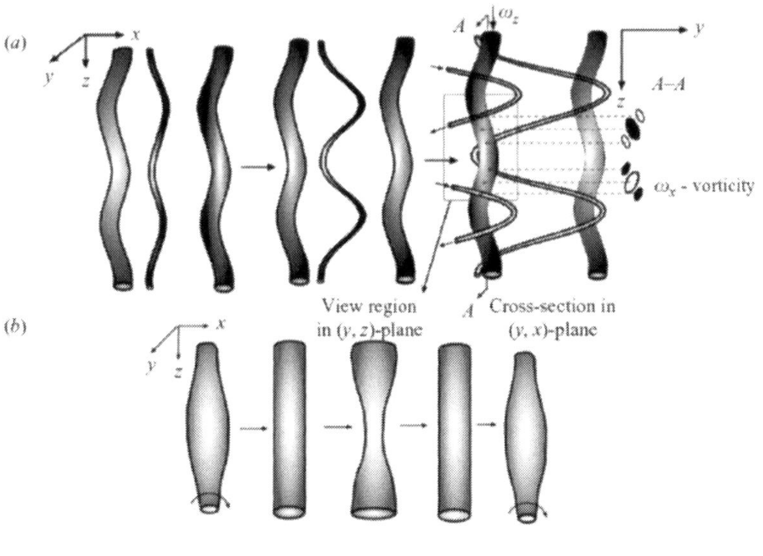

Fig. 1. Sketch of spatial development of different three-dimensional disturbances: (a) bending, (b) bulging.

In our recent experimental work [12,18,19] a mixing layer was perturbed in three-dimensions by using oscillating 'chevron'-shape flapper, thus introducing spanwise disturbances along with the streamwise ones. The measurements revealed formation of either bending or bulging of KH rolls. An unexpected result was observed when flapper fluctuations magnitude was doubled from A=1.5mm to A=3mm. This did not lead to a faster growth of the mixing layer thickness due to larger perturbation amplitude, as we expected. On the contrary, the amplitude growth was slowed down. At the same time, the mixing layer characteristics, such as cross-flow position of the average mean velocity value y_0 (y_0 is the cross-flow coordinate where the mean velocity is equal to a half of the sum of free stream velocities in the lower and upper borders of the mixing layer) and the momentum thickness θ, exhibited qualitatively different dependence on the spanwise coordinate. At lower flapper amplitude the y_0 position oscillated while the momentum thickness remained almost a constant. At larger amplitude we observed oscillations of θ in the spanwise direction and almost a constant value of y_0. We assumed that larger flapper amplitude triggers the bulging instability instead of the bending one. Further observations revealed additionally that when the flapper was static the spanwise oscillations of y_0 were even stronger than in the case of 1.5mm amplitude of the flapper oscillations, while the momentum thickness did not change across the span. Due to its geometry, steady Chevron flapper generates strong static spanwise perturbations responsible for formation of translational instability, while its oscillations result in additional oblique initial perturbation generating bulging. Further increase of the oscillations amplitude from 1.5 mm to 3 mm leads to a prevailing role of bulging generated perturbations, which results in observed variations of y_0 simultaneously with approximate homogeneity of the momentum thickness in spanwise direction. Similar observations were reported by Bell & Mehta [8], but no explanation was provided.

In our computational modeling we studied development of 3D instabilities by introducing a 3D perturbation at the stages of under saturated and saturated KH billow, but before starting of a rollup or pairing. According to Pierrehumbert & Widnall [1], a perturbation responsible for bending should have a spanwise vorticity component antisymmetric in the cross-flow direction, and a symmetric one for bulging. Basing on this prediction we studied evolution of the symmetric and antisymmetric perturbations and had found that they really lead to bending or bulging of KH rollers. However, the result, i.e., bending or bulging, depends not only on the symmetry in cross-flow direction, but also on the phase shift of the initial perturbation in the streamwise direction when an oblique wave is generated, and on the choice of which Fourier mode is perturbed, i.e., purely spanwise or oblique. Thus, by switching between spanwise or oblique initially perturbed wave, or by placing the perturbation amplitude maximum inside the roller or inside the braid, we can switch between the bending and bulging instabilities. It should be emphasized that dependence of three-dimensional evolution of the mixing layer flow on initial disturbances is well documented (see, e.g., [8,9,19]). Here we show how a controlled initial disturbance can be used to switch between the bulging and bending modes.

Above observations were made for the fully 3D formulation where KH rollers were developing in time. Therefore it remained unclear whether growth of either bending or bulging results from a linear instability, in a complete analogy with the Stuart vortices, or time evolution of the billows and nonlinear interactions should be taken into account for explanation of the phenomena. To answer this question we repeat the above numerical modeling taking (i) small perturbation amplitudes to ensure linear initial growth, and (ii) KH billows frozen at two time moments corresponding to under saturated and saturated roller. This approach resembles one used [20]. The simulation with frozen billows shows growth of the bending and bulging modes in complete consistence with the fully non-linear modeling. This shows that bending and bulging can be interpreted as a result of linear instability of a combined flow consisting of mean flow and frozen KH billow. Moreover, this numerical exercise and monitoring the Fourier modes amplitudes allow us to conclude that development of both instabilities results from the triple interaction of the Fourier modes (1,0), (0,1) and (1,1), where mode (1,0) represents the finite amplitude frozen KH billow and the two-dimensional base flow is a combination of the above mode (1,0) and the mean parallel flow that can be generalized as mode (0,0). The finite amplitude of the frozen KH billow is of order *O(1)* as the mean flow, while two others modes (0,1) and (1,1) are of order $O(\varepsilon)$ at least at initial stages of the three-dimensional perturbations growth.

The following bulk of the paper is organized as follows. Formulation of the problem and a short description of numerical technique are given in Section 2. Numerical results obtained for the full and frozen-billow models are described in Section 3. All the results are discussed in Section 4, after which the concluding remarks are given.

2. Formulation of the problem and numerical method

We consider flow of Newtonian incompressible fluid described by the dimensionless momentum and continuity equations

$$\frac{\partial \boldsymbol{v}}{\partial t} + (\boldsymbol{v}\cdot\nabla)\boldsymbol{v} = -\nabla p + \frac{1}{Re}\Delta\boldsymbol{v}, \qquad \nabla\cdot\boldsymbol{v} = 0 \qquad (1,2)$$

in a three-dimensional rectangular box $0\leq x\leq\lambda_x$, $-H\leq y\leq H$, $0\leq z\leq\lambda_z$, where $\boldsymbol{v} = (u, v, w\ )$ is velocity and p is the pressure. The initial dimensional velocity distribution is defined as $\tilde{u} = U_{max}tanh(y/\delta)$, where δ is the mixing layer thickness. Equations (1) and (2) are rendered dimensionless using δ, U_{max}, δ/U_{max} and ρU_{max}^2 as the scales of length, velocity, time and pressure, respectively. We assume periodicity in the x- and z- directions and vanishing of shear stresses at $y = \pm H$. Since the base initial flow is directed along the x-coordinate we refer to the velocity components $u, v,$ and w and the corresponding directions as streamwise, normal and spanwise, respectively.

To visualize the evolution of mixing layer flow and similarly to [3] we add also a passive scalar T, whose evolution is described by a transport equation

$$\frac{\partial T}{\partial t} + (\boldsymbol{v}\cdot\nabla)T = \frac{1}{Pe}\Delta T, \qquad (3)$$

where Pe is the Peclet number. The initial distribution of the passive scalar is given by the hyperbolic tangent $T = tanh(y/\delta_T)$, where δ_T is the temperature layer thickness. In the following $\delta_T = \delta$. Similarly to the velocity filed we prescribe periodicity conditions in the x- and z- directions and vanishing of $\partial T/\partial y$ at $y = \pm H$.

The numerical solution of the problem is carried out by mixed spectral/finite-difference method of [21]. Each scalar variable $f(x, y, z, t)$ is represented as truncated Fourier series

$$f = \sum_{|m|<M}\sum_{n=0}^{N-1}\left[f_{nm}^{c}(t,y)cos(2\pi n\frac{z}{\lambda_z}) + f_{nm}^{s}(t,y)\sin\left(2\pi n\frac{z}{\lambda_z}\right)\right]exp\left(2\pi im\frac{x}{\lambda_x}\right) \qquad (4)$$

Since the result of summation in eq. (4) must be real we impose an additional constraint on the coefficients f:

$$f_{n,m}^{c} = \left(f_{n,-m}^{c}\right)^{*}, \quad f_{n,m}^{s} = \left(f_{n,-m}^{s}\right)^{*} \quad , \qquad (5)$$

where asterisk denotes complex conjugation. Substitution of the series (4) in the equations (1)-(3) and calculation of the corresponding Galerkin projections in the x- and z- directions yield a system of non-linear PDEs for calculation of the y- and time-dependent coefficients f. We apply the finite difference method with the hyperbolic tangent mapping function to resolve in the y-direction. The time integration is carried out using a semi-implicit Runge-Kutta scheme [21].

In the following we refer to every term of the series (4) as a (m,n) mode $f_{m,n}$ of a function f. For example, the (m,n) mode of the streamwise velocity is defined as

$$u_{m,n}(x,y,z,t) = \left[u_{nm}^{c}(t,y)cos(2\pi n\frac{z}{\lambda_z}) + u_{nm}^{s}(t,y)\sin\left(2\pi n\frac{z}{\lambda_z}\right)\right]exp\left(2\pi im\frac{x}{\lambda_x}\right) + c.c. \quad (6)$$

with *c.c.* standing for the complex conjugated terms. Clearly, the coefficients f^c_{nm} and f^s_{nm}, and in particular the coefficients u^c_{nm} and u^s_{nm} in eq. (9), are real, while each of the (*m*,*n*) modes is a complex function if $m \neq 0$. The modes corresponding to $m = 0$ describe perturbations that change in the spanwise plane and are uniform along the streamwise direction.

The amplitude F_{mn} of a (*m*,*n*) mode of a function *f* is defined as

$$F^2_{mn} = \frac{1}{\lambda_x \lambda_z} \int_0^{\lambda_x} dx \int_0^{\lambda_z} dz \int_{-H}^{H} f^2_{mn}(x,y,z,t) dy \qquad (7)$$

In addition we define the total amplitude of modes having a particular spanwise wavenumber F^S_n and the total amplitude of all modes F^{3D} as

$$(F^S_n)^2 = \sum_{m=-M}^{M} F^2_{mn}, \qquad (F^{3D})^2 = \sum_{n=0}^{N} (F^S_n)^2 \qquad (8)$$

Borders of the computational domain are chosen according to experiments of [22] and are set to λ_x=15, λ_z=10, and *H*=20. This yields the ratio between spanwise and streamwise wave lengths λ_z/λ_x = 10/15 = 0.66, which corresponds to the optimal value obtained in [1] for Stuart vortices. In most of numerical runs the solution was computed using 64 spectral modes in streamwise and spanwise directions and 128 grid points in *y*-direction. In some limited number of runs 128 spectral modes and 192 grid points are used. This increase of the resolution drastically changed the time of the run but did not affect significantly the final results.

To characterize developed mixing layer flow we use also its momentum thickness θ and position of the zero streamwise velocity y_0. Taking into account that minimal and maximal dimensionless velocities are equal to ±1, the momentum thickness is defined as

$$\theta(x,z,t) = \int_{-H}^{H} \frac{u - U_{min}}{U_{max} - U_{min}} \left[1 - \frac{u - U_{min}}{U_{max} - U_{min}}\right] dy = \frac{1}{4} \int_{-H}^{H} (1 - u')(1 + u') dy \qquad (9)$$

Where u' is the mean velocity profile shifted relatively to $(U_{min} + U_{max})/2$ and normalized by velocity difference $(U_{max} - U_{min})/2$ [22]. The experimental velocity profile in the vicinity of the splitter edge fits satisfactory the hyperbolic tangent function used in the computations as the initial velocity profile. The position y_0 is defined by the equation

$$u'[x, y_0(x,z,t), z, t] = 0 \qquad (10)$$

Following former and recent experiments [12,22] where perturbation was introduced at the edge of the splitter plate by a periodically oscillating flapper, we introduce small-amplitude perturbations of normal velocity v in the basic plane-parallel flow. In the computations described below we perturb (1,0) mode of v to obtain an evolution of primary Kelvin-Helmholtz billows, and then perturb additionally the modes (0,1) and (1,1). Amplitudes of these modes were chosen to be symmetric or antisymmetric with respect to the midplane $y = 0$ and were defined by an even or odd function of *y*, respectively:

$$v_{sym}(y) = \begin{cases} A[1 - y^2] sech(y), & y \le 1 \\ 0, & y > 1 \end{cases} \qquad (11)$$

$$v_{asym}(y) = \begin{cases} A[1 - y^2] sech(y) tanh(y), & y \le 1 \\ 0, & y > 1 \end{cases} \qquad (12)$$

The amplitude *A* in eqs. (11) and (12) is defined as $A = A_0 e^{i\alpha}$, where A_0 is a small positive and real value, α defines a phase shift in the *x*-direction. The phase shift is used to place maximum of a streamwise wave either in the region of a billow or a braid.

All the numerical experiments described below were carried out in the following way. First, we started with a perturbed two-dimensional (1,0) mode to obtain the KH rollup in the (*x*,*y*) plane of the mixing layer. Evolution of a 2D KH billow is illustrated in Figs. 2-4. Then all the two-dimensional modes $(m, 0)$ were frozen at a certain time, thus fixing a snapshot of the developing KH billow as a base state for the further investigation. Also all the non-linear interactions of 2D modes $(m, 0)$ with other 3D modes were forcefully canceled. Then three-dimensional perturbations of modes (0,1) or (1,1) were introduced and the evolution of three-dimensional perturbations was studied. This evolution involves both the exponential growth of perturbation amplitudes and all possible non-linear interactions of all three-dimensional modes. This approach allows us to study the response of the mixing layer flow with developed primary rolls, both under- and over-saturated, to different three dimensional initial perturbations. Apparently, it must be started from a small-amplitude initial perturbation and remains meaningful as long as amplitudes of three-dimensional modes do not reach the order of amplitude of primary rolls.

3. Results

Non-linear evolution of the primary KH instability is illustrated in Figs. 2-4. Evolution of the amplitude of the primary 2D KH mode is shown in Fig. 2. Snapshots of isolines of the passive scalar T and spanwise vorticity component are shown in Figs. 3 and 4, respectively. Being the most unstable linear mode the primary KH billow grows exponentially during a certain time interval. When the roll up of the mixing layer increases (Figs. 3 and 4) the velocity inside the roll becomes of the same order of magnitude as the mean velocity of the base flow. At this stage the roll amplification terminates, and the amplitudes start to decrease. One observes that the KH rollup reaches saturation. At the same time the thickness of the mixing layer increases and its value becomes such that the base hyperbolic tangent base flow profile becomes linearly stable. Further development of the roll can be characterized by changes in its shape and a slow decrease of the velocity amplitudes (Fig.2). Actually, the roll energy decreases and there might be even energy transfer from the roll to the mean flow which is often observed in experiments and is also confirmed in this work (Fig. 2).

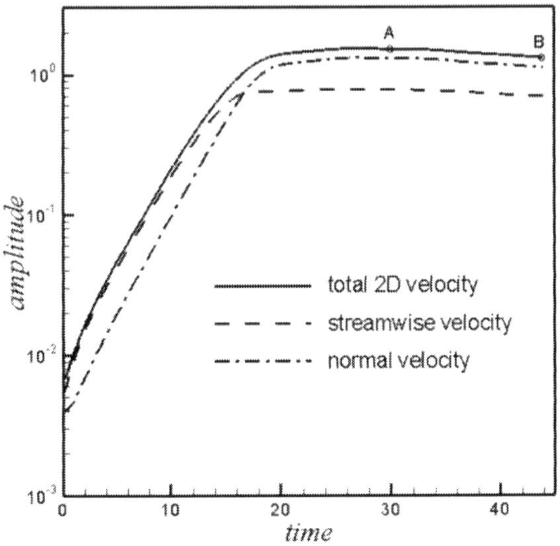

Fig. 2. Evolution of the amplitude of 2D perturbation of the primary base state for $Re = 250$.

The exponential growth of perturbations, characterized by the parameter β, can be separated into different time ranges corresponding to different instability stages. At the very beginning of the run (~ 2

Journal of Physics: Conference Series **216** (2010) 012019 doi:10.1088/1742-6596/216/1/012019

time units) the growth is very steep followed by a rapid expansion of the layer thickness. The slope is characterized by $\beta \approx 30\%$. This value has been reported previously in [3]. When entering the later stage the growth rate drops by approximately a half ($\beta \approx 15\%$) and remains almost constant for a considerable time until the primary roll forms completely. In the calculations described below the 3D-perturbations are added at two different times: 30 and 44 time units. At $t = 30$ the primary rolls do not overlap, while at $t = 44$ the subsequent rolls overlap and oversaturation takes place (see Figs. 3 and 4). Comparing of the growth rate of 3D modes at different times we conclude that they do not depend the saturation stage of the primary roll.

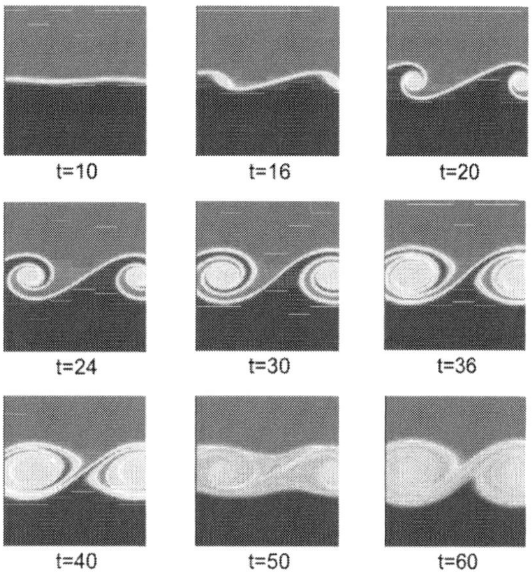

Fig. 3. Snapshots of isolines of the passive scalar T corresponding to a developing 2D flow at different times for $Re = 250$.

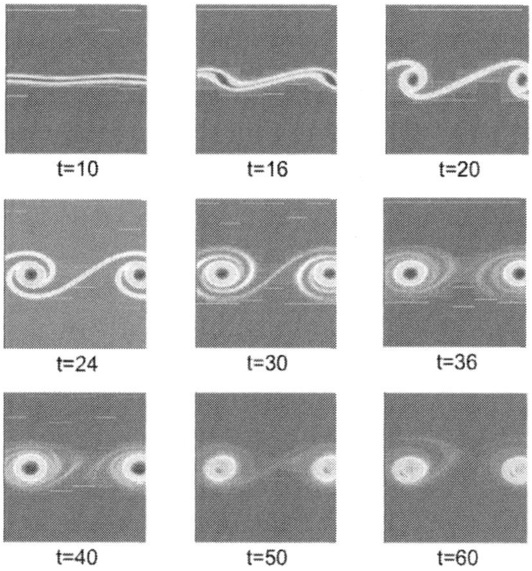

Fig. 4. Snapshots of isolines of z-vorticity of the developing 2D flow at different times for $Re = 250$.

Time moments, at which the primary roll was frozen, perturbed 3D modes, and the perturbation profiles (symmetric or asymmetric) for all the runs described below are summarized in Table 1. Figures 5 and 6 show time evolution of amplitudes of the components (0,1) and (1,1) for the 7 runs defined in the Table 1, in which either symmetrical (0,1) or antisymmetric initial 3D perturbation (1,1) were introduced at different times. The initial perturbation amplitude was 10^{-3} in all runs.

First we describe runs corresponding to the developed, but not saturated primary KH roll at t=30. When the initial perturbation is defined by the mode (0,1) and its y-profile is *symmetric*, growth of the secondary disturbance is very fast (Fig. 5, run 1) and the primary roll is subject to bending. Another run with the same initial (0,1) mode but with an *antisymmetric* y-profile leads to a significantly slower growth (Fig. 5, run 2) and bulging of the primary roll is observed. Difference in the growth rates is indicated also by the values of β in Table 1. Also, the antisymmetric y-profile causes strong undulations of the amplitudes. Note also, that the excitation of the mode (0,1) causes a similar growth of the mode (1,1) (Fig.

6, runs 1 and 2), which is a result of the triad interaction of the modes (0,1) and (1,1) of the order $O(\varepsilon)$ with the frozen mode (1,0) of order $O(1)$.

Table 1. Summary of runs

Number of run	Time of switching on of 3D perturbation	type of perturbation	initially perturbed modes	Phase shift, α	β	Resulting structure
1	30	sym	0,1	0	5%	bending
2	30	asym	0,1	0	2%	bulging
3	30	asym	1,1	0	5%	bending
4	30	sym	1,1	0	2%	bulging
5	30	asym	1,1	90°	2%	bulging
6	44	sym	0,1	0	5%	bending
7	44	asym	0,1	0	2%	bulging

When the mode (1,1) is excited first, contrarily to the (0,1) mode excitation, we observe a rapid amplitude growth in the case of the *antisymmetric* y-profile (Table 1 and Fig. 6, run 3). Similarly to the above observations, this rapid growth leads to the bending of primary roll. When the initial perturbation profile is symmetric (Table1, run 4), we again observe a slower amplitude growth with undulations (Fig. 6, run 4) that leads to the bulging. As in the previous case we observe a similar growth of the mode (0,1) resulting from the triad interaction of the mode (1,1) and (0,1) of the order $O(\varepsilon)$ with the frozen mode (1,0) of order $O(1)$.

When the initial perturbation of the mode (1,1) is shifted by $\alpha = 90°$ in the x-direction the behavior of the above initial perturbations switches in the case of *antisymmetric* y-profile. Namely, the shifted mode with $\alpha = 90°$ grows slowly, exhibit undulations, and leads to the bulging (Table 1, run 5), unlikely the same excitation with $\alpha = 0°$, that exhibits a fast growth and leads to the bending (Table 1, run 3).

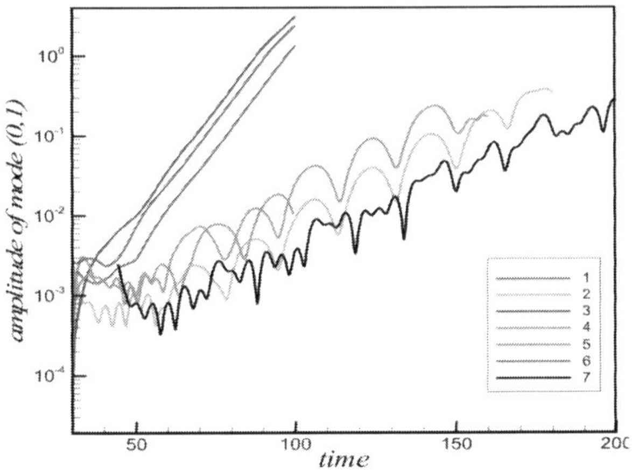

Fig. 5. Time evolution of the mode (0,1) for the 7 runs defined in Table 1.

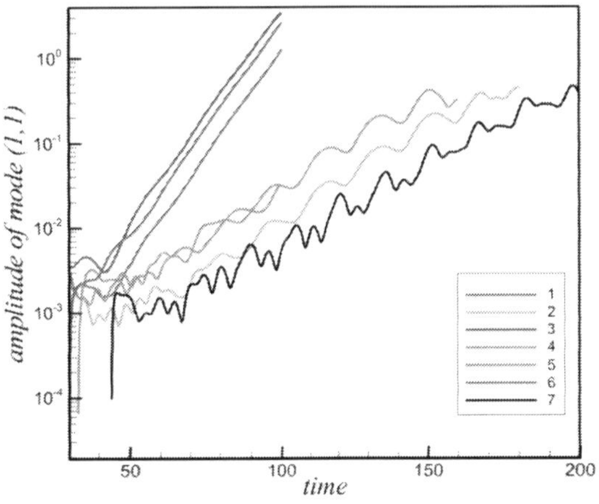

Fig. 6. Time evolution of the mode (1 ,1) for the 7 runs defined in Table 1.

Runs 6 and 7 in Table 1 correspond to the oversaturated primary KH roll frozen at t=44. Here we report only two runs with the symmetric and antisymmetric excitation of the mode (0,1), that also lead to either fast growth with bending or slow growth with bulging

The bending and bulging instabilities are illustrated by snapshots of isolines of the passive scalar T and the spanwise vorticity component Ω_z in Figs. 7-10. In these figures we show results obtained with the excitation of (0,1) mode only, since all other cases exhibit similar behavior. Note that the Prandtl number is equal to 1 and therefore the governing equations for Ω_z and T are similar. At the same time contours shown in Figs. 7-10 differ significantly, which is a result of different initial profiles and boundary conditions. The development of bending is clearly seen at the snapshots corresponding to shorter times in Figs. 7 and 8. Similarly, the bulging shape can be observed in Figs. 9 and 10. When the energy of 3D-modes becomes large, the contours become strongly distorted (Figs. 7-10). It can be observed also that the contours change drastically with the change of the relative location of a cross section in the streamwise direction. In spatially developing mixing layer, which is typical for experimental studies of free shear flows, it would mean a strong variation of these variables at different phases of the flapper fluctuations.

Isosurfaces of the streamwise vorticity ($\Omega_x = \partial w/\partial y - \partial v/\partial z$) are shown in Figs. 11 and 12 for the bending and bulging instabilities, respectively. The primary KH roll is illustrated in these figures by isolines of the scalar T plotted in the midplane $z = 0$. In the case of bending instability the streamwise vorticity starts to grow at the roll edges and is absent in the center of the roll. At earlier times the vorticity is antisymmetric with respect to the midplane $z = 0$, however at later times, e.g., at t=94 (Fig. 11), the antisymmetry breaks. Note that at the early stages (t=32) the streamwise vorticity does not develop at the mid-braid region and at the roll center. The large values of streamwise vorticity located near the roll edge and at the mid-braid emphasize a relative thickness of the braid during this stage of evolution. Evolution of the mixing layer leads to a fast growth of vorticity intensity and to evolving of strong vortices within the roll and in the braid region, as is it seen at t=56 (Fig. 11).

Fast increase of streamwise vorticity intensity can be observed at later times, t=78 and 94 (Fig. 11). The growth takes place both in the central plane of the roll and of the braid. For the time interval of 22 time units the intensity has been magnified almost in order of magnitude, and then gained another order of magnitude after 18 next time units. Note that the amplification of the streamwise vorticity takes place not only at the centers of primary roll but also at their periphery. We can speculate that a rib vortex, which is initiated in the braid plane reaches the primary roll edges. Since the primary roll is frozen the development of the secondary streamwise vorticity at short times is essentially a linear process. It includes the triad interactions of the primary roll, which can be represented as the mode (1,0) of order 1 and two secondary modes, (0,1) and (1,1) that are of order ε, which is equal to 10^{-3} in our calculations. The interactions of each of the secondary modes with the primary mode are the only ones essential. The other secondary mode is generated through this interaction when the first one is initially excited. The cross-interaction of the two secondary modes is not of any importance at the linear stage since it is of the order of ε^2.

At larger times, when the intensity of secondary 3D vortices become approximately of the order of the primary 2D roll (t=94, Fig. 11) the isosurfaces of the secondary streamwise vorticity shrink in the spanwise direction and their size considerably decreases. Their cross-section reminds more and more cross-section of rib vortices. The secondary streamwise vorticity arriving from the braid region and that developing within the primary roll become tilted, which can be attributed already to the development of non-linear effects.

Journal of Physics: Conference Series **216** (2010) 012019 doi:10.1088/1742-6596/216/1/012019

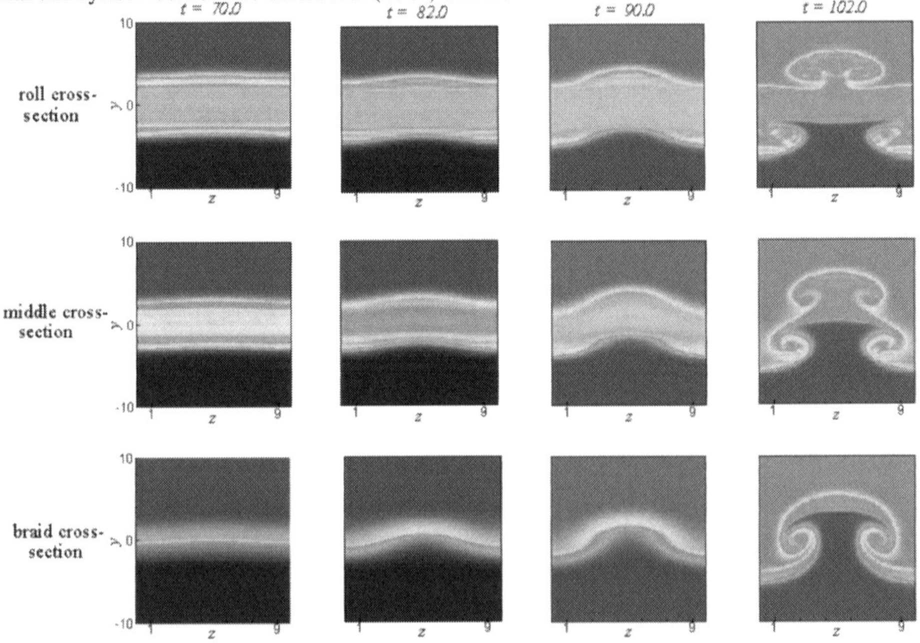

Fig. 7. Isolines of the passive scalar T in characteristic spanwise planes. Snapshots at $t = 70$ and 82 correspond to the initial bending instabiluity. Snapshots at $t = 90$ and 102 correspond to the developing collapse of a 3D structure.

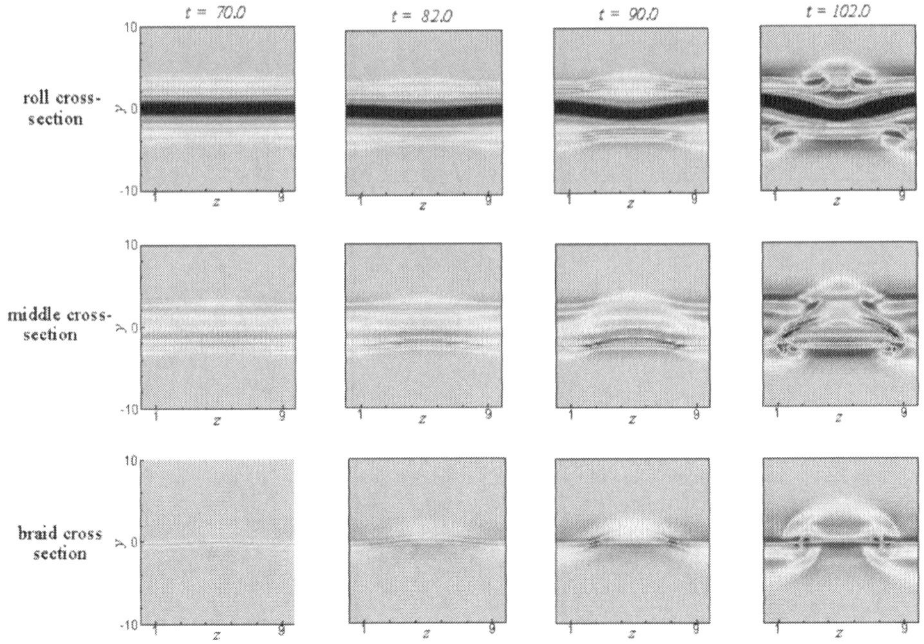

Fig. 8. Isolines of the spanwise vorticity component in characteristic spanwise planes. Snapshots at $t = 70$ and 82 correspond to the initial bending instabiluity. Snapshots at $t = 90$ and 102 correspond to the developing collapse of a 3D structure.

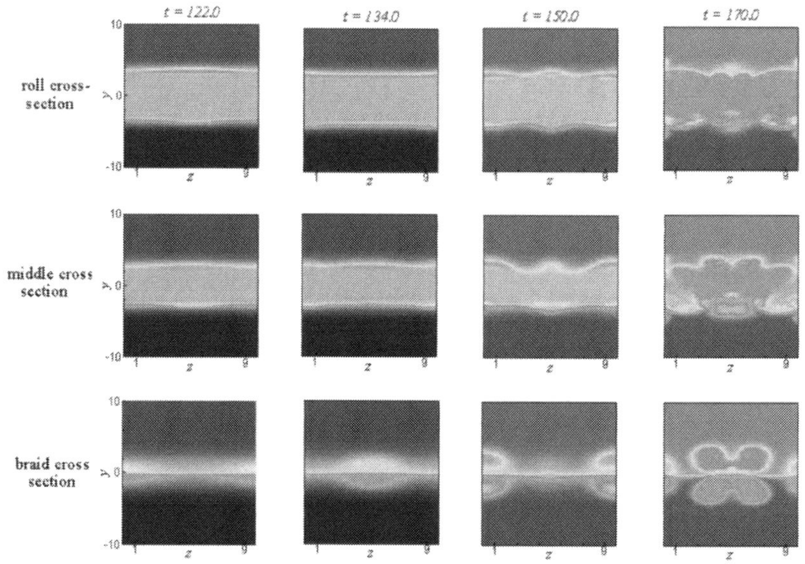

Fig.9. Isolines of the passive scalar T in characteristic spanwise planes. Snapshots at $t = 122$ and 134 correspond to the initial bulging instability. Snapshots at $t = 150$ and 170 correspond to the developing collapse of a 3D structure.

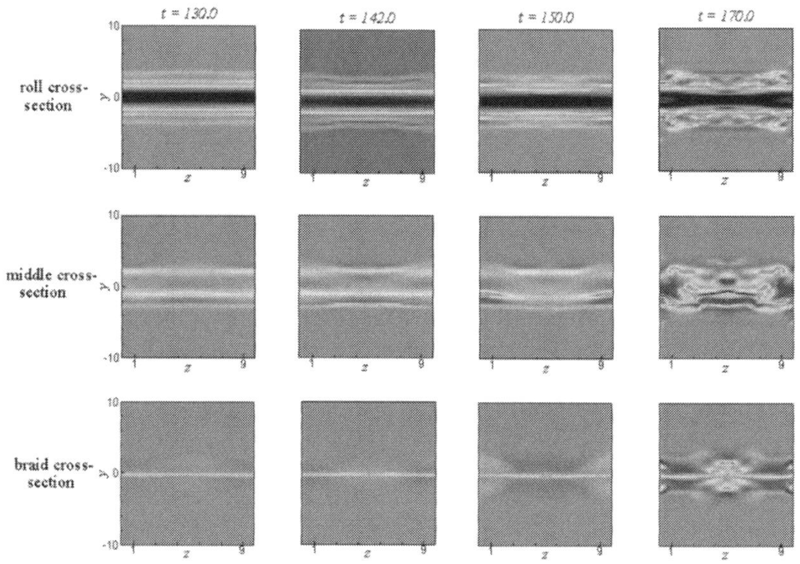

Fig.10. Isolines of the spanwise vorticity component in characteristic spanwise planes. Snapshots at $t = 130$ and 142 correspond to the initial bulging instability. Snapshots at $t = 150$ and 170 correspond to the developing collapse of a 3D structure.

Journal of Physics: Conference Series **216** (2010) 012019 doi:10.1088/1742-6596/216/1/012019

The evolution of the streamwise vorticity differs significantly from that shown in Fig. 11 if the secondary instability is excited by a slow bulging mechanism. At relatively short times, e.g., t=72 in Fig. 12, a large number of vorticity minima and maxima in the primary roll are related to the streamwise vorticity growth within well-distinguished layers developing in the process of rolling up. As in the previous case the streamwise vorticity is antisymmetric with respect to the midplane $z = 0$. These antisymmetric pairs are observed for all cross-sections including the center of the braid section. The latter is of special importance since it indicates that no stretching of the vorticity lines develops in the linear regime. Thus, development of rib vortices typical for translation bending mode of secondary instability does not take place.

As in the previous case the streamwise vorticity grows with time (Fig. 12). At the same time the shape of the isolines does not change significantly during a rather long time period (compare t=100 and 140), and then is destroyed at later times by non-linear mechanisms (t=172). Obviously, when the bulging mode develops the streamwise vorticity grows significantly slower than in previously considered case when the translational bending mode and rib vortices are being developed.

Stretching and folding of vorticity lines are shown in Figs. 13 and 14. The first frame of Fig. 13 illustrates the initial locations of the lines determined by the largest vorticity intensity. These lines are located at the primary roll center and at the center of the braid. Further stretching of vorticity lines in the case of bending (Fig. 13) leads to a development of primary roll bending and rib vortices. Hair pin structures appear at larger times. The red color indicates their penetration into the upper semi-volume, $y > 0$, and blue into the lower.

A typical development of bulging is illustrated in Fig. 14. When the vorticity lines at the roll center attain a bulged shape their counterparts in braid region remain practically unchanged. At later times the non-linear effects become important and collapse of the primary roll occurs, generating rings. The vorticity lines in the braid region are stretched and bended in a very complicated manner with a symmetry breaking already observed for the streamwise vorticity. Note that in the case of bending the latter structure appears mostly in the lower semi-volume.

Figure 15 presents spanwise distribution of the momentum thickness θ and the lateral coordinate y_0 corresponding to the zero value of mean streamwise velocity at different times of mixing layer evolution in the case of bending instability. These values have been computed after the averaging over the streamwise coordinate has been made. As follows from Fig. 15 no variation of θ and y_0 is observed till $t \approx 50$. At larger times both values θ and y_0 start to vary. At t=58 one does not observe strong spanwise variations of θ while y_0 attains a spanwise oscillation with the same spanwise spatial period as the initial perturbation. At larger times (t=82), the amplitude of y_0 oscillations increases significantly. The momentum thickness θ starts to vary as well, seemingly, at double frequency. At even higher values of time, t=94, the non-linear effects become strong and both variables vary in spanwise direction but not at the fundamental frequency. Very similar results were observed in the experiments [12] in the presence of oscillating Chevron flapper with amplitude of 1.5mm. At larger distances, the developing turbulent fluctuations in the experiments, which are not accounted for in the numerical simulations, change the spanwise distributions of both variables.

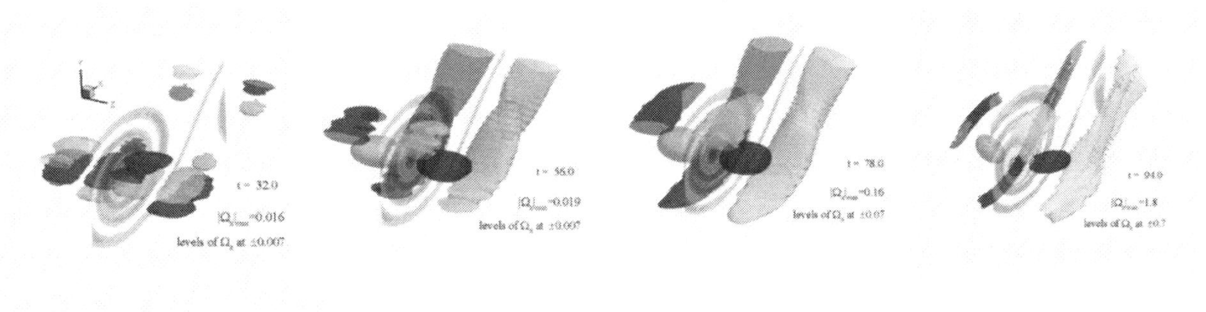

Fig. 11. Isosurfaces of the streamwise vorticity for bending instability. Parameters of run 1 (Table 1).

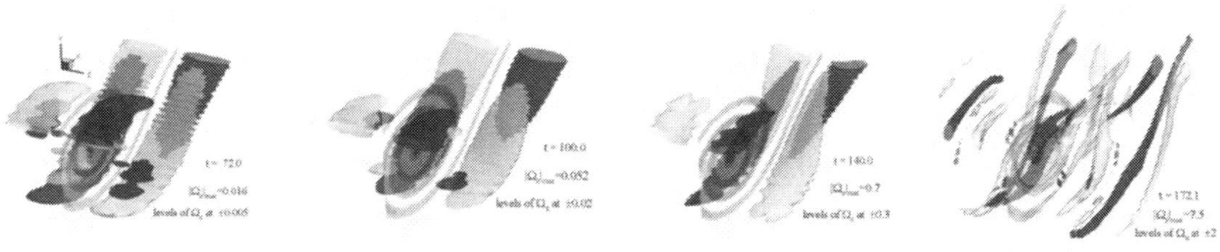

Fig. 12. Isosurfaces of the streamwise vorticity for bulging instability. Parameters of run 2 (Table 1).

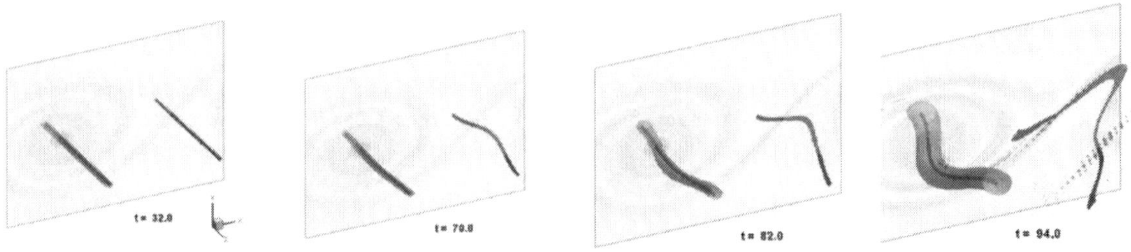

Fig. 13. Isosurfaces of the spanwise vorticity for bending instability. Parameters of run 1 (Table 1).

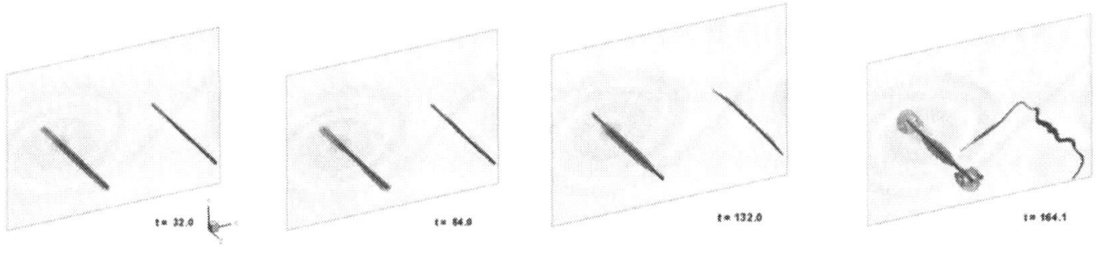

Fig. 14. Isosurfaces of the spanwise vorticity for bulging instability. Parameters of run 2 (Table 1).

Distributions of the momentum thickness θ obtained for bulging mode at several different times are shown in Fig. 16. As can be expected for the bulging mode, the position y_0 remains unchanged at all times and is not shown. We observe that, compared with the bending mode (Fig. 15), the momentum thickness noticeably oscillates in the spanwise direction at large times (t=170). In general, the bulging processes develop slower than the bending ones. In the experiments with chevron flapper it is impossible to obtain a separated bulging mode. Nevertheless, the increase of chevron flapper perturbation amplitude up to 3mm increased the relative contribution of bulging mode, which lead to a relative increase of the momentum thickness compared to the variation of y_0 [12].

The mean velocity and temperature profiles at the mid spanwise position z=5 are shown in Figs 17 and 18 for bending and bulging instabilities, respectively. The averaging was conducted over the entire streamwise spatial period. The vertical y-coordinate in these figures is normalized by velocity momentum thickness θ computed for the corresponding time. Since in these computations $Ri = 0$, there is no stratification and the temperature behaves as a passive scalar. The temperature is affected by velocity field but not vice versa. The velocity profiles remain smooth and the initial hyperbolic tangent profile keeps more or less its shape during the evolution. The profile of passive scalar varies significantly even during development of two-dimensional perturbations, t=30 (Figs. 17 and 18). At this time strong temperature gradients are observed at the mixing layer edges. Also in the mixing zone the temperature noticeably, so that local overturning of the profile can be noticed. Within the primary vorticity core these effects are expected to be stronger. Figure 3 indicates appearance of low temperature layers above layers with higher temperature. As expected with initial perturbations generating the bending mode, the mean profiles change faster (cf. Figs. 17 and 18). The lack of symmetry observed at greater times is most probably related to fast development of three-dimensional non-linear effects with time evolution.

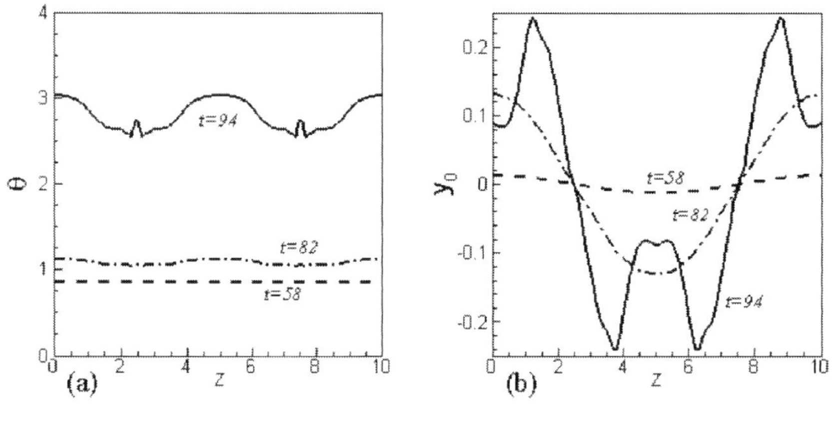

Fig. 15. Spanwise distributions of momentum thickness θ and the lateral coordinate y_0 corresponding to the zero value of mean streamwise velocity for bending instability (run 1 of Table 1).

The RMS distributions of amplitudes of all the three velocity components, temperature and spanwise vorticity are presented in Figs. 19 and 20 for the bending and bulging instabilities, respectively. The most interesting result in both cases is the observation of relatively strong streamwise and cross-stream velocity amplitudes penetrating into the potential velocity region while the temperature and streamwise vorticity are limited to the mixing layer region. The variation of temperature fluctuations is almost abrupt: it changes from relatively high values to almost zero at very short lateral distances. The spanwise velocity component emerges at greater time indicating evolution of three dimensional effects. At this specific spanwise coordinate, z=5, its lateral distribution is more symmetric for the bulging mode case

(Fig. 20). This is perhaps the reason for more symmetrical shape of temperature and spanwise fluctuations distributions.

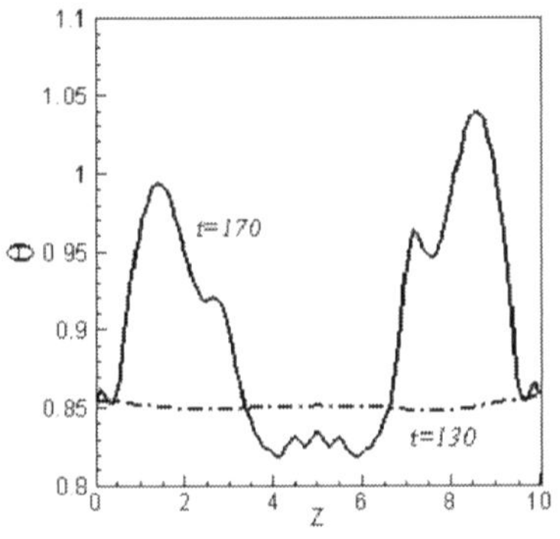

Fig. 16. Spanwise distributions of momentum thickness θ for bulging instability (run 2 of Table 1).

Velocity and temperature (averaged across XY plans) correlations along spanwise direction are presented in Figs. 21 and 22. The shape of these correlations is indicative for understanding of the secondary structures. In particular, when these correlations are close to unity the flow structure is two-dimensional, e.g., as obtained at a short time t=30. At longer times, this value deviates from unity, thus, demonstrating development of secondary 3D-structures. While the underlying processes are linear, the correlations exhibit sine-like variations (t=58 in Fig 21 and t=102 and t=130 in Fig. 22). Strong evolvement of non-linear processes results in non-harmonic shapes of the correlations as can be noticed in Fig. 21 at times, t=82 and t=94 and in Fig. 22 at t=170. Two velocity components, streamwise and cross-stream, are *symmetric* in spanwise direction while the spanwise component is *antisymmetric*.

At longer times the spanwise periodicity remains unchanged for the bending mode but varies significantly in the bulging mode case (t=170, Fig. 22). This result may be important when correlations of coherent velocity components obtained in the experiments are considered. In particular, it can be used to determine the mode obtained in these experiments. The spanwise variation of correlation of streamwise velocity component is significantly larger than that of lateral velocity component. A similar result was observed also in experiments [12].

4. Discussion and conclusions

The computational approach adopted in the current study with two-dimensional and three-dimensional perturbations introduced into the mixing layer at different times allowed us to distinguish between evolution of primary two-dimensional roll and development of secondary three-dimensional structures. Although this distinction is somewhat artificial it helps to understand better the mechanisms of development of secondary 3D structures. In particular, this approach is justified by a faster amplification of 2D-primary roll and lower level of initial 3D perturbations. Due to these effects, the primary roll becomes dominant even in case when all perturbations *are excited simultaneously*. Also, it helped to understand how the symmetry features of initial perturbation affect evolution of secondary 3D structures. It must be emphasized that the actual perturbations considered here are composite due to triad interaction of the two initial secondary modes: (0,1) and (1,1) and the primary, independent on the spanwise coordinate, two-dimensional mode (1,0). The latter develops into the primary roll during the initial stage of evolution.

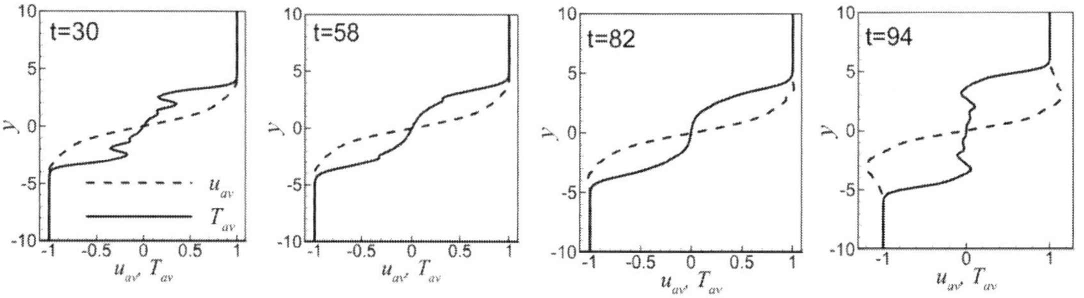

Fig. 17. Normalized mean velocity and temperature profiles at the mid spanwise position z=5 for bending instability (run 1 of Table 1).

As can be observed from Figs. 5 and 6 symmetry properties of secondary perturbations determine uniquely the mode of 3D instability: bending or bulging. The oversaturation of primary roll affects the rate of growth and in case of bulging mode the frequency of periodic oscillations during the growth, but secondary disturbances patterns remain qualitatively unchanged. When the bending mode develops, the rib and core vortices evolve while bulging mode results in a complicated structure of core vortices. No rib vortices are observed in the case of bulging.

When time is long enough the size of primary roll increases (oversaturation) and the successive primary rolls overlap. In this case velocity and shear in the middle of the braid are affected by the primary rolls (Figs. 3 and 4). The mushroom structure is observed at large times, e.g., starting from $t \approx 94$ for the case of Fig. 7. Similar mushroom structures were observed in some experimental studies (see, e.g.,[7]). Also in measurements with SPIV system these structures were obtained when the smoke was not homogeneously spread across the plane. The continuous bulging structure breaks to separate rings at larger times (Fig. 8, t=150), similar to what was observed in [3].

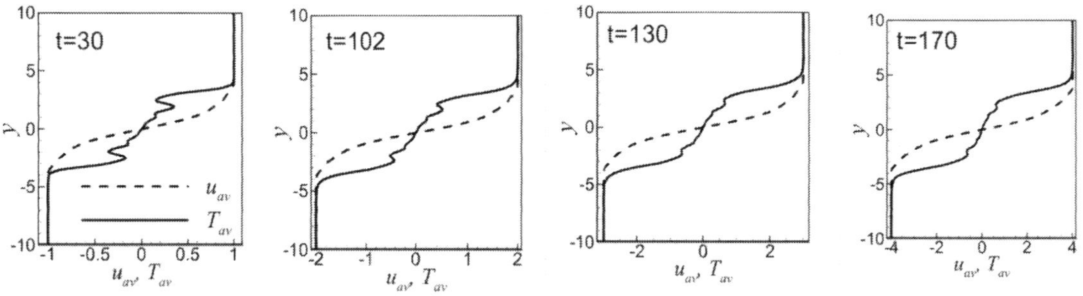

Fig. 18. Normalized mean velocity and temperature profiles at the mid spanwise position z=5 for bulging instability (run 2 of Table 1).

Evolution of streamwise vortices clearly shows generation of hair-pin vortices. It follows from Fig. 13 that stretching of braid lines, which are deprived of vorticity due to roll up process, is relatively fast and represents mechanism leading to formation of hair-pin vortices. These structures eventually interact with the primary rolls and results in their collapse and developing of strong turbulence within these cores. One can speculate that in the bending regime the 3D-turbulence develops in the peripheral region. In the bulging regime, where rib vortices do not develop and the instability occurs mainly in the core, the turbulence should develop within the core.

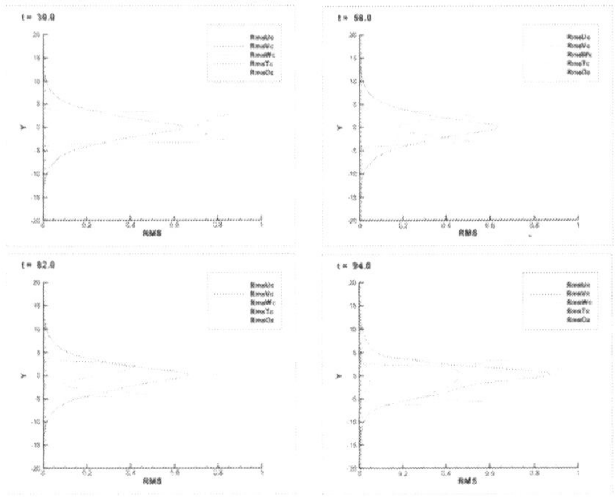

Fig. 19. RMS amplitudes of velocity components, temperature and spanwise vorticity for the bending instability, z=5.

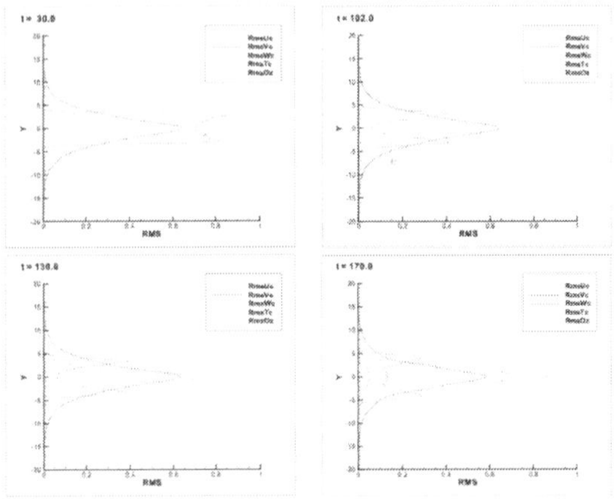

Fig. 20. RMS amplitudes of velocity components temperature and spanwise vorticity for the bulging instability, z=5.

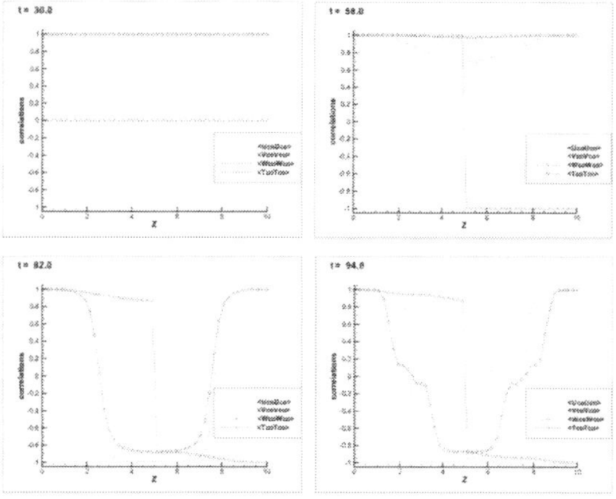

Fig. 21. Average velocity and temperature correlations in the spanwise direction. Bending instability.

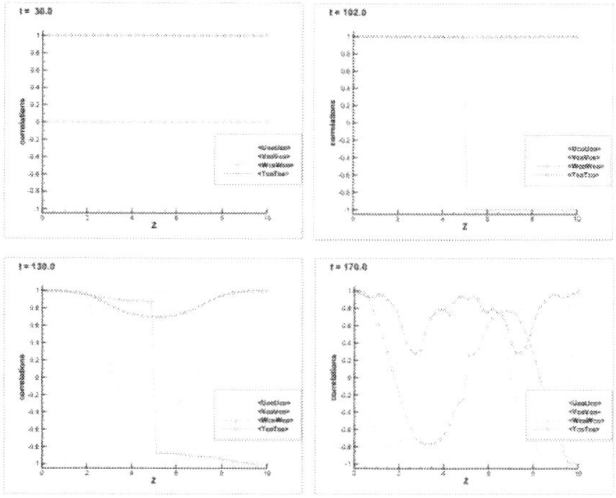

Fig. 22. Average velocity and temperature correlations in the spanwise direction. Bulging instability

Computational modeling of the fully nonlinear model with both two- and three-dimensional perturbations introduced in the flow reveals qualitatively similar instabilities development. At smaller times, the flow is mainly two-dimensional and the spanwise variations are relatively small. Depending on the kind of the three-dimensional perturbation, at longer time (t= 82 and t=130 for bending and bulging, respectively) the flow becomes essentially three-dimensional. At even larger times (t=94 and t=170) when non-linear tree-dimensional effects prevail the growth of perturbations becomes steeper (especially for the bending/translation mode) and the mixing layer diverges faster. It should be stressed that these computations did not account for the pairing, which can be of great importance at more advanced stages of mixing layer evolution. The whole non-linear three-dimensional evolution of the mixing layer flow takes place in a very good agreement with the experimental results [12]. More details on these calculations can be found in [23].

The main conclusion of the present study is that three-dimensional perturbations, harmonic in the streamwise and spanwise directions, enhance the primary (KH) instability and modulate the ensuing eddies along the span. The three-dimensional disturbances can be divided into two sets that differ by their cross-flow symmetry and streamwise phase, and lead to development of either bending or bulging instability. Perturbations belonging to the first set enhance the translative instability that bends the (initially spanwise-uniform) KH vortices and develop streamwise rib vortices. The second set of initial perturbations generates the bulging mode that leads to spanwise alteration of the KH billows width. Both sets create coherent streamwise vorticity within the rollers and braid regions, which lead to the development of streamwise vortex tubes (rib vortices and bent or bulging spanwise vortices) farther downstream. Both types of three-dimensional distortions of the primary spanwise rolls, bending and bulging, can be distinguished by spanwise variations of the momentum thickness θ and vertical position of the mean velocity y_0. In the case of bending one observes almost constant θ and spanwise oscillations of y_0. Conversely, the bulging leads to spanwise oscillations of θ and almost unaltered y_0 position.

The streamwise vortices become less coherent with time increasing. At longer times, when the non-linear effects become essential, they seem to disappear in the straining field existing between adjacent KH rolls (braid region), as well as becoming less regular and weaker within the rolls themselves. This might be an outcome of a strong interaction between spanwise KH type of rolls and streamwise vortex tubes. Before collapse of the rib vortices, very clear structures reminding hair-pin or lambda vortices are observed. This result might explain observation of such structures in real physical experiments.

The velocity profiles and all the parameters, such as momentum thickness and location of the mixing layer center (where streamwise velocity becomes zero) obtained using these profiles in the current experiments are in very good qualitative agreement with experimentally observed coherent structures in spite of the fact that computations are conducted for the temporally evolving mixing layer while the measurement were carried out for spatially developing one.

An interesting observation from this work is the strong linkage between the lateral perturbation nature (by means of symmetry) and the spanwise & streamwise modes (2D or 3D) to the instability type. This behavior has not been explained within the current study and might be worth further investigation.

Acknowledgements This work is supported by the Bi-National Science Foundation (Grant 2004087) and in part by the Israel Science Foundation (Grant 964/05).

References

[1] Pierrehumbert R T and Widnall S E. 1982 The two- and three-dimensional instabilities of a spatially periodic shear layer *J. Fluid, Mech.*, **114** 59

[2] Schoppa W, Hussain F and Metcalfe R W 1995 A new mechanism of small-scale transition in a plane mixing layer: core dynamics of spanwise vortices *J. Fluid Mech.* **298** 23

[3] Rogers, M M and Moser R D 1992 The three-dimensional evolution of a plane mixing layer: the Kelvin-Helmholtz rollup *J. Fluid Mech.* **243** 183

[4] Schoppa W, Hussain H S, and Hussain F 1993 Nonlinear instabilities of free shear layers: subharmonic resonance and three-dimensional vortex dynamics In *Nonlinear Instability of Nonparallel Flows* (ed. S.P. Lin et al.), pp. 251-280, Springer.

[5] Jimenez J 1983 A spanwise structure in the plane shear layer *J. Fluid Mech.* **132** 319

[6] Bernal L P and Roshko A. 1986 Streamwise vortex structure in plane mixing layers *J. Fluid. Mech.* **170** 499

[7] Lasheras J C and Chol H 1988 Three-dimensional instability of a plane free shear layer: an experimental study of the formation and evolution of streamwise vortices *J. Fluid Mech.* **189** 53

[8] Bell J H and Mehta R D, 1993 Effects of imposed spanwise perturbations on plane mixing layer structure *J. Fluid Mech.* **257** 33-63.

[9] Nygaard K J and Glezer A 1994 The effect of phase variations and ceoss-shear on vortical structures in a plane mixing layer *J. Fluid Mech.* **276** 21

[10] Tung S and Kleis S J 1996 Initial streamwise vorticity formation in a two-stream mixing layer *J. Fluid Mech.* **319** 251

[11] Estevadeordal J and Kleis S J 2002. Influence of vortex-pairing location on the three-dimensional evolution of plane mixing layer *J. Fluid Mech.* **462** 143

[12] Kit E, Wygnansky I, Friedman D, Krivonosova O and Zhilenko D 2007. On the periodically excited plane turbulent mixing layer, emenating from a jagged partitition *J. Fluid Mech.* **589** 479

[13] Metcalfe R W, Orszag S A, Brachet M E, Menon S and Riley J J 1987 Secondary instability of a temporally growing mixing layer *J. Fluid Mech.* **184** 207

[14] Ashurst W T, and Meiburg E 1988 Three-dimensional shear layers via vortex dynamics *J. Fluid. Mech.* **189** 87

[15] Caulfield C P and Peltier W R 1994 Three dimensionalization of the stratified mixing layer *Phys. Fluids* **6** 3803

[16] Khac M D, Basdevant C, Hiêp-Lê T, and Dang-Tran K 1998. Topological and dynamical analysis of a three-dimensional mixing layer *Aeroespace Sci. Technol.* **1** 13

[17] Kit E, Krivonosova O, Zhilenko D and Friedman D 2005 Reconstruction of large coherent structures from SPIV measurements in a forced turbulent mixing layer *Experiments in Fluids* **39** 761

[18] Kit E and Wygnanski I 2008 On periodically excited turbulent mixing layer created downstream of a plane chevron partition *Physica Scripta T*, **doi**:10.1088/0031-8949/2008/T132/014008.

[19] Scott Collis S, Lele S K., Moser R D. and Rogers M M 1994 The evolution of a plane mixing layer with non-uniform forcing *Phys. Fluids* **6** 381

[20] Klaassen G P and Peltier W R 1985 The onset of turbulence in finite-amplitude Kelvin-Helmholz billows *J. Fluid Mech.* **155** 1

[21] Nikitin N V 1994 A spectral finite-difference method of calculating turbulent flows of an incompressible fluid in pipes and channels *Comp. Maths. Math. Phys,* **34** 785

[22] Gaster M, Kit E and Wygnanski I 1985 Large-scale structures in a forced turbulent mixing layer *J. Fluid Mech.* **150** 23

[23] Gerstenfeld D 2007 3D simulation of secondary instability modes of kelvin-helmholtz billows in plane mixing layer, M. Sc. Thesis, Faculty of Engineering, Tel-Aviv University, 77pp.

AUTHOR INDEX

Institute of Physics
Dirac House, Temple Back
Bristol BS1 6BE UK